From Lexington to Baghdad and Beyond

From Lexington to Baghdad and Beyond

War and Politics
in the American Experience

Third Edition

Donald M. Snow and Dennis M. Drew

Routledge
Taylor & Francis Group
LONDON AND NEW YORK

First published 2010 by M.E. Sharpe

Published 2015 by Routledge
2 Park Square, Milton Park, Abingdon, Oxon OX14 4RN
711 Third Avenue, New York, NY 10017, USA

Routledge is an imprint of the Taylor & Francis Group, an informa business

Library of Congress Cataloging-in-Publication Data

Snow, Donald M., 1943–
  From Lexington to Baghdad and beyond : war and politics in the American experience /
by Donald M. Snow and Dennis M. Drew. — 3rd ed.
     p. cm.
  Rev. ed. of: From Lexington to Desert Storm and beyond. 2nd ed. 2000.
  Includes bibliographical references and index.
  ISBN 978-0-7656-2402-4 (cloth : alk. paper)—ISBN 978-0-7656-2403-1 (pbk. : alk. paper)
  1. Politics and war. 2. Sociology, Military—United States. 3. Military art and science—United
States—History. 4. United States—History, Military. 5. Civil-military relations—United States
—History. I. Drew, Dennis M. II. Snow, Donald M., 1943– From Lexington to Desert Storm
and beyond. III. Title.

  UA23.S5247 2009
  306.2′70973—dc22   2009012390

ISBN 13: 9780765624031 (pbk)
ISBN 13: 9780765624024 (hbk)

Contents

Preface

This volume is a revision and expansion of *From Lexington to Desert Storm and Beyond: War and Politics in the American Experience*, which in turn was an expansion of *The Eagle's Talons*, published by Air University Press in 1988. This new edition came about after conversations with Patricia Kolb, our editor at M.E. Sharpe, and has two primary purposes.

The first is to expand the coverage to include critical changes that have occurred since the book was last produced. Writing for that edition was completed in 2000, and obviously things have changed since then: the United States has entered into long military involvements in Afghanistan and Iraq.

This third edition reflects those new military experiences. The American intervention in Afghanistan was a direct response to the 11 September 2001 attacks on the United States, aimed at destroying the bases from which those assaults were planned and executed. Sadly, that effort has been incomplete and inconclusive, for reasons explored in Chapter 9. In 2003, the United States invaded Iraq with stated motives that were similarly—if perhaps speciously—related to terrorism. The Iraq War is examined in Chapter 10. It is the authors' conclusion that both the Afghanistan and Iraq experiences can be better understood in the context of the framework laid out in the pages that follow and that, had something like the framework offered here been considered in making those decisions, the results might have been different.

Donald M. Snow
Dennis M. Drew

Introduction

As the first decade of the twenty-first century winds down, the United States finds itself revisiting the question of the appropriate role of military force as an instrument of power to support the objectives of foreign policy. That debate is part of a much larger disagreement over America's place in the world. The poles of the larger debate about power range from the expanded responsibilities the United States must assume in a still dangerous, terrorist-infected world to the dangers of overextension of American resources.

Future applications of American military might are clearly important elements of the post–cold war world because military force is, and has been throughout history, a major means societies have used to settle their differences. One can decry or celebrate that observation, but one can ignore it only by the considerable application of selective perception. The overriding questions for Americans are where, if anywhere, and over what, should they be willing to shed their blood and expend their treasure.

Three factors stand out and collectively define the issues underlying the debate. The first is the issue of limited war, a problem of adjustment to fighting less than World War II–sized conflicts, and a problem the currency of which surrounds American limited intervention in places from Somalia and Bosnia and Herzegovina to Iraq and Afghanistan. The second is the effects of the end of the cold war and hence of the missions and ways of doing things that flowed from that competition. The third is the lingering legacy of America's most traumatic and unsuccessful adventure at arms, the Vietnam War and its aftermath.

The period since the end of World War II represents a watershed in the reasons and purposes for which the United States goes to war. For a period beginning with the American Civil War and ending with World War II, the significant uses of American force were in support of total political purposes: the overthrow of the enemy's regime, a purpose adequate to justify the large-scale mobilization of American society.

Such wars are not history's norm, but in popular (including the military's) thinking they became the norm. Total wars are simple and straightforward: the enemy must be vanquished and whatever means are necessary will be employed to that end.

In a world where total war could mean nuclear holocaust (which is still physically possible, even with the demise of the Soviet Union), such war has become inconceivable. Indeed, the only instance since World War II where the United States exercised the unlimited goal of overthrowing an enemy regime was in the Iraq War, where the United States overthrew and replaced the Iraqi government.

America and the American military have had to adjust to an era when force is used for more limited purposes, and it has been difficult. Wars with limited objectives are often fought with less than the total means available, which is frustrating to the military. Limited goals are also generally less morally lofty and uplifting; the enemy is not so evil that he must be exorcised. Particularly when the war is confined to a single country, differentiating good from evil is difficult. Moreover, the purposes underlying limited war are often vague and poorly articulated and understood.

During the cold war, U.S.–Soviet competition necessitated limits on any wars in which the United States—or for that matter, the Soviet Union—was involved. Since military engagements generally pitted the United States against a Soviet ally—North Korea, North Vietnam—a concern always had to be whether some action might cause escalation to direct confrontation that could lead to nuclear war. Certainly such reasoning affected the way both the Korean and Vietnam conflicts were fought.

Since the end of the cold war, that inhibition no longer exists. American involvements, however, remain limited, but the cause of limitation has changed. Before 9/11, the conflicts were small, mostly internal affairs in areas where American interests were minimal by traditional definition. These wars, as in Bosnia and Herzegovina, Somalia, Kosovo, and the Kurdish enclaves in Iraq, did not demand maximum force, nor were they situations where large-scale force could achieve defined objectives. Reactions to 9/11 broadened the purposes for using force, but the means and ends remained constrained.

The second factor is the end of the cold war itself. The central reality for the American military during the forty-four years between the end of World War II and the beginning of the end of the Soviet empire in 1989 was the menace posed by Soviet-led communism. As a military problem, it had two sides: the conventional balance between the North Atlantic Treaty Organization (NATO) and the Warsaw Pact countries in central Europe, and the thermonuclear balance between the Soviets and the Americans.

This confrontation provided the conceptual framework from which all other thinking and planning derived. The purposes of military force were straightforward: to deter either kind of attack by convincing the opponent he could not succeed (deterrence) or to defeat an attack should deterrence fail.

Nuclear weapons played a special, evolving role in this regard. As nuclear arsenals grew and the calamity of their use became appreciated, they increasingly became the limiting factor for the whole competition. Everyone on both sides came to understand that a nuclear war would be catastrophic and could not be allowed to occur. Since any direct conflict between the superpowers had within it the potential to escalate to nuclear exchange, all confrontations had to be avoided.

The result was a "necessary peace" between the superpowers where the avoidance of mutual annihilation was the conceptual glue. In that circumstance, the whole competition gradually became more ritualistic than real: we planned, we organized, and we practiced for a war we knew could never occur. When that ritualistic nature was combined with the economic crisis of the Soviet system, the fate of the cold war was sealed.

The end of the cold war has also meant the increasing irrelevance of the concepts on which it was premised. The new international system that is the successor to the cold war system is still not clearly in place. One component of that system will be a set of criteria defining when the United States will and will not use military force. One factor that continues to condition the evolution of thinking about that use is what remains of the Vietnam legacy.

The third concern is a product of the limited-war problem in general, and more specifically, the American experience in Vietnam. The war in Southeast Asia was a national trauma. In many important ways, it triggered the entire debate over the American role in the world and the utility of military force, a debate that has been rekindled by U.S. involvement in the Persian Gulf.

The Vietnam War was bewildering. The U.S. effort inundated Indochina with a flood of American equipment, advisers, technology, and combat troops. Applied American military power dwarfed the physical efforts of the Vietcong and the North Vietnamese. Militarily, the American battlefield effort yielded an almost unbroken string of victories. American forces commanded the air and sea and could operate anywhere with ground forces. On the one occasion when the enemy stood and fought American forces in a conventional style, the Tet offensive in 1968, the enemy was so badly mauled that it could not launch another major offensive for four years. But despite all this "success," the United States could not prevail.

The Vietnam War was the first war in which none of the American political objectives were attained. Despite an enormous military effort, continual success on the battlefield, and the sacrifice of more than 50,000 American lives, the United States was unable to translate apparent military success into success in the larger war. Perhaps worse, the controversy that eventuated over U.S. participation in the war ripped apart the fabric of American society as private passions about the war erupted into massive demonstrations and occasional violence. Many became disillusioned with the country's military and political leadership. In addition, the fiscal consequences of the war were still being felt in the American economy more than a decade after the end of U.S. involvement. Finally, the struggle's end was as difficult as its conduct. According to one point of view, the United States shamefully abandoned its South Vietnamese ally, thus making a mockery of the American effort and of those who were sacrificed in that effort.

That the Vietnam experience was traumatic is unquestionable; the question that remains is what can be usefully learned from the experience—and again, Americans have not achieved consensus. Purported lessons have included simplistic cries for "no more Vietnams" to equally simplistic pleas demanding that the military be unfettered when prosecuting such struggles in the future. Other "lessons" span the gamut from geopolitical to technical and have been equally diverse and contradictory. In short, the passions caused by the Vietnam trauma have generated considerable heat but little light by which Americans can guide future actions. The "ghosts" of Vietnam continue to haunt the country in Iraq and Afghanistan.

Confusion, bewilderment, anger, and simplistic solutions are symptomatic of the vague and often myopic historical view of most citizens, including, unfortunately, many civilian and military leaders charged with important national security responsibilities. The authors contend that current concerns about the effects of the end of the cold war, limited war, and the Vietnam experience are but threads in the much larger historical tapestry of American politico-military experience. We believe that just as individual threads should be viewed as only part of a whole tapestry, current politico-military concerns can be evaluated accurately only if viewed in their historical context.

Simplistic approaches to politico-military problems are also indications that Americans have not needed to deal comprehensively with the role of force. History, and especially military history, has treated the United States kindly. Few countries share the American experience of carving a new state from a vast wilderness only sparsely populated by aboriginal tribes and then transforming that bountiful wilderness into a great democratic experiment. Americans have a legacy of optimism: despite obstacles, Americans expect to achieve their goals. The "can do" spirit has been an indelible element in our collective psyche.

Much of this American optimism is supported by the country's reprieve from history's darker side, but history has shown another face to Europeans. Plowshares have been beaten into swords as often as the reverse and conquerors have regularly scourged the land. The result has been to breed caution, reserve, suspicion, and a belief that peace and prosperity may be only temporary interludes. In short, Europeans have a long tradition in which war is an integral, if not central, part of political activity.

The luxury of long isolation from the internecine struggles of Europe has molded the American view that war is an aberration, an unfortunate diversion from the normal course of events. Rather than a political instrument, to Americans war represents the failure of political policy—the failure to deal successfully with a direct threat to the essential virtue of the American experiment. When forced to arms, Americans tend to view warfare as a great crusade to overcome a well-defined evil.

If war has historically touched Americans with less frequency and effect than others, it has also left Americans with a legacy of military success. That this legacy is partly mythology built on selective memory is almost beside the point. Thus ,Americans cherish the tradition that the United States is not only "slow to anger" and enters into war only with great reluctance but also wins when the crusade is mounted. Despite such contrary evidence as the American military performance in the War of 1812 and some isolated unseemly or embarrassing episodes, applied force and victory have been inextricably linked in the American recollection. For Americans who believe in their country's infallibility, the tarnish from recent experience (whether Vietnam or lesser debacles such as Beirut or Desert One) is all the more disturbing.

Because fortune has shielded Americans from some of the nastier realities of military force, most Americans have been able to avoid coming to grips with the central role military force has had and continues to play in an international setting where the recourse to force remains the "court of last resort" for achieving national ends. More specifically, American innocence has allowed us to avoid confronting war as a political act and learning the often harsh relationship between politics and military force. In a relatively uncomplicated world in which the United States was not a central player, that innocence was affordable. In a world of increasing mutual dependency in which the United States is center stage, that innocence is a too expensive luxury.

That it is now time to decide when, where, and why the United States should be willing or unwilling to use force in the future is obvious and compelling. The debate has been joined but remains jaundiced by proximate concerns and exacerbated by historical myopia. We believe significant insights into why and how the United States could and should use force in the future lie in understanding why and how America has done so in the past. At a minimum, an excursion into American military history may dilute the tendencies to separate political and military affairs and to treat all events as unique and discrete.

In some important ways, this new edition represents an extrapolation beyond the historical record by including an examination of American efforts in Afghanistan and Iraq. These two wars, which are the subjects of the two new chapters in this third edition, are ongoing conflicts the outcomes of which are unknown at the time of this writing. There is little objective, dispassionate analysis of either, and what is available tends to be clouded by lack of perspective and distorted by disagreements about their worthiness and eventual outcomes. The success or failure of any military action is ultimately decided by the postwar conditions that take hold after the fighting is over (the better state of the peace), and the contours of the outcome usually require time to take shape. Afghanistan and Iraq are still war zones with more-or-less continuing levels of violent contention, with differing and contentious prospects for resolutions, and uncertainties over outcomes. As a result, these conflicts must be treated more tentatively than when analyzing completed wars. Nonetheless, the authors believe that there is intellectual merit in applying the framework developed for this book to aid understanding of these conflicts.

Given the complexity and breadth of the subject, the organization of this volume takes on considerable importance. Following an introductory chapter to provide the framework for our analysis, we have devoted an entire chapter to each of America's major wars and one to three minor wars. Each case study is organized in sections titled (in order): Issues and Events, Political Objective, Military Objectives and Strategy, Political Considerations, Military Technology and Technique, Military Conduct, and Better State of the Peace. We have attempted to write each chapter so that it can stand alone and, at the same time, flow together with other chapters to form an integrated whole. Within each chapter, each section is written to stand alone and yet contribute to a coherent analysis. These objectives are somewhat mutually exclusive and could only be accomplished by a limited degree of repetition in the text. The result is a survey history of the American experience in war and individual surveys of political objectives, military strategies, military technology, and the other subjects that are treated discretely in each chapter and successively from chapter to chapter. The final chapter draws conclusions and delineates important trends evident in the broad sweep of the American experience.

From Lexington to Baghdad and Beyond

1

WAR AND POLITICAL PURPOSE

"War is a continuation of political activity by other means," the great Prussian strategist Carl von Clausewitz wrote about 180 years ago. His famous dictum, so disarmingly simple and straightforward, is mimicked constantly in discussions about the role of military force in accomplishing the goals of groups and states. Despite its obvious truth and power, it is a statement shallowly comprehended and constantly forgotten.

What Clausewitz meant, and what is at the very heart of understanding why countries go to war, is that military force is a tool, one among many, by which states (or groups of states or groups within states) seek to accomplish their ends. Those ends are defined politically in terms of imposing the policies of one group on another. Force is certainly not the only means by which countries seek to accomplish their political ends, but because it inevitably involves the taking of human life, it is the most extreme of the so-called instruments of national power. Other instruments of power are conventionally described as the economic and diplomatic instruments: the use of various forms of economic reward or deprivation and of persuasion to achieve ends. What should never be forgotten is that the instruments of power are ultimately judged and gain their entire meaning by the extent to which they serve national policies.

Despite its bestial and grotesque nature, war continues to be a tool of national policy. Americans must understand war and its purposes as clearly as possible to choose most intelligently when to use and when not to use the military instrument of power. That is the purpose of this volume.

The bulk of the concern is why Americans go to war. One must begin by looking at why people generally have gone to war as context for looking at why Americans have found and will find the use of military force necessary. To begin to unravel that relationship, one must begin with two general questions. The first deals with the environment in which we find ourselves: How and why does the international system permit circumstances in which opposing states determine that only the use of armed violence will allow them to settle their differences? Once that question has been answered, the second question can be addressed: What is the role of military force in solving political differences?

The key concept in understanding how and why the structure of the international system permits and even sometimes encourages the use of armed force is *sovereignty.* Sovereignty means supreme and independent political authority, and it is a quality possessed not by the system itself but by its constituent members, the states. What this means in practice is that within the territorial boundaries of a country, the authority of the national government is supreme and knows no superior source of authority. In the relations among states, the implication is that there is no higher source of authority to regulate those relations and to resolve policy differences when they arise.

This situation is utterly unlike the relations among individuals and groups within states (at least where national political authority is effective), because in that instance there is an arbiter, the state. All states have rules established to regulate internal conflicts of interest and, in the ultimate, the mechanisms of state (e.g., the judicial and legislative systems) provide forums for the authoritative settlement of policy disagreements short of the use of violence (which is uniformly proscribed in word if not in deed). A sovereign exists as the ultimate settler of differences.

There is no equivalent in the relations among states because the members of the international system are themselves the sovereigns. There is no authority superior to the state that can be called on to resolve the differences between the states. When states come into disagreement over policy, they cannot take the matter to court to gain a resolution, simply because there is no court with that kind of authority.

Why is this the case? The answer flows from the notion of sovereignty and finds expression in the idea of "vital national interests." Vital national interests are those interests about which the state is unwilling to compromise, will not submit to arbitration, and hence will seek to protect by all available means. The most basic of those interests are the territorial integrity of the state itself and the maintenance of sovereign control over that territory.

Should, for instance, Mexico decide to reassert its claims to the American Southwest, the United States would be unwilling to take the matter to the World Court (which gains authority over cases only when the states who are party to a dispute specifically give it jurisdiction for that particular matter). Why? The answer is simple in a world of sovereign states. The Southwest is a vital interest of the United States, and we would clearly be unwilling to relinquish sovereign control over it. If we went to court, we might lose. Since we would not honor the verdict, the simplest way to handle the situation is to avoid having a mechanism capable of rendering unfavorable authoritative judgments. And that is the way the system is.

In such a system, given that disagreements over policy will inevitably arise, how are policy differences resolved? The answer, once again, is straightforward: states can favorably resolve policy differences to the extent that they can impose their will on others. The principle is known as *self-help*, and it means that international politics are fundamentally an exercise in power. Power, in turn, can be defined as the ability to get people to do something they would not otherwise do, in this case to accept policies in opposition to those preferred.

Take the hypothetical case of Mexican irredentist claims on the Southwest as an example. Should such claims exist, there would be a clear policy disagreement between the United States and Mexico, with American policy based in continued sovereign control of the Southwest and Mexican policy demanding its return. The policy disagreement is total: only the United States or Mexico can exercise sovereign control over the territory. Since the current situation reflects American policy, the problem for Mexico is how to get the United States to change its policy. In the absence of authoritative mechanisms to resolve the dispute, the problem for Mexico thus becomes one of self-help, the effective exercise of power to achieve its political ends. This brings us back to the question of the instruments of national power and the ability to apply them effectively.

As stated earlier the instruments of power are conventionally divided into the three categories of diplomatic, economic, and military power. Diplomatically, the Mexican government might seek to engage in negotiations, using its most persuasive diplomats and framing its argument in historical or demographic terms, to convince us voluntarily to cede the territory because of a superior Mexican claim. Failing in that, the Mexicans might threaten or carry out economic sanctions or promise rewards if we would agree to the cession of the Southwest. They could, for instance, threaten to deny American access to Mexican oil reserves or, using a more positive approach, they could offer unlimited access to those reserves in return for the territory. The degree to which such a strategy might be effective depends on American dependence on Mexican oil. If we were highly dependent, the Mexican government might have an effective lever that would compel us to accept its policy. If not, the economic instrument of power would be ineffectual.

Should all else fail there is always the military instrument of power. Should Mexican claims be serious enough (considered a vital national interest) and should other instruments fail to achieve the purpose, then Mexicans might consider the use of military force to seize and control the Southwest. That may not be the way one likes to think of things, but it is sometimes the way things are.

Force is thus a tool of political authority, and its purpose is either to guarantee that the inimical policies of others are not imposed on the political unit or to impose one's own policies on a recalcitrant adversary. Seen this way, military power gains its meaning as an agent for realizing the political purposes and objectives of the state (or whatever designation the political unit has). Unless this subjugation of military force to the political authority from which it flows is fully comprehended, the role of force cannot be adequately understood. Unless policy is made for military force starting from this ordering, the result is likely to be inappropriate policy and unnecessary friction between political authority and the military. Their roles may be distinct, but the military is an agent that implements the decisions of political authority.

Interestingly, it was Clausewitz who best understood this relationship. One level of this understanding is the Prussian dictum with which this chapter began: "War is a continuation of political activity by other means." The dictum is not an advocacy for using force to resolve political differences. Clausewitz, as a military man, understood that the decision to use force resides with political authorities; his role was to implement those decisions should that determination be made. The dictum merely states the relationship between war and politics. When the policies of two or more states become so incompatible that they cannot be pursued simultaneously, some means to resolve those differences must be found. Military force is one means to resolve those differences—it is another means to continue the political process of conflict resolution.

The relationship can be seen in another light captured by Clausewitz in an equally true but less cited observation that war has its own grammar but not its own logic. What he meant was that once the decision to go to war has been reached, the nature of conducting warfare—the so-called military art and science—dictates how war should be fought on the battlefield (the grammar, or as most people would say today, the language of war). The reason for going to war and the political objectives for which war is fought do not flow from that language, but derive from the overall political objectives (the logic of war) to which they are subordinate. In the heat of campaign that subordination is often blurred by the passion of the moment, but the Prussian was quite explicit that one should never lose sight of the relationship.

The language of war, quite clearly, is written in blood, and it is man's most extreme means of resolving differences. Because its consequences include the expenditure of human life and the destruction of the things people value, it is a political remedy *in extremis.* The use of force is the means chosen when the objective is vital and where other, nonviolent instruments of power have been ineffective in resolving political differences.

The purpose of force, thus, is to exercise power. "War is," as Clausewitz notes, "an act of force to compel our adversary to do our will." Doing "our will" is, however, a more complex matter than the quotation may suggest. A good deal of the misunderstanding about the role of force arises from oversimplifying how political and military aspects of war contribute to achieving the imposition of will.

In more contemporary terms observers often refer to the objectives of overcoming hostile will and ability. Hostile will contains at least two distinct parts. On one hand, hostile will consists of the willingness to continue to resist the imposition of hostile policies. What levels of cost, in terms of deprivation and suffering, are a people willing to endure, and at what point is the price of accepting the adversary's policies less than the cost of continuing to resist? Hostile will as willingness to continue to resist is well captured in the term *cost-tolerance:* what levels of cost are you willing to accept? On the other hand, hostile will also, and ultimately, is defined in terms of the willing acceptance or embrace of the originally objectionable policies. How does one go about convincing an adversary that the policies one seeks to impose are right and to the benefit of those who opposed them? The notion of hostile ability is more straightforward, referring to the physical ability of an adversary's armed forces and society to continue resistance.

Many practitioners and theorists have underemphasized the distinction between the two forms of hostile will and have consequently distorted the degree to which political authority and the military instrument contribute to achieving the ends of overcoming hostile will and ability. The assumption, implicit or explicit, has been that once the decision to use force has been made, it is the appropriate task of the military instrument to overcome hostile will and ability. We contend that it is more complicated than that.

Because hostile ability is represented by an adversary's armed forces, the military instrument is most clearly useful in removing that source of opposition. The classic method of defeating an enemy is to destroy his army, which is to say his hostile ability, although this is a realistic objective only part of the time. If one's armed forces are inferior to those of the enemy, then destroying those forces is usually an impossible way to achieve one's goals. In that case one may be forced to attack hostile will (cost-tolerance) by forcing the enemy to endure more suffering than his goals are worth. Sometimes one can pursue both objectives simultaneously; that is, break the enemy's will while destroying his army.

Overcoming hostile ability is clearly a military imperative and hostile willingness to persist an ambiguous military or political goal. Overcoming hostile will (defined as acceptance of originally odious policies) is a political problem solvable only in the peace that follows hostilities. Obviously, the military aspect plays a part and there is a sequential relationship: until either hostile will or ability is overcome, one can neither impose nor convince the adversary to accept one's policies. At the same time, military victory does not ensure the later psychological acceptance of the outcome by the vanquished. Military victory may allow one to "compel our adversary to do our will," but in the long run, it is acceptance at the psychological level that renders the outcome totally successful.

This is a subtle but very important and often overlooked point. Victory or defeat in war has two distinct definitions. The most obvious is military victory because that aspect is easiest to view. The other, and ultimately more important, definition is the achievement of the political purposes for which war is fought in the first place, and that means acceptance by the adversary of the political objectives for which the war was fought. In turn adversaries must be convinced that the objectives for which they fought were wrong and that those for which you fought were correct. Military force may be able to enforce the terms of peace, but convincing the enemy population to embrace the peace is a political task of persuasion for which military force may be irrelevant or counterproductive. It is indeed possible to "win the war and lose the peace" if one assumes that once hostilities are concluded victory is complete. The lesson of World War I, when a punitive peace virtually ensured that the German people would not embrace the peace treaty, is only the most obvious example.

The purpose of this discussion is to establish the intimate, complex relationships between war and its political purposes. Americans tend to think of war primarily in its military aspects, but that is clearly not enough if we are to comprehend fully the dynamics of military conflict and where military force can and cannot be applied intelligently and effectively, especially in the contemporary context. Rather, the complex interaction between military and political affairs needs to be viewed systematically, and it is our purpose in the rest of this chapter to lay out a framework for organizing that relationship, which we will then apply to the American military experience in subsequent chapters.

The first element in that framework is what is often referred to as the causes of wars: those underlying issues that make the recourse to war an apparent solution and the proximate events that lead to the decision to go to war. The political objective that directs the war effort and gives it meaning emerges from these issues and events. Political objectives, in turn, lead to the determination of military objectives to achieve the political objective and military strategies that will accomplish the military task. We will then turn to the purely political considerations that affect the conduct and outcome of hostilities and in that context examine selectively the actual conduct of each conflict. Because technology has been such an enormous influence on the evolution of war, we will look at technological innovations—how they were or were not applied effectively, and how they affected the conduct and outcomes of wars. Finally, we will examine whether or how the political purposes were achieved, using Sir Basil Liddell Hart's "better state of the peace" and the notions of overcoming hostile will and ability as yardsticks.

Issues and Events

The decision to go to war is seldom a casual matter. The road to war is generally a long one, and with the considerable assistance of hindsight, one can normally detect a gradual deterioration in the relations between what became warring units over underlying issues or sets of issues that were not resolved peacefully. Those underlying issues were transformed into events that serve as lightning rods that made the end result seem inevitable.

A caveat is in order here. One of the important concerns of historians and other social scientists is to speculate on the true "causes" of war and to devise elaborate theories about why there is war. This volume does not propose to add to that body of thought in the sense of proposing any grand scheme or overarching grand design to explain why men go to war. The concern here is more limited and descriptive. It begins from the more modest premise, supportable by evidence, that Americans from time to time make the political decision that armed violence is the way they must settle disputes. From that premise, it is the authors' purpose to look at those instances and to see if there was commonality and to see how the decision chain led to and directed the political and military objectives. The decisions to make war will in all likelihood be made again. It is our hope that those determinations will be made wisely and will be translated into appropriate, supportable, and achievable political and military aims and objectives.

With that context, one can divide the road to war into two analytical distinctions. The first deals with the underlying issues (or causes) in the preceding peace that eventually led to war. What kinds of incompatibilities in policy fester to the point that differences appear solvable only by the sword? How did these come about, and were they resolvable by other forms of action? Were the issues fundamental, or did they simply devolve because of inattention or the inability of men to resolve them? How did these issues evolve into the political objectives for which the war would be fought (which is really the most important question of all)?

The second distinction arising from those underlying issues is the specific events (or proximate causes) that normally emerge to hasten the process toward war. Clearly, one is not always in control of these events because they can be precipitated by either antagonist. It is, however, those proximate events that either galvanize popular opinion behind the decision to go to war or fail to create that support. The important factor is the dynamic relationship between the underlying issues and the proximate events.

The distinction may best be demonstrated by example, contrasting two conflicts from the American experience—the Civil War and Vietnam. As will be argued in Chapter 3, the underlying issue from which the Civil War arose was a fundamental clash of cultures. The North had gradually evolved from an agrarian to an industrial society while the South remained an agricultural society based around the plantation and the cultivation of cotton. The United States had become not a society but two distinct societies. The issue was fundamental and pervasive; two distinct socioeconomic systems could not coexist within one political framework indefinitely, and the differences gradually consumed more and more of the social and political fabric of the country.

The specific events that led to war flowed from this underlying incompatibility. Whether it was the debate over protectionist tariffs, the extension of slavery to the territories, or the election of Abraham Lincoln in 1860, all these events can be seen as reflections of the underlying issues. Within that context, the country could not survive as it was evolving. There were only two solutions available: either disunion that would allow each society to be represented by its own political system or union wherein one society triumphed and imposed its will.

Contrast the clarity and profundity of the issues and events leading to the Civil War with the parallels that led to American involvement in Southeast Asia. If there was a clear underlying issue (a point that remains contentious), it only indirectly involved the United States and North Vietnam. Rather, there were asymmetric issues: the American commitment to the policy of containment and the North Vietnamese desire to unite the country, by force if necessary. Ho Chi Minh cared little about the American policy, and the United States had little direct stake in the

North Vietnamese objective. Rather, the Southeast Asian peninsula simply became a forum wherein quite different concerns would clash. Moreover, the underlying issues were of a different nature: the U.S. policy was abstract (containing Communist expansion), whereas the adversary's goal was concrete (unification of a divided nation).

If the underlying issue was vague and less than pervasive from an American standpoint, the translation into specific events was also less than crystalline. At the beginning the events that would lead to American combat involvement had relatively little to do with Vietnam per se, but were instead part of a generalized response precipitated by North Korea's invasion of South Korea and a concern that the French economy was not recovering adequately due to the drain of the war against the Vietminh. Moreover, the road to war had an incremental flavor: relatively discrete decisions by a series of American presidents eventuated in a war they all hoped to avoid.

Political Objective

The basic reason for war, which should provide the definitive guide for its conduct, is to attain its political objective. The definition of the political objective is normally framed in terms of the peace that ensues after war is complete, and is well captured in Basil H. Liddell Hart's concept that the object of war is to produce a better state of the peace. (The *better state* being defined in the victor's terms.) Broadly speaking, those political objectives can be either total or limited, depending on the extent of policy incompatibility between the antagonists. Although the instigator of war may have a clearer vision of the better state of the peace at the outset, both (or all) parties in a war ultimately justify their efforts in terms of what is and is not a satisfactory ensuing condition of peace.

The political objective serves two basic functions, at least in a democratic society such as the United States. The first function, to which allusion has already been made, is to provide a framework for directing the war effort. The political objective provides guidance for the proper conduct of hostilities, which should be aimed at attaining the political objective. The second function is to provide a rallying cry for public support of the war. Because modern war involves, to some extent, societal commitment and sacrifice, war can be conducted by democratic societies only if it has explicit and continuing support. The failure adequately to galvanize support through the political objective almost inevitably leads to a lagging willingness to continue the effort (or the exceeding of cost-tolerance).

This critical role of the political objective suggests some criteria for how the objective must be framed if it is to have broad support. At the risk of some oversimplification, the American experience suggests a "good" political objective should have all or most of four characteristics. The more of these that are met, the more strongly supported the war is likely to be. Conversely, the more they are violated, the more unpopular (and hence unsustainable) the effort is likely to be.

The first characteristic of a good political objective is that it is simple, straight-forward, and unambiguous. Given that war is an inherently complex business, the public needs a readily understood reason for supporting it. At best the objective should be reducible to a catchphrase that is widely acceptable. "Independence" (the political objective of the American Revolution) or "destroying the Hitler monster" met that criterion.

Second, the objective should be morally and politically lofty. This need is particularly important to Americans, who have always considered themselves a special, even morally superior, people. In the American experience support has always been most unwavering when the purpose resembled a crusade. "Making the world safe for democracy" had the kind of loftiness that gained broad American support; restoring the status quo in Korea did not.

The third and fourth criteria overlap somewhat. The third is that attaining the objective must be seen as vital to the interests of the United States. This, of course, is a difficult criterion to get a precise grasp upon and is difficult partially because of the subjective nature of what is vital to the United States. For instance, was the ending of impressment (War of 1812) vital enough to go to war over? The War Hawks thought so; others disagreed. The same question can be raised about Iraqi supposed possession of weapons of mass destruction (WMD) and their alleged connections to terrorists in 2003. Moreover, the expeditionary nature of American military adventures invariably creates a debate about whether vital interests are involved. Precisely which overseas threats are threats to core American interests will always be a point of debate. Reexamining what constitutes a vital interest will have a major impact on defining when to use force in the post–cold war world, since obviously vital American interests rarely seemed at risk in places like Haiti and Kosovo during the 1990s, where American arms were summoned.

The fourth criterion is that the interests of most Americans must appear to be served by the decision to go to war. This criterion was most problematical in the wars of the nineteenth century when sectionalism was an important concern. Support for the War of 1812, for instance, was far greater in some parts of the country than it was in others, and the Mexican War had very little appeal in either New England or the South.

The American experience in Iraq suggests a fifth criterion: the truth value of the threat to American interests that are posed by an adversary. The stated political objectives of the United States in Iraq included the unacceptability of Iraq's possession of WMD. After the invasion, no Iraqi WMD were ever discovered, leading many to conclude their elimination was a false objective. The realization of that falsity has contributed to decreased support for the war.

Although wars of limited political purpose have been by far the more frequent through history, Americans have tended to show the greatest support for unlimited wars. This is the case partly because total war most obviously meets the criteria for a "good" political objective: total defeat of the foe is a simple and unambiguous goal; total defeat must be necessary because an inherent evil requires eradication (loftiness); the vital interests of the United States must be threatened or a total effort would not be necessary; and most Americans can agree that the outcome is necessary. The destruction of fascism and its symbols in World War II is the most obvious case.

Limited political objectives, on the other hand, are more likely to violate one or more of the criteria. The objectives may not be simple and understandable, as was the objective of containment in Vietnam. There may be moral ambiguity in the cause (seizing the Southwest United States in the Mexican War can be viewed as either manifest destiny of the American people or as naked imperialism). In a limited action vital interests of the United States may or may not be involved. (If they are, why would we not go all out to win?) Limited ends can divide the American people (regionally as it did in the nineteenth century, or as the limited political objectives in Korea or Vietnam did in the second half of the twentieth century). Virtually all the political involvements of the United States in post–cold war conflicts will confront this problem to some extent.

A final point about the limited–total distinction should be made. Clearly, both sides in war have their own political objectives, which are in opposition (both cannot be achieved simultaneously). That does not mean, however, that the objectives are necessarily symmetrical in limited–total terms. It can and does happen that one side may have limited political objectives while the other has total objectives. In the American case two examples come to mind. In the American Revolution, the purpose of independence was total and indivisible: you cannot be partly independent. The British objective of restoring control was, as it evolved, not total. Ultimately, the restoration of British authority was justified as exemplary to other parts of the empire (if the Americans won, other parts of the empire might get seditious ideas). The even clearer case is Vietnam. The American objective of containing Communist expansion and thus allowing the South Vietnamese to engage in self-determination was clearly a limited one. We did not seek the overthrow of the North Vietnamese government (although we would not have objected to that outcome). North Vietnam had the total objective of overthrowing the government of the Republic of Vietnam and of uniting the country by force.

Military Objectives and Strategy

If war is politics carried on by other means, then the fundamental objective of all military operations in wartime is quite simple and straightforward. The military's basic task is to overcome the enemy's ability to resist our policies militarily. Although straightforward, this fundamental objective is so broad that it provides little practical meaning or useful guidance for military planners. It is, however, instructive to keep this fundamental objective in mind for two different reasons.

First, the fundamental military objective excludes certain specific objectives as legitimate pursuits for armed forces. For example, the battle for the "hearts and minds" of an enemy population would be an inappropriate undertaking for military forces. Such a battle is better reserved for civilian authorities who can make political, economic, and other nonmilitary policy decisions that will have a direct impact on the perceptions and attitudes of hostile populations. This is not to deny that successful military operations are often necessary prerequisites to winning "hearts and minds." Such was certainly the case in World War II. Complete military victory allowed the imposition of nonmilitary policies that has resulted in over 65 years of friendship and support from former enemies. Nor does this deny that the military can often be the executive agent to implement these nonmilitary policies. Perhaps the most memorable instance of the military functioning as the executive agent is found in Japan following World War II. General of the Army Douglas MacArthur was the de facto dictator of occupied Japan responsible for imposing the enlightened nonmilitary policy decisions that resulted in a resurgent Japan friendly to American policy objectives.

Second, the enemy's ability to resist militarily is directly affected by the enemy's will to resist. Thus victory on the battlefield does not necessarily translate into victory in the war. Defeat on the battlefield does not necessarily mean that the cause is lost. America's experience in the Vietnam conflict provides conclusive evidence that one belligerent can win virtually all of the significant military engagements and yet lose the war. Conversely, America's enemies in the Vietnam conflict demonstrated that there are occasions when simply avoiding catastrophic defeat while exacting a high price in blood from the enemy can make a decisive contribution to the destruction of the enemy's will (and thus his ability) to resist militarily.

It cannot be overemphasized that the objective of war is not military victory. Rather, the objective of war is to attain the political objectives which spawned the war. Military victory is merely one means to political ends. In some wars, it may be a necessary means to the end, but it is never an end in itself.

One might assume that specific military objectives would flow naturally from political objectives, but such is not always the case. Although what follows is not an encyclopedic list, four situations have commonly caused conflict between political and military objectives.

First, one can confuse means and ends, particularly when deeply mired in bloody conflict. Such was the case for the belligerents in World War I, particularly those struggling in the trenches on the Western Front. As the casualties mounted along with the frustrations of a stalemated war, the declared objectives on all sides disappeared into the mud of Verdun and Flanders and were replaced with simple hatred and the desire for retribution. In many respects, the object of war became the war itself rather than the peace that followed. The result was unsatisfactory to all sides and formed the breeding ground for an even greater conflagration.

Second, political objectives can clash with military expediency. Sherman's famous march from Atlanta to the sea was a military expedient that surely shortened the American Civil War. The wanton destruction caused by his rampaging troops gutted the heart of the Confederacy and led to serious morale problems among Confederate troops. On the other hand, it led to long-lasting and deep-seated bitterness among the vanquished and postponed true reunion between the North and the South. This drives home the point that the manner in which a war is fought can have a significant effect on the peace that follows. Moreover, political objectives can frustrate prudent military operations. In the Vietnam conflict military operations were banned in certain areas for political reasons. These sanctuaries, however, ensured that the military security required to win the "hearts and minds" of the civilian population could never be achieved. In a sense, political objectives were hoisted by their own petard.

Third, political objectives can be so abstract that the military is left with little on which to base its objectives. For Americans, the political objectives for most wars have been concrete (although they may also have been simplistic and shortsighted) and military objectives followed easily. Such was the case in the Civil War and both world wars. Each was a great crusade against a clearly defined "evil." In Korea the political objectives were more obscure but were still definable because of an overt invasion. In Vietnam political objectives remained abstract and the resulting military objectives were nebulous and muddled. As a result, military strategy was often confused and inappropriate, interservice rivalry flourished as parochial interests came to the fore, and morale crumbled. In situations such as Somalia, it is difficult to determine what the objective was beyond interrupting the starvation. In contemporary situations like Bosnia and Kosovo, the military objective may be no more than suppressing the violence. The same may be true for Afghanistan and Iraq, at least suggesting the possibility that vague political objectives may simply be a characteristic of modern war.

Finally, the military may be given the task of accomplishing political objectives that are inappropriate for military means. The Vietnam experience may be the classic case. Given the unrest in third world areas and the importance of those areas to the industrialized nations, this situation may be more common in the future. As will be described in later chapters, the principal American problem in Vietnam was not military. Rather, the problem was one of state building, an objective that requires vigorous nonmilitary action. Military actions in such a situation could only provide the security needed for other actions to succeed. And yet the military was the principal power instrument used by both the South Vietnamese and the Americans, while nonmilitary actions were given far less attention. The result was predictable. The South Vietnamese state was never built and, ironically, the proximate cause of its downfall was the failure of the South Vietnamese military. Many contemporary problems are comparable. Some, for instance, have suggested a parallel dynamic in Afghanistan.

Although the most fundamental military task in war has remained relatively constant (at least in nonnuclear war), specific military strategy—the technique of developing, deploying, and employing military forces—has evolved. The two centuries of the American experience have witnessed rapid and fundamental changes in military strategy. The eighteenth century was the era of limited war, limited by the nature of the political objectives sought by the interrelated absolute monarchies that dominated all of Europe (the exception being Great Britain). They waged war for a province here, a city there, or control over royal succession in another kingdom, but rarely to overthrow a brother monarch. There was little passion involved, the objectives being the royal rather than popular interest. The armies that fought for these objectives were composed of the dregs of society and mercenary soldiers recruited from throughout Europe. In essence, the bulk of the population was isolated from both the objectives of war and those who waged it.

The linear tactics developed to use the limited-firepower technology of the day led to bloody but indecisive battles. Rather than risk their expensive and hard-to-replace armies in pitched battle, eighteenth-century generals sought to gain advantage through maneuver to cut the enemy's line of supply. The elaborate depot and magazine system required to support an army in the field presented a convenient vulnerability that could decide a campaign with minimum risk of pitched battle.

Near the end of the eighteenth century, the idealism of the American Revolution returned ideology to warfare. The common man had a political objective for which he would voluntarily fight and die. The trend continued during the French Revolution and Napoleonic Wars. Ideological objectives and the democratization of war eventually paved the way for mass popular armies, more flexible tactics and supply systems, and, finally, wars fought for unlimited objectives. By the time of the American Civil War, this drift toward total war was well under way.

In many ways, the American Civil War was America's first total war. For the Union, at least, the political objectives were unlimited. No compromise could be accepted; militarily, the Confederacy itself had to be crushed and forced back into the Union fold. Confederate political objectives were also unlimited (independence). However, its military objectives were limited as it sought only to repel Union invaders and maintain its independence. From these objectives flowed the strategy of each side. For the Union an annihilation strategy was in order: it needed to destroy the Confederate armies and overthrow the Confederacy's government to achieve reunion. Confederate strategy remained largely that of attrition, seeking not to destroy the Union Army and overthrow the Northern government, but to defend the Confederacy and inflict enough pain on the Union forces to discourage further attacks on the new Confederate state. The North had to destroy Confederate hostile ability; the South's objective was to exceed Northern cost-tolerance.

The Civil War was also a war of unlimited means. Mass armies took to the field. Maneuver was critically important but bloody battle was the decisive factor. Both sides attempted to mobilize their civilian populations and industrial bases for war. Finally, civilian populations and economies became military targets, at least for the Union forces. Thus the Union blockaded Southern ports in an attempt to shatter the Confederate economy and starve the Rebel population. The rationale for Gen. William T. Sherman's march through Georgia and Gen. Franz Sigel's less well-known but equally devastating attacks in the Shenandoah Valley were directed to the same end.

The Civil War also demonstrated for the first time the importance of mechanization in warfare. It was the first American war in which railroads played a major role. Mass armies were transported over vast distances with great speed and were kept well supplied. Rail transportation expanded the scope of the war, which was conducted simultaneously in widely separated theaters of operation. But with the opportunities of mechanization came limiting factors. Much of the military strategy revolved around rail lines themselves. Cutting the enemy's rail lines or protecting one's own became a major preoccupation of Civil War generals.

As the trend toward mechanization continued into the twentieth century, changes in military strategy continued apace. The internal combustion engine added greatly to military flexibility and the speed of maneuver. Combined with other technological gadgets, the internal combustion engine gave rise to armored war on land, undersea warfare, and war in the air. Each of these developments provided new opportunities for the military strategist, particularly when combined with the unlimited objectives of modern total war. Rapid and fluid maneuver, deep penetration, and increased firepower characterized modern warfare, culminating in the campaigns of World War II.

But increased mechanization presented vulnerabilities as well as opportunities. Supply lines became even more important (one cannot forage for spare parts and fuel) and thus were a prime target for attack. The same was true for the industrial base that supported mechanized forces in the field. Traditional naval blockade (and blockade by submarine) remained an exceptionally important tactic. The airplane offered the opportunity to attack the civilian industrial base directly and with more immediate effect than blockade. Thus direct combat operations ranged from the front lines to the skies over civilian industrial centers. Total war was all-encompassing, sweeping up civilian and soldier alike.

The nuclear weapons developed and used at the end of World War II significantly changed the way we think about war, particularly after other countries also developed nuclear weapons. Even the incredible cost of total war could, until then, be justified by unlimited objectives that seemed to be of greater value. But the advent of nuclear weapons raised the specter of total war involving total cost—the possible annihilation of civilization itself. Surely no political objective could be worth the all too real risk of mankind's extinction. Thus the concept of deterrence came to the fore along with its arcane language and arabesque logic.

The Clausewitzian dictum, it seemed, no longer applied, or at least it did not apply to nuclear war. Nuclear war, it appeared, could not be an extension of politics because the possible death of civilization served no rational political purpose. The fundamental objective of nuclear deterrence was and is to ensure that nuclear war never will serve any rational political purpose, that no one would ever risk the possible catastrophic consequences. And thus the paradoxical situation came about where incredibly powerful weapons are developed and fielded, their sole purpose being to ensure that neither they nor similar weapons possessed by any enemy will ever be used.

The nuclear stalemate has not been matched at lower and less threatening levels of warfare. Although both superpowers carefully avoided any direct military confrontation, even at lower levels of conflict, both fought lesser foes at these lower levels. Additionally, many countries in the third world have taken up arms against one another, often supported by one or the other of the superpowers. What has happened, in effect, is that we have returned to eighteenth-century limited war—but with two significant differences. First, the political objectives of these conflicts became ideologically based. They tend to arouse impassioned support and a considerable degree of fanaticism.

Second, and as mentioned earlier, at least when the United States has been involved in these conflicts, the political objectives of the belligerents have been asymmetrical. While America has waged limited war in these instances (limited objectives, limited means), its smaller opponents have waged unlimited war. Their objectives were unlimited in the sense they dealt with perceived vital interests that could not be compromised, and they used all means at their disposal to conduct the war. Such a situation places the United States at a distinct disadvantage.

American strategy has been only partially successful in this new situation. The Korean War was little more than a World War II–style conflict on a limited scale. American objectives at the time we intervened were clear and understandable. American attitudes were helped by the fact that this war involved an outright invasion without provocation. Although Americans (both military and civilian) chafed at fighting a war with less than the total means available, the war was brought to a conclusion that achieved our originally stated national objectives. The same cannot be said for the American venture into Southeast Asia. Protracted insurgent warfare using guerrilla-style tactics continues to frustrate American political and military strategists, and since guerrillas assume military inferiority, it would seem impossible to deter such conflicts.

Total war was more the product of unlimited objectives than the product of modern weaponry. In fact, it has been the terrible impact of modern weapons of mass destruction that forced a return to limited war by the superpowers. The limited wars of the post–World War II era, despite the availability of modern weapons of mass destruction, prove the point. American political objectives have been limited during this period, and the means by which America has fought have been constrained. Clearly, the primary influence on military objectives and strategy is the political objective, which is as it should be. It is also true that military technology and technique have also strongly influenced military strategy. In truth, military strategy is the result of the interplay of numerous factors. All of these interrelated factors play a significant role in determining how a war is fought and in doing so affect the peace that follows.

Political Considerations

Just as political concerns lead to war and the political objective defines its scope and purpose, there is a dynamic relationship between military and political considerations before and during the conduct of hostilities. Both domestic and international political concerns affect the way war is fought and, conversely, the ebb and flow of warfare influence political forces at each level.

Especially in a democratic society, when the use of the military instrument is contemplated, a major consideration must be the likely level of public support for the enterprise. In some cases, of course, the decision may be thrust upon one, as with Pearl Harbor. For a country whose primary military engagements are expeditionary, however, there will generally be some meaningful opportunity to consider the question. Public reaction to Vietnam paralyzed the use of force for a time, but a string of at least apparent successes in the last few years has reenergized support for the military instrument of power. Current concerns over Afghanistan and Iraq have reopened some of the questions originally raised over Vietnam.

The question of public support has become increasingly important. While not attempting to exhaust all confounding influences on public support for American military adventures, at least four can be mentioned, each of which will arise as the odyssey through the American experience at war unfolds. They are presented in no particular order of importance.

The first, and perhaps least well-understood factor is the impact of the media, and especially the electronic media. At the most obvious level, the electronic revolution permits coverage of military operations at a speed and with an intimacy heretofore impossible. Certainly, media coverage of the battlefield is not particularly new. In the American experience, close and rapid coverage of war goes back to the Civil War and the introduction of the telegraph, which allowed next-day reportage of engagements. What is unique about the electronic media, however, is that they bring a vivid, visual quality to coverage.

One obvious effect has been to deglamorize war. The blood, maimed bodies, and corpses that constitute the tribute of combat cannot be hidden from the camera's eye. It is an image difficult to ignore. Moreover, the television camera is most effectively used in capturing the discrete and dramatic event. Television coverage does not focus effectively on the long lulls between combat that make up the vast majority of war; instead, it trains on the spectacular—the maimed child, the firefight, the exploding bombs. At the same time, television reportage of violence and atrocity may have the opposite effect of creating public support for using force to relieve misery as in Somalia. As mass atrocity became commonplace during the 1990s in places like Bosnia and Rwanda, public tolerance of carnage may have increased as audiences become desensitized to the sights of war.

A second factor is the burgeoning expense of war. This expense is largely due to the increased sophistication of weaponry that makes equipment both more costly and more deadly, a phenomenon discussed in the section on technology. The result is that even fairly minor military involvements place great burdens on resource bases on which there are multiple competing claims. The Vietnam War, while certainly not a minor engagement, is estimated to have cost the United States $150 billion. The enormous costs of the Iraq War, all of which have yet to be tallied but which will have had a strong impact on the country's economy, are a contemporary example. The question is just what economic sacrifices will people be willing to make in the future?

A third problem is that likely future scenarios involve conflicts for limited political objectives in third world countries where American interests are minimal. As discussed earlier and as will be amplified in the pages that follow, developing and sustaining support for these kinds of wars is often difficult. The ramifications of that observation are explored in the final chapter.

Fourth and finally, there is the historical American aversion to things military—and particularly to the costs of military forces—that goes back to the birth of the country. Part of the American Anglo-Saxon heritage is to suspect military force and to look toward other instruments of power to solve problems. When the dollar ruled supreme in the world during roughly the quarter century after the end of World War II, the economic instrument could be used effectively to achieve American ends. That relative advantage has eroded, however, and Americans must contemplate the use of alternate instruments, one of which is military power.

The interplay of politics and military force does not end once the decision to go to war has been reached. Rather, in some ways that relationship intensifies at essentially two levels, what one may call the high politics and low politics of war.

The high politics of war refers to the direction of military efforts to achieve political objectives and to ensure that the application of force does not alter the objectives beyond politically acceptable bounds. This latter problem is particularly acute in wars of limited political objective because purely military imperatives and political objectives most often come into conflict in these situations.

Clausewitz, once again, recognized the problem. The tendency in war, he observed, is to intensify and broaden the scope of action, a dynamic against which he warned political authorities to be constantly vigilant. The dynamic can occur whether one is winning or losing. If one is successful in achieving the objective through military means, there is a powerful temptation to broaden the objective and attempt to achieve even more. The decision to cross the 38th parallel in Korea in late 1950 exemplifies the point. At the same time, losing may cause one to intensify the effort, to up the ante, to avoid defeat. The gradual buildup in Vietnam and Gen. William C. Westmoreland's ceaseless pleas for more troops represent this case.

The low politics of war refers to the direction and exploitation of military situations for personal, political, or other gain. In its seamier aspect, it may be the use of military activity to serve the ends of political figures. The image of President James Polk diligently watching the Washington press to see if either of his military commanders (both of whom were potential political opponents in the 1848 presidential election) was gaining too many headlines and then directing their campaigns to minimize such publicity comes to mind. Conversely, military victory or defeat can influence political outcomes. Lincoln, for instance, might well have lost the 1864 election had Atlanta not fallen to Sherman in the nick of time, just as stalemates in Korea and Vietnam helped drive two presidents into retirement.

The conduct of war affects and is affected by international concerns as well. Although there is really no such thing as international public opinion, the general opposition of America's NATO allies (largely reflecting public opinion in Europe) to our involvement in Vietnam weighed heavily on U.S. decision makers. Likewise, military events may influence foreign powers in ways that can alter the situation. For instance, the American victory at Saratoga in 1777 (our first significant battle-field success) convinced the French to intervene openly and probably allowed the success of the American Revolution.

It is this dynamic interaction between military and political affairs during wartime that is most nettlesome to the military, and forms the basis for charges of political interference or derogations about political wars. Once again, the problem is greatest in limited-war situations, where concerns about remaining within the limited political objective may call for limitations on military actions that impede the effective application of force. Military art and science teach the virtue of maximum force to achieve the destruction of enemy forces wherever they may be. Within the confines of limited objectives, it may not be possible to unleash the full fury of military capability without running the risk of broadening the war and, implicitly, its political objectives. The granting of sanctuaries in Korea and Vietnam are examples, and they are the kinds of constraints under which the military is likely to be forced to labor in any future conflict.

Military Technology and Technique

Military technology is only one of many factors that influence the conduct and the character of war. However, it is the primary factor that determines how battles within a war are fought. The history of battle is, to a great extent, the story of military men struggling to cope with technology. The outcome of battle is often determined by the ability of one antagonist or the other to make the best use of available military technology. Thus to the extent that it actually influences the outcome of battles, technology directly influences the course of war and, more indirectly, the peace that follows.

Since the ancients first took up arms, one clearly identifiable trend has been constant. The power and destructive efficiency of weapons have become ever greater, as reflected in their explosive power, accuracy, and range. In the twentieth century, we may have reached the ultimate extension of this trend with thermo-nuclear weapons riding atop intercontinental ballistic missiles. Today weapons of previously unimagined power are only minutes away from any spot on the face of the earth. In essence, it has been changes in power, accuracy, and range of weapons that have caused the nature and technique of war to change over the two-hundred-odd years of the American experience. These changes are reflected in the scale, intensity, tempo, organization, scope, and cost of battle.

The first change has been to increase the physical scale of conflict. The size of the individual battlefield began to expand in the eighteenth century as generals attempted to maximize smoothbore musket firepower by packing more men armed with these weapons onto the battlefield. However, instead of the densely packed formations of earlier ages, eighteenth-century warfare was characterized by linear formations. Formations spread laterally and were generally only three ranks deep. This allowed for fire by all three ranks while minimizing the possibility of muzzle-blast damage to one's own troops. Thus the frontage of individual units was vastly expanded as was the size of the battlefield.

In the nineteenth century, advances in transportation and communications technology expanded warfare far beyond the already swollen eighteenth century battlefield. The railroad allowed mass armies to be transported rapidly over great distances and permitted the efficient resupply of these far-flung armies. The importance of rail transport first became obvious during the American Civil War. In that conflict separate armies operated throughout different theaters of war and yet did so with considerable coordination. Railroads made it possible to operate effectively in distant theaters and the telegraph made coordination of these operations both possible and practical.

In the twentieth century, globe-girdling warfare has become the norm. All modes of land and water transportation increased in both speed and carrying capacity, making overseas force deployments relatively commonplace. By mid-century, air transport added a new dimension to deployment speed while the advent of radio provided instantaneous communications without reliance on fragile wires. In addition to the worldwide breadth of twentieth-century military operations, the development of the airplane and the submarine expanded the battlefield vertically—beneath the sea and into the sky.

In essence, technology expanded the battlefield to proportions unimagined in the eighteenth century. Advances in transportation and communication solved the twin problems of logistic support and coordinated command. Modern military commanders have been forced to expand their horizons far beyond the confines of the immediate battlefield. During the same time period, however, technology also brought about rapid increases in the intensity of warfare and the speed of maneuver on and between battlefields. The ongoing applications of the fruits of the telecommunications revolution to warfare promise an ever-accelerating rate of technological change on the battlefield.

In the eighteenth century, warfare was conducted at a rather leisurely pace. Military campaigning seasons—when the weather was good and gunpowder could be kept dry—were relatively short, and armies regularly went into winter quarters during the cold and wet months of the year. Even during the campaigning seasons, little fighting or maneuvering occurred during hours of darkness. By the twentieth century, keeping one's powder dry was no longer a significant problem because of cartridge ammunition. The incandescent lamp facilitated night operations, as did electronic marvels such as radar and light amplification equipment. Thus, by the last half of the twentieth century, war had become an intense activity waged around the clock throughout the year.

The tempo of war increased along with its year-round intensity. In the twentieth century, wheeled and tracked self-propelled vehicles sped the maneuver of men and guns on the battlefield, just as railroads, fast ships, and even faster aircraft speed the transport of men and weapons to the battlefield.

The organization and scope of combat has multiplied as well. At the same time that technology expanded the scale, intensity, and tempo of military operations, it also created significant problems. First, it complicated the structure of armed forces and placed a premium on quality staff work. The infrastructure required to command, control, and support a technologically sophisticated military force has expanded at a geometric rate and the logistic support required has assumed momentous proportions. Second, the coordination of ground, air, and sea forces in modern three-dimensional war requires large, complex, multiservice staff structures. These factors have combined to change radically the ratio of combat to noncombat troops in modern armed forces over the last 200 years.

However, it is more significant for present purposes to realize that technological sophistication has made the industrial base that supports armed forces in the battlefield vitally important. The production of modern military weapons, vehicles, aircraft, and ships requires a robust industrial base (or access to one). Naturally, this same industrial base has become an important target for military operations. Civilian populations that support a country's industrial base have also become targets for attack either directly (physical attack, psychological operations, and deprivation) or as a by-product of attacks on industrial plants (so-called collateral damage). As technology made the home front an important target, that same technology also made it possible to attack home-front targets without first defeating armies and navies.

Last, but not least, is the economic cost factor, which is both a military and political concern. The cost of war has dramatically escalated during the American experience in terms of both blood and treasure. Although the ratio of killed and wounded to the total in uniform during a war has decreased since the eighteenth century, the absolute number of casualties has vastly increased. This is a natural product of larger armed forces and far more lethal weapons. At the same time, the survival and complete recovery rate for those wounded in action has increased dramatically because of rapid advances in medical science and military efforts to bring medical care to soldiers on the front line.

The treasure expended in war has escalated to mind-boggling proportions. In the eighteenth century, the village smithy could produce most of the kinds of weapons required by an army and could do so at a relatively low cost. Modern armies require weapons of great complexity and incredible cost. The lethality of these weapons means that battlefield attrition and consequent replacement demands add enormously to the total cost of a modern high-intensity war.

In essence, technology expanded the battlefield horizontally, vertically, and finally in depth to include the home front. Modern total war increased in intensity, in tempo, and in scope to include everyone, not just the soldiers on the front lines.

Armed forces that had originally protected the home front can no longer do so. Factory workers have essentially become frontline soldiers without guns. Thus in many ways the unlimited involvement characteristic of modern total war has matched the unlimited objectives of ideologically-based modern war.

The most recent development in warfare flies in the face of the trend of developments since the eighteenth century. In the face of total war with totally destructive nuclear weapons, the second half of the twentieth century saw a resurgence of limited warfare reminiscent of the eighteenth century. Protracted insurgent and partisan (sometimes called asymmetrical) warfare using guerrilla tactics allow the weak to compete on the battlefield with the strong. Forces using these tactics pick the time and place of battle, refuse to stand and fight unless they desire to do so, and melt away into a friendly (or at least neutral) civilian population when they do not. These irregular, asymmetrical forces make no pretense of protecting civilian populations on which they depend for their survival. Rather, the guerrilla reverses the relationship and uses the civilian population as a shield to hide from the enemy. As a result, the terrible destructiveness of modern weaponry has only limited utility in combating the asymmetrical warrior. Where success in "conventional" warfare depends on the ability to kill people, success in combating guerrilla forces depends on the control of people. Only if the population is controlled and secure can guerrillas be ferreted out of the general population they use for protection.

Military Conduct

Much has already been said during the discussions of military objectives, strategy, technology, and technique about how wars have been and are fought. In the eighteenth century, war was, in effect, a battle of masses. Although the American Revolution was an exception, war in that era often held to Voltaire's observation that "God is always for the big battalions." By the time of the American Civil War, however, the situation began to change. More complex weapons and the use of mechanically powered transportation systems put a premium on industrial capability and capacity. By the time of the two world wars, this trend matured and war became a battle of factories. Industrial plants were so important that the factories themselves became military targets rivaling deployed forces in importance. Finally, the technological explosion during the last half of the twentieth century made warfare a battle of brains. It would seem that Voltaire has been turned on his ear as God now seems to favor the best technology. The exception to the modern trend, asymmetrical warfare using guerrilla tactics, makes much modern military technology irrelevant. Perhaps God is on the side of the smart battalions.

Although the evolution of warfare from a battle of masses to a battle of brains encapsulates the evolution of war fighting during the American experience, one important point remains to be made if one is to understand the nature of modern warfare and if one is to understand the difficulties of winning the peace.

Viewed broadly, warfare became a progressively more desperate undertaking during the nineteenth and twentieth centuries. War has always been a desperate affair for the individual warrior. However, the increasingly ideological basis of warfare and its consequently unlimited political objectives have turned much of modern warfare into a death struggle between rival societies. Failure in such a war is catastrophic, and as a result, the struggle is fought with bitterness and desperation. War is no longer the glorious adventure to which men march with bands playing and flags waving. The glorification of war was a tradition buried beneath the mud of Flanders fields if not earlier. Any sense of chivalry has all but disappeared from battle as the unlimited ends sought seem to justify unlimited means. Naval blockades indiscriminately starve civilian and soldier alike. Civilian population centers have been routinely attacked and weapons of mass destruction have been used. Defenders have often used scorched-earth tactics and one finds many instances of fierce fighting when all hope of victory has clearly been lost.

One might be led to believe that the "limited" and "low-intensity" wars characteristic of the nuclear age would have a less desperate nature. However, one must remember that such terms as *limited* and *low intensity* are given meaning by one's perspective. For those on the battlefield, no war is limited and all wars are intense. On a larger scale, all of America's opponents in these types of conflicts (North Korea, China, and North Vietnam) were essentially waging unlimited wars. Only their relative military weakness (when compared with a superpower) and the secondary importance of our objectives in these conflicts have allowed us to characterize these conflicts as limited or low intensity.

The modern desperation in war produces a bitter legacy. Few, if any, are untouched by the horror of modern warfare. All sides harbor bitter feelings because of widespread death and destruction. The losing side agonizes over how much it gave and how much it lost. The winner resents the suffering endured in relation to the objectives achieved, which often seem hollow in the harsh light of war's aftermath. The bitter legacy makes the task of the peacemaker far more difficult than in any other age. Winning a better state of peace after a modern war may be the most difficult of all tasks.

Better State of the Peace

"There is no substitute for victory," Gen. Douglas MacArthur said in testimony about American conduct of the Korean War. It is a beguiling statement and so straightforward as to appear unimpeachable. But what does it mean? What exactly constitutes victory in war?

The most obvious answer, and the one to which Americans (including General MacArthur) are drawn, is that winning the war means military victory, but that is an incomplete answer. Military victory is only a part of winning wars, and although one tends to associate military triumph with victory, such is not always the case. Victory has another, more profound meaning. The other, ultimately more important, sense of victory is the attainment of the political objectives for which war is fought in the first place.

These two aspects of victory should not be confused. Although military victory on the battlefield is usually prerequisite to "imposing our will" on the adversary, such is not always the case. It is possible, for instance, to fight to a militarily inconclusive ending but still accomplish the political objective of war, as the American colonies did in the Revolution. At the same time, one can actually lose the military campaign and achieve at least part of the political objective, as the United States did in the War of 1812. Finally, it is also possible to win all the military campaigns in war and yet lose politically, a distinction demonstrated by the American experience in Vietnam.

To comprehend the two senses of victory, one needs to return to the notions of hostile will and ability. As was argued earlier, the distinct task of the military is to overcome hostile ability, and military victory (or its absence) is normally determined on that basis. Overcoming hostile ability in the American experience has usually been equated (at least since the campaigns of Grant and Sherman in the Civil War) with the destruction of the enemy's armed forces and that is the most obvious, measurable, and observable outcome of war. The most vivid expression was the unconditional surrender of Axis armed forces at the end of World War II. But there is more to victory than that. One may also win wars militarily by overcoming hostile will, defined as exceeding cost-tolerance, which is the way the colonies defeated Great Britain and the North Vietnamese overcame the United States.

Victory in a political sense, however, is best equated with the notion of political will defined as acceptance by the vanquished of the political objectives of the victors. The political objective includes imposition and acceptance of policies and that forms the reason for going to war in the first place. Clearly at the outset, the visions of a better state of the peace are diametrically opposed, or there would be inadequate reasons to engage in war. In the end, one vision of the better state of the peace must prevail and the other must vanish (unless, of course, they are modified or compromised).

The question is how the better state of the peace is accomplished. The answer must begin with the realization that acceptance of policies is itself a political process, and one that must be accomplished by political authorities during the ensuing peace. It is a political, not a military, task.

Broadly speaking, there are two ways of implementing desired policies. The first way is through the simple imposition of the victor's vision of the better state on the vanquished in the form of a *punitive peace*. The terms of the peace can simply be imposed, and if the former enemy objects or tries to alter those terms, coercive means can be used to guarantee continued compliance. Punitive peace was the model for the settlement of World War I and, to a lesser degree, the American Civil War.

The other means is to engage in active efforts to convince former foes that one's policies are more enlightened than theirs were and that they are better off embracing this vision of the better state of the peace. This form is known as a *reconciliatory peace* and requires intense political and psychological efforts to overcome the residue of hostile feelings. The process of reconciliation is inherently political and can be accomplished only by political authorities through political processes (although the military may serve as the agents for political authorities, as in the postwar occupation of Japan). The clearest case of a reconciliatory peace accomplishing the political objective occurred in Germany and Japan after World War II, and it was the path that Abraham Lincoln had charted for the Confederacy before his assassination.

As a general rule, a reconciliatory peace is preferable to a punitive peace because it involves acceptance and embracing of political objectives rather than sullen acceptance of imposition. As a result, the residue of hostile will (resistance to policies) is more likely to be overcome. Unfortunately, a reconciliatory peace is the most difficult to attain for at least two reasons.

First, the nature of war itself makes reconciliation difficult. War leaves physical and emotional scars in the forms of death, maiming, and the destruction of property. The result is a natural inclination, on the victorious side, toward vindictiveness as the caskets and maimed veterans come home and the rubble is sorted through. In those circumstances the impulse to punish is a powerful feeling that is difficult to overcome. The carnage and physical destruction of World War I had a powerful influence on French attitudes toward the peace negotiations. At the same time, the way war is fought may exacerbate the hostile feelings of the vanquished. Sherman's march to the sea and through the Carolinas, while militarily justifiable, undoubtedly inflamed lingering bitterness in the defeated South. The expanded nature and lethality of modern war exacerbate these problems.

Second, and this is particularly true in the American experience, there is a tendency to ignore (or at least inadequately consider) the task that remains after physical hostilities have concluded. This tendency is part of the classic American mobilization-demobilization pattern that says once the war is won (militarily), it is time to return as quickly as possible to the more normal condition of peace. It is a tendency against which one needs to guard. The physical conduct of war, as Clausewitz correctly noted, is instrumental, a means toward an end. The end is attainment of the political objective and that is a "battle" that can be won only in the ensuing peace. If we are to avoid winning the war but losing the peace in the future, that is a lesson that must be well learned.

2

AMERICAN REVOLUTION

The American Revolution was this country's great formative act. Along with the Civil War, the Revolution stands as the most influential military event in American history in terms of its political purposes and political impact. Although it was not particularly recognized at the time it was being fought, the Revolution also marked an important turning point in international politics: it was both the first major war of independence against European colonial domination and the conflict that reintroduced ideology into the underlying causes and sustainers of warfare.

In purely military terms, the Revolution pales in comparison to the great struggles that began with the Civil War and continued into the twentieth century. The armies that contested the war were tiny by modern standards; at its apex, the British had a force numbering about 32,000 on colonial soil, and George Washington's Continental Army never numbered more than 20,000. Battles were relatively few and comparatively bloodless, reflecting the eighteenth-century style of warfare. In the climactic Battle of Yorktown (which is more properly the Siege at Yorktown) an army of 16,000 Continentals and French marines in about equal numbers faced a British force of about 7,500 under Lord Cornwallis. When the British surrendered, the battlefield toll was slightly more than 200 killed, of whom only 20 were American.

The Revolution also continued the American mythology about the American military tradition begun in the French and Indian War. Yorktown, Saratoga, Trenton, Princeton, King's Mountain, and other fields of battle have become symbols of the tradition of military prowess and the American self-image as winners. That most of these battles were not as successful as we remember them, nor as militarily significant, and that they were interspersed with a series of military reverses that were nearly decisive, (e.g., the battles of New York and Brandywine Creek) have faded from the popular mind. An objective evaluation of the war itself shows that militarily it was no better than a draw for the Americans, who were aided by mediocre British generalship.

In addition, the revolutionary experience did much to create the myth of the militia tradition in America. When the war began almost accidentally at Lexington and Concord, the only military forces available to the rebellion were militia units. Throughout the war's conduct, militia units played an important role (especially the so-called revolutionary militia, about which more will be said later). Colonial success created the myth that militia was effective against regular troops and that the United States did not need a standing armed force of any size during peacetime. The militia could be quickly mobilized and hold its own until a regular army could be fielded to carry the day. Once the war was over, demobilization could be rapidly accomplished (in 1784, a year after the peace treaty was signed, the standing army consisted of 80 regulars whose sole purpose was to guard military supplies). The performance of militia units in the Revolution, viewed carefully, hardly justifies that level of faith, but it created a tradition that still leaves Americans at least slightly uneasy about maintaining a large standing force during peacetime.

That the war was not exactly what we choose to remember does not depreciate the importance of the struggle. Although the American Revolution may not have been a major military struggle, it was a major political event. There were military lessons to be learned. Europeans would again be faced with the problem of attempting to retain control of colonial empires against determined indigenous resistance, and Americans would be confronted with parallel difficulties in Southeast Asia.

In some ways, our revolution was the harbinger of things to come for the European powers, first the Spanish and Portuguese in Central and South America and later other European powers in Asia and Africa. The British military problem of snuffing out a determined rebellion fought by unconventional methods far from home was one of those precursors. The cost and political unpopularity of such endeavors which, as much as the success of the Continental Army, drove the British from the field would be revisited by the French in Indochina and Algeria. That the parallel was well appreciated and the lesson well learned is doubtful.

The great irony and tragedy of the revolutionary experience is in its parallel with the American morass in Southeast Asia 190 years later. As one sifts through the Revolution, the parallels draw closer and closer, and the reader is encouraged to look for them. Great Britain was attempting to quell a rebellion far from home against a force and population largely hostile to it, just as the United States did in Vietnam. The war was fought by an enemy who usually refused to stand and fight in the accepted manner, preferring instead guerrilla tactics. Moreover, the British had to fight two wars, one against the revolutionary militia guerrillas who attacked from ambush and who suppressed loyalist support (leading one British commander to refer to the activity as "the dirty little war of terror and murder") and the other against Washington's regular army. The parallels with the Vietcong and the North Vietnamese army are striking. Finally, public opinion turned against the British cause at home as success eluded them, just as American support waned for the struggle in Vietnam. The British problem in America was the American problem in Vietnam. That the outcome for Britain in the Revolution and for the United States in Southeast Asia should have been the same is less than surprising. That the parallel was not seen nor the analogy drawn is tragic.

Issues and Events

Like most momentous political events, the underlying issues that led to the first great modern revolution built up over a period of time. At heart the underlying issues were relatively moderate by comparison to a contemporary world that thinks in terms of the Iranian revolution of Ayatollah Khomeini or the Russian or Chinese Revolutions' bloodbath of the twentieth century. Many historians view the American Revolution as an essentially conservative revolt.

The most pervasive issue was the relationship between the Crown in London and the citizens of the colonies. The heart of the matter was the standing of the colonials: virtually all Americans of the 1760s and early 1770s considered themselves loyal British subjects. (With the exception of inhabitants of enclaves like formerly Dutch New York and settlements of persecuted religious groups from the European continent, most Americans were of British stock.) Because they viewed themselves as coequal British subjects, they believed they should have the same political rights as other English citizens. The core of their grievance was not that most wanted to be treated distinctly as Americans nor that they desired autonomy; rather, most aggrieved Americans (and by no means did all Americans feel aggrieved) simply wanted to be treated more like any other "Britisher."

The matter boiled down to a disagreement over rights versus obligations. From the colonial viewpoint, the basic issue was the political rights adhering to them as British subjects from which obligations logically flowed. Most Americans did not so much mind obligations in the form of taxes and the like so long as they went hand-in-hand with the full rights of English citizenry. It was the imposition of duties without parallel rights that troubled them.

The Crown disagreed. From the royal viewpoint, the subjects of the colonies were, after all, *colonial* subjects, and the primary role and purposes of colonial subjects were to support the Crown. The fact that these particular colonials were of British descent might make one more sympathetic with them than with subjects one had conquered, but the fact did not alter the basic relationship between colony and mother country. Colonies had, in a word, obligations, not rights.

Constructed this way, the whole matter has the abstract quality of a debate over political philosophy, and through a good bit of colonial history the matter remained at that level. Political rights might have symbolic importance but were not basic as long as the mother country was in essence leaving the colonies alone, thus failing to impose obligations and granting rights indirectly. The debate became concrete and lively when the Crown began to impose obligations on colonial subjects without granting matching rights. The precipitating event was the French and Indian War and the debts that Great Britain ran up prosecuting it. The lightning rods after 1763 became the dual and interrelated issues of taxation of colonials and the permanent stationing of British troops on colonial soil.

Fairness dictates that one look at the issues from both sides. The Crown's position was straightforward. The British Crown had contributed heavily both in terms of manpower and treasure to defending the colonies during the French and Indian War (as well, of course, as kicking the French out of Canada and adding Canada to the British Empire). Moreover, after the war Great Britain was forced to maintain a garrison of about 6,000 troops along the frontier to protect against the Indians (who were often French inspired). All of that cost money and someone had to pay the bills. From the royal vantage point, it seemed entirely logical that those who were benefiting (the citizens of the colonies) should help pick up the tab.

In the Crown's view there were two ways that the colonials could contribute. The first, and most generally vexing to the colonials, was through taxation. Originally, most of the tax revenues were to be generated through import and export taxes (the taxes on tea and sugar), but other forms of taxation like the stamp tax were also included. Second, since the colonies were British colonies, it seemed entirely reasonable to the Crown that the home islands should have a special trading relationship with the colonies from which British private enterprise, and ultimately governmental coffers, would be beneficiaries. The result was restrictions on whom and with what the colonies could trade. The restrictions excluded colonial trade in some items with anyone but Britain (rum was an example), gave preferential access and treatment for goods produced in the colonies to the British, and required that a great deal of commerce be shipped on British ships.

Those colonial subjects who would ultimately lead the Revolution disagreed with this logic and the consequences it produced. The colonials appreciated neither the idea of permanent garrisons of British troops on their soil nor the idea of being taxed without their explicit consent. Obviously, the two issues were related since the costs of maintaining the troops created both the perceived need and the rationalization for taxation. Despite this linkage one can separate the two for analytical purposes.

The problem with having British troops on American soil in peacetime was twofold. First, it was an imposition that British subjects on the home islands would themselves not have tolerated, and hence was a reminder that the Crown did not consider the colonials as equals. Second, part of the Anglo-Saxon tradition is a deep and abiding suspicion of standing armed forces in peacetime, because British history had taught that such forces could be used for political repression. The Glorious Revolution of the seventeenth century had had as one of its major outcomes a specific banning of standing armies during times of peace. Despite the threat of Indian attacks, most colonials chafed at the idea of having these troops around. There was a suspicion that potential political repression was the real reason that the troops were there. The colonial militia, after all, was capable of defending against the Indians, so why were the Redcoats there? To many the question was rhetorical.

The second issue was the imposition of the various sets of taxes upon the colonials to help defray the costs associated with defense of the frontier. The colonials were unfavorably disposed to taxation in any form and for any reason, partially because they were unused to the idea and partly because it hit them in the pocketbook. The taxes added to the cost of goods, and restrictions on trade forced them to buy from British suppliers, generally at a higher price than they could obtain elsewhere. More specifically, however, they objected to the imposition of those taxes in an arbitrary manner without being consulted. Taxation, the imposition of burden, became the lightning rod that enlivened the debate about rights and obligations, and the rallying cry became "no taxation without representation."

As causes that would eventuate in revolution, these issues were relatively mild. The colonists were not crying "off with their heads," as the truly radical French revolutionaries would do more than a decade later. Rather, the heart of the matter was the request to be treated as British citizens. Had there been even nominal direct colonial representation in London (possibly even nonvoting observers in the Parliament) and had British troops not been called in from the frontier, there likely would not have been a revolution of any consequence.

It should be added that the disgruntlement of Americans with the Crown's acts was largely regional and that it affected different groups within the society differently. Objections were strongest in New England because that region was the most strongly affected. Most of the British soldiers on the continent were stationed in New England, and the merchant class most disadvantaged by the trade restrictions was largely based in that region, especially around Boston. As one moved farther south through the middle Atlantic and southern colonies, there was considerably less British military presence and a smaller trading class as well. Although the British restrictions were generally disadvantageous to colonies, they were more onerous to some than others, and some sectors, such as the indigo plantations in South Carolina, actually benefited from the Intolerable Acts. These differentiations in grievance would affect the general appeal of the Revolution regionally, as well as help to dictate the strategies of the combatants once the Revolution entered its military phase.

Thus the underlying causes of the Revolution were less than what we would now think of as radical, and the proximate events that led to the first shots at Lexington and Concord were gradual and more nearly accidental than carefully preplanned. The situation basically accumulated until a comparatively minor event triggered "the shot heard 'round the world." The evolution of each underlying cause reflects this gradual nature of the accumulation of events.

The various tax measures imposed on the colonies included the Sugar Act of 1764, the Stamp Act of 1765, and the new Tea Act in 1772. The taxes were levied sequentially and became a gradually irritating factor. The reason that new taxes had to be added was that each tax, which was supposed to meet British revenue needs, failed to produce the amount anticipated. The reason for the shortfalls was colonial resistance to paying the taxes.

The inefficiency of tax collection created, in the minds of the British monarchy, the need for more effective means for royal tax collection. The instruments selected were the very British forces the taxes were supposed to underwrite, and gradually those forces (who, it will be remembered, were there to protect the frontiers against the Indians) were moved from the frontier to the urban areas. Their presence, previously realized but unseen, became visible, particularly when troop contingents were moved into the urban areas, notably Boston. Matters were made even worse when the Quartering Act of June 1765 required that citizens open their homes to these troops. Now the Redcoats were not only in town, they were in private homes, and the cry against "quartering" of soldiers was heard across New England.

Tensions gradually increased and there were instances of hostility on both sides. As early as the winter of 1770, a group of Boston youths pelted a contingent of British troops with rock-filled snowballs. The Redcoats panicked and fired into the crowd, and the result was the famous Boston Massacre. In 1773 a group of the Sons of Liberty, disguised as Indians, sneaked aboard a British merchant ship in Boston Harbor and dumped its cargo in defiance of the Tea Act (the Boston Tea Party).

As acts of hostility and sedition increased and relations between colonial authorities and the citizenry became more strained, the Crown and its representatives became increasingly more concerned about the possibility of open rebellion. As these thoughts emerged, the size of the colonial militia, the only organized armed forces that could oppose the Crown, increased as well. Although the fighting capabilities of the militias were suspect, they did have considerable stores of arms in their armories without which armed resistance would be impossible. Disarming the militias became an appealing way to nip a potentially nasty problem in the bud.

It was this motivation that caused the British governor of Massachusetts to order British troops to march out of Boston and to seize arms caches at Lexington and Concord. As word spread (carried by people like Paul Revere in his famous ride) of what the British were up to, militia units began to form along the road as the British column marched toward its destination. Tempers flared, names were called, and someone (no one knows on which side) fired the shots that signaled the beginning of the American Revolution. In the fighting that ensued, the British retreated into Boston, and the militia units, almost instinctively, followed them and took up posts at Bunker (Breed's) Hill on the Charlestown Neck, overlooking Boston. The British counterattacked at what is usually called the Battle of Bunker Hill, and the militia (which was gradually reinforced by units from adjacent areas) began the siege of Boston. The Revolution was joined, with very few having any real idea of what would follow.

Political Objective

Largely because of the largely accidental way in which it began, the American Revolution belies the neat depiction of a clear political objective defining the reasons for the outbreak of the war and then determining how the war would be fought.

Rather, the Revolution began without a clear notion of a political objective shared by the majority who initiated it, and nearly a year of violence occurred before the larger purpose for which the Revolution was fought emerged.

At the time of the skirmishes at Lexington and Concord and the subsequent siege of Boston, the idea of political independence was not a widely shared vision. Indeed, even after the siege was laid, the Continental Congress was unwilling to make independence the goal of the colonies. Rather, in August 1775, four months after the initial battles, the Congress instead sent the so-called Olive Branch petition to King George III asking him to intervene in the colonials' behalf to protect them from the "tyrannies" (represented by the Intolerable Acts) of the Parliament. The Crown, however, rejected this offer of conciliation, instead declaring the colonies to be in a rebellion that would be put down. It is at least arguable that had the king responded favorably to the Continental Congress's request, the Revolution would have either died or become a small, isolated movement.

The military situation forced a defining of political purpose, rather than the other way around. The key event was the evacuation of British forces from Boston on 17 March 1776 to Halifax, New Brunswick. When the British garrison left Boston, there remained no British military presence in the colonies. De facto political independence had been achieved, and it was then a matter of declaring formally that independence was the goal of the revolution.

Even with independence militarily asserted, translating that circumstance into a formal political purpose was no easy task. Three months elapsed between the evacuation of Boston and the formal promulgation of the Declaration of Independence on 4 July 1776. At that, the Declaration, as an ideological statement, was a comparatively mild document that had as its basic features statements of grievance directed at violations of rights by the king and the assertion of largely commercial, mercantile rights. Once again, had the British monarchy embraced those principles and agreed that the colonials were entitled to them, it is questionable whether broad-based support for the objective would have been forthcoming.

Once political independence had been asserted as the goal and that purpose had been rejected by the Crown, the issue was joined and was central to the objectives for both sides in the following seven years before a peace treaty was signed. Political independence within the context of a free and, by the standards of the time, democratic polity became the objective of the revolutionary cause, reintroducing political ideology into warfare for the first time in over a century and setting the precedent for ideologically based warfare ever since. The objective was not embraced by all or even a majority of Americans. The Declaration did, however, galvanize enough support to recruit an army and to sustain the military effort to its conclusion.

The major feature of the Declaration of Independence as a political objective was that it placed an absolute purpose on hostilities. The issue of independence was indivisible: it is impossible to be partially independent. With the issue so defined, compromise was unlikely; the rebellion cither had to succeed or be crushed. There were no alternatives.

The British objective, of course, was the obverse of the American purpose. Initially, the Crown viewed the purpose as a discrete matter of quelling a rebellion and reasserting British authority in the colonies.

This objective did not meet with great popular or parliamentary support, as there were sizable numbers of Britons who felt that the colonials' demands were either reasonable or at least not worth fighting over. Given this lack of enthusiasm, the Crown eventually evolved its own version of the "domino theory" by asserting that resistance to rebellion was a matter of precedent. If the Crown did not forcefully react to sedition in the American colonies, other parts of the empire would try the same thing.

Military Objectives and Strategy

In the simplest terms, the fundamental British objective during the American Revolution was to regain sovereignty over the American colonies. As such, the British military objective was to provide the circumstances in which that sovereignty could be reclaimed. The problem was how to accomplish this goal.

By July 1776 British authority had been effectively removed from the rebellious colonies, although Canada remained firmly under British control. Looking at the situation at hand, the British saw the colonies as divided over the issue of independence and believed (rightly) that strong loyalist sentiment remained. They saw a rebel military force composed of ragtag militia, officered by a group of men with limited military experience and training. They viewed the colonies as a long, thin, disconnected string of outposts clinging to the edge of a great wilderness and, as such, vulnerable at innumerable points to dominant British sea power. The British also realized they could not occupy all of the colonies. Britain's traditionally small land army was not designed for such a task, especially considering that the American colonies were not Britain's only responsibility—or even her most important colonial possession.

What options did the British have? Three options presented themselves, to be used individually or in combination. The first was to destroy the rebel military force, which would eliminate the instrument of the rebellion—the means of military resistance. The second option was to occupy the decisive places that were the bases of the rebellion and thus to choke the rebellion to death. The third was to win over the "hearts and minds" of the uncommitted colonists who, combined with the Tory loyalists, would themselves put an end to the revolt.

The British results were also less than successful. Gen. William Howe, leading the first British counterthrust at New York, attempted to combine all three possibilities. He desired to defeat Washington's small army while attempting to negotiate, since he was empowered to make concessions to the rebels. At the same time he would seize New York, which the British assumed to be a vital center, and was prepared to launch a two-pronged campaign to separate New England from the rest of the colonies. Unfortunately for Howe, he quickly discovered that Washington would not stand and fight a decisive battle. Rather, Washington fought and then retreated to preserve his army to fight another day. Howe also found the American rebels totally unwilling to negotiate; and although New York made an excellent base of operations, it was anything but a vital center in the European tradition.

This pattern was to repeat itself throughout the war. Rebels would not negotiate, Tory sympathizers were kept in check by the revolutionary militias, British victories came often but were indecisive, and there were no vital centers to capture.

American military objectives were much more complex. As Dave R. Palmer pointed out in his brilliant exposition on the subject, American military objectives were time dependent. The Revolution was actually in four phases, each with its own military objectives. Phase one, lasting from the battles at Lexington and Concord in April 1775 until July 1776, was the Revolution itself. The military objective was quite simple—throw the British out. Even if the long-term political objective was some sort of political settlement with the British retaining sovereignty, the rebels could not negotiate with the British in control or occupying a threatening position. The rebels could only negotiate from a position of strength. Such an objective dictated an offensive military strategy. Thus came the capture of Fort Ticonderoga (with its cannons), the siege of Boston, and the invasion of Canada in hopes of sparking an uprising by French Canadians.

The invasion of Canada was a failure, but by the summer of 1776 the Revolution was complete. The British had been thrown out of the 13 rebellious colonies. Now the rebels were almost prisoners of their successes. The revolt had been so successful and so complete that some sort of precipitous political action was a foregone conclusion. Thus came the decision by the Continental Congress for complete political independence.

The Declaration of Independence marked the beginning of phase two of the war and a new military objective. The rebel army now had to defend its newly won independence against the invasion by the British that was sure to come. With independence declared but not yet won, the Americans suddenly had everything to lose. It was clear at this point that Washington could not defeat a determined professional British army in a decisive battle. Thus Washington had two objectives. The first, and more important, was to prevent a decisive defeat. As long as he could preserve his army and keep it in the field, the Revolution probably could remain alive. The second objective was to make the British pay a high price for their invasion of the newly independent colonies. He must wage a war of attrition, hoping to wear the British down, hoping they would tire, and hoping the cost of imperialism would become too high. Then, too, there was always the hope that Britain's enemies would come to the aid of the struggling rebels.

Phase three began in February 1778 with the signing of a treaty between the rebels and the French. With the French as allies more troops and arms became available. More important, the French navy challenged British control of the American coast. The British navy no longer had a free hand in moving troops from one point to another. Washington could afford to take more risks. With new troops and newfound sea power he might, with luck, actually inflict a decisive defeat upon the British. Clearly, phase three called for an offensive strategy just as the circumstances of phase two had dictated a defensive strategy. The siege and surrender of Yorktown ended phase three with the decisive victory Washington sought.

Phase four lasted from the victory at Yorktown until the signing of the peace treaty recognizing American independence. Although the focus of the war shifted away from the colonies to a wider theater, Washington had to concentrate on a better state of peace. His objective was to keep his victorious army intact despite great pressures to disband the force. The Americans had everything to lose once again. Washington was successful and the favorable terms of the peace treaty were due in no small part to the fact that the Continental Army was still in existence and was still a force of significant ability.

The success of the American Revolution was in large part due to Washington's accurate reading of the proper military objective at the proper time. Reckless offensive action during phase two would surely have led to decisive defeat and the probable collapse of the rebellion. Lack of aggressiveness in phase three might have left the British with a threatening position to exploit at their leisure. Disbanding the army to conform to the popular will in phase four might have had disastrous consequences for the treaty negotiations in Paris.

Political Considerations

Throughout the Revolution purely political concerns helped to influence what happened on the battlefield. At the level of domestic politics within the colonies, there was a continual contest to nurture and sustain public support for the rebellion, and the war effort was hampered by the lack of formal legitimacy of the Continental Congress until the Articles of Confederation were ratified in 1781. Internationally, the need for outside support for the Revolution was a pressing concern until the alliance with France was completed in 1778, and significant effort was always directed at the prospect of turning British public and parliamentary opinion against the war.

As the first ideologically motivated war of the modern era, the American Revolution had to win the hearts and minds of the American people if it were to succeed. As stated earlier, the cause was not universally embraced. A sizable portion of the population remained loyal to the Crown (the Tories), and many of these people emigrated to Canada and elsewhere after independence was achieved. At the same time, a large portion of the population was indifferent to the whole affair, especially in the South (before the war moved to the southern theater) and on the frontier.

Public support was absolutely necessary in raising and sustaining the army that stood as the principal obstacle to British reassertion for political control. As a practical matter, this created two problems for the revolutionary cause: The Revolution had to appear to have a reasonable prospect of success and Tory sentiment had to be suppressed. The conduct of hostilities can be understood only with these considerations in mind.

First, Washington faced the considerable task of projecting revolutionary forces as winners. The problem was greatest when British forces arrived in 1776 and the Continental Army was routed in the battles around New York. It has been argued that had General Howe pursued and destroyed what was left of that army—about 2,500 effectives at the low point—the British could have crushed the organized rebellion. Howe did not, apparently believing that the Continental Army would dissolve on its own and that he could mop up what was left of the rebellion in the spring. Faced with the real prospect that the army would simply vanish, Washington was motivated to attack the isolated garrisons at Trenton and Princeton, not because of their military significance but to show the American people that he could in fact win (and to gain needed military stores to survive the winter).

These same kinds of symbolic concerns forced Washington to engage in two battles that he could not win. As noted, Washington opposed the British occupation of New York and was soundly defeated (at least partially because he overestimated the fighting abilities of his troops, based on their successes around Boston). The reason for the engagement, simply put, was to show that the army was of consequence: if it would not defend the nation's major port city, what good was it? For the same reason, Washington positioned his army in front of the British force moving to occupy the colonial capital of Philadelphia in 1777, and the army only narrowly avoided envelopment and destruction at Brandywine Creek. The defense of Philadelphia was not of great military importance, but defending the seat of the Continental Congress was of great symbolic importance. In both cases, the destruction of the Continental Army, the vital force of the Revolution, nearly occurred, but political rather than military considerations forced the risk.

The second problem, the job of suppressing Tory support, fell to the revolutionary militias. These militias were irregular, locally based forces who occasionally engaged in direct combat with the British (as at Saratoga), but who served the additional purposes of maintaining revolutionary control of areas not occupied by the British and of providing a recruitment base for Washington's army.

The militias adopted classic guerrilla methods for these additional roles. When they fought, they surprised their adversaries, engaged in hit-and-run engagements, and then faded away (techniques which, at the time, were considered cowardly in the European military tradition). To suppress Tory sentiment in local communities, they harassed and intimidated the loyalists, including burning their property.

The other domestic political problem was the status of the Continental Congress, which, until the Articles of Confederation were ratified in 1781, had a dubious legal status. Although the body convened regularly and made policy affecting the operation of the war, it was little more than an advisory committee since it was empowered by no legal constitutional act. Rather, its members were representatives of the various state legislatures, and any and all authority the Congress had arose from those legislatures agreeing to carry out congressional policies.

At least part of the reason the Continental Congress was given no authority was the colonial distrust of strong central government. Many in colonial legislatures feared that endowing the Congress with independent and superior authority would result in the same kind of tyranny against which they were revolting. Even when the Articles of Confederation came into force, the powers of Congress derived strictly from the state legislatures (members of Congress were representatives of the legislatures and received their instructions from those bodies), and the powers to implement policy (for example, imposing taxes) came from the state governments and not the central government.

This kind of institutional arrangement, which would later plague the Davis administration during the American Civil War, greatly impeded the war effort. The Continental Congress did not have the authority to conscript or otherwise raise troops; rather, the Congress established quotas for each colony, which the legislatures were free to meet or not. The degree of compliance is indicated by the fact that the Congress "authorized" an army of 76,000 but the fighting strength of the Continental Army never exceeded 20,000 and was generally smaller than that. Provision of such basic supplies as food, clothing, and weaponry was a perpetual problem since the Congress could not levy taxes to pay for these items, having instead to rely on the largesse of the colonies. A great deal of Washington's time was spent trying to cajole the Congress to support him more adequately, but a combination of congressional impotence and suspicion of the army itself (the fear it might be used to suppress the Congress) meant there were continuing problems.

There were international political concerns as well. The greatest concern was securing foreign support to counter obvious British advantages in military equipment (the colonies had no organized armaments industry), in control of the oceans, and in military manpower.

There were several candidates for the assistance role. The most obvious was France, seething over her defeat in the Seven Years' War, which had removed all French colonies from North America. Less likely candidates were Spain, which could always be relied on to oppose Britain's design in the new world, and such commercial rivals as the Dutch Republic and Russia. All four countries, directly or indirectly, contributed to the establishment of American independence.

None of them did so, however, out of any sense of affinity with the American cause. The Spanish, who refused throughout the war to receive American emissaries, hoped the colonials would eventually lose the struggle because they feared an independent North American state would cast a possessive eye on Spanish holdings in North America. The Dutch cast their lot with the League of Armed Neutrality in retribution for the British seizure of the islands of Saint Eustatius and Saint Martin in the Caribbean, which had been transfer points for Dutch military contraband being shipped to the colonials. The common thread in foreign motivation was, in a few words, revenge against Great Britain. For these countries creation of an independent United States was at most a by-product of European power politics and not necessarily a desirable outcome at that.

The problem for the colonials was that no foreign power was about to come to the aid of the Americans unless it could be made to believe that the colonials had a chance of winning. The French were particularly interested in helping, but supporting a losing cause would not produce the desired retribution. The problem was made more difficult by the early lack of success of the Continental Army. Following the successful siege at Boston, the army had suffered a string of defeats during 1776 and 1777 that came close to breaking the back of the rebellion, and its minor military successes at Trenton and Princeton did not compensate for them.

The victory that tipped the scales was Saratoga, where Gen. Sir John "Gentleman Johnny" Burgoyne surrendered his army to Gen. Horatio Gates. In a matter of months a formal alliance was signed between the colonies and France. With that agreement more adequate supplies of war materiel became available to the revolutionary cause, the French navy became available, at least part of the time, to menace and harass the Royal Navy, and, in time, French forces would fight alongside the Continentals. French assistance was particularly crucial at the final colonial military victory of the war, the Siege of Yorktown, where, as noted, almost equal numbers of Continentals and French marines formed the siege lines and the French Caribbean fleet blockaded the coast.

The other, and ultimately pivotal, international concern was the battle for British public opinion. The Americans knew from the beginning that there was considerable opposition to the war, led by William Pitt the Elder, Edmund Burke, and others. Realizing he lacked the military muscle to defeat a determined British army, Washington chose a strategy of attrition aimed at increasing British war weariness (the fact that Britain had been involved in a series of expensive wars in the years preceding the Revolution contributed to war opposition). If the war dragged on long enough in an inconclusive way, British opinion might become a factor that would lead to independence. It was a matter of hanging on long enough to exceed Britain's "cost-tolerance," and, aided by the stunning victory at Yorktown, the strategy worked in the end.

Military Technology and Technique

The character of battle in the late eighteenth century was largely determined by the basic infantry weapon available at the time. The standard weapon was a large smoothbore musket that was difficult to load and had a short range. The British infantry weapon, nicknamed the Brown Bess, was typical of the muskets of the various European armies. Brown Bess was over five feet in length, weighed 12 pounds, and had a three-quarter-inch-diameter muzzle. Although 250 yards was its maximum range, it was extremely inaccurate. Against man-sized targets, 50 yards was considered the maximum effective range, and its rate of fire was about two rounds per minute.

A key to success on the battlefield was to increase firepower, but how does one increase firepower using such weapons? The standard solution was to put more muskets on the battlefield and pack the troops tightly together. This practice, however, led to additional difficulties. The muzzle blast from the large-caliber weapons could easily rupture eardrums unless troops were placed in proper formation. The answer was rigid linear formations with the men placed shoulder to shoulder in long lines, generally three ranks deep. After one rank fired, a second stepped forward (or the first rank retired to the rear) and fired while the first rank began to reload. After the second rank fired, the third took its place and fired. If all went well, the first rank would have reloaded its muskets and be ready to fire when the third rank had discharged its weapons. To increase the shock effect of the weapons, each line would fire in volleys on the command of its officers. The effect was a curtain of lead, smoke, and noise, certain to terrify all but the most highly disciplined soldiers.

As frightening as such massed volley fire was, the real terror may have been in waiting for the enemy to come within range. When attacking, a linear formation had to march with great precision to maintain its rigid alignment. (Prussian officers were known to stop their troops in order to realign them.) Slowly the distance to the enemy, also drawn up in a packed linear formation, closed. Finally, when the two sides were very close, one side or the other fired a volley that would shatter the other's line. The volley was quickly followed by a bayonet charge, which most generals believed would actually decide the battle.

Such rigid tactics, terrifying even to think about, required enormous discipline. Many European generals followed the maxim of Frederick the Great, who observed that soldiers had to fear their own officers even more than the enemy; otherwise, the ordinary soldier would break in battle very quickly. Training for such rigid tactics was long and involved. If more than one volley was to be fired, reloading had to be mastered and conducted in unison. Troops had to march with exacting cadence and pace to keep their formation properly aligned.

Both linear techniques and the available technology made eighteenth-century warfare leisurely by modern standards. It took considerable time for commanders to arrange their formations for battle, and, in effect, both opponents had to agree tacitly to fight before a battle could commence. Because the flintlock weapons of the era were extremely unreliable when wet, few battles were fought in winter when rain and snow were common. The general result was limited warfare at a leisurely pace, but warfare that was deadly once joined.

The British brought to their American colonies a professional army with a heavy concentration of mercenaries hired from other countries, a common practice in the eighteenth century. They were skilled in the exactions of linear warfare and ready to do battle with the best European armies. Doing battle with the Americans, however, led to peculiar problems for which their training provided few answers.

The Americans had the same basic weaponry as the British, except that they lacked artillery in the early stages of the war. Sufficient artillery was quickly captured and later supplied by overseas allies. At the beginning of hostilities the Americans also did not have a well-trained force, professional or otherwise. To defend the colonies, the British had relied on a small contingent of regular army troops supplemented by militias raised and trained by each colony. The quality of these militia units varied widely; for example, the Massachusetts militia required only four days of drill per year, hardly enough to engage a well-trained enemy in linear warfare.

The American solution to these disadvantages was threefold. First, Washington built a small professional army that, by the end of the war, could acquit itself favorably. Help from various European professional soldiers who fought with the Americans was invaluable in this effort.

Second, because the Continental Army was never large enough to be a decisive force by itself, relatively untrained militias were used either as skirmishers or to expand the Continental "line." As skirmishers the militia could harass the British line and inflict considerable casualties while presenting fleeting targets inappropriate for massed British volley fire. When used to expand the Continental line, untrained militia units often broke, but they could initially increase the firepower available.

Third, militias were used effectively on the defense by "going to ground" behind strong breastworks. This technique was demonstrated at Breed's Hill, where militia units behind strong breastworks inflicted grievous casualties on British regulars advancing in linear formation. In the best European tradition, the Americans held their fire "until they could see the whites of their eyes," which means the British were within effective range. Meanwhile the British were frustrated because the Americans were hiding in relative safety from British volley fire instead of standing up and fighting. The technique of going to ground would not be limited to militia armies in the future; as weapons became even more deadly, an infantryman's best method of survival was to take cover, whether he was a professional or an amateur.

Some analysts have argued that the Americans actually had a technological advantage because some of the rebels used rifles that had much longer range and far more accuracy than British smoothbore muskets. This, however, is a misconception. Relatively few rifled weapons were used, and certainly not enough were employed to make a decisive difference. Although rifles had advantages in range and accuracy, they also had two telling disadvantages. First, they had a much slower rate of fire because they were much more difficult to load. Second, rifled weapons were not equipped with a bayonet. This lack led to a disaster at the battle of Brooklyn Heights, when American riflemen were bayoneted by charging British troops before they could get their weapons reloaded. Even Daniel Morgan (whose troops used rifled weapons) admitted that the rifle was effective only when supported by muskets with bayonets.

Military Conduct

As has been pointed out, the decisive revolutionary act occurred almost by accident. The spark that set the tinder aflame was a British expedition from Boston to seize militia arms stored at Concord. Aroused local militiamen met the British troops on

Lexington Green, where someone fired the "shot heard 'round the world." Neither the gunfire at Lexington nor the running battle with the British forces during their return to Boston was the decisive revolutionary act, however. For years many of the colonists had resisted British policies and attempts to enforce those policies. The real revolutionary act came when the irate Americans sealed the British in Boston and put the city under siege. The shots fired at Lexington, Concord, and en route to Boston could be considered the acts of overwrought subjects. The "army" surrounding Boston was a clear challenge to the authority of the Crown—the makings of a true revolution.

Army is a charitable term. The besiegers were a motley group of militia: ill trained, poorly led, undisciplined, and with no real legal standing. But events quickly transpired to begin the process of turning this group into what would eventually become an effective fighting force. On 14 June 1775 the Continental Congress moved to take advantage of the situation by "adopting" the force surrounding Boston and appointing George Washington, a Virginian with some military experience, as the army's commander. Meanwhile, another rebel group captured the small British garrison at Fort Ticonderoga (along the river-lakes route to Canada). The booty from this victory included artillery pieces which, when hauled to Boston, would eventually convince the British to evacuate.

There was little real fighting during the siege. The only serious confrontation took place on Breed's Hill. The British sought to oust the Americans from the hill because from that position the Americans could bring the British forces under direct artillery fire. Anticipating a British countermove, American forces built a considerable redoubt on the hill where they hid waiting for the approach of the scarlet-clad British. After an unsuccessful attempt to turn the American flank, the British launched a frontal assault on the redoubt in the finest tradition of eighteenth-century warfare. The Americans waited in relative safety behind their breastworks. American volley fire took a terrible toll, but the British tried three times to break the American line. The Americans were finally forced to withdraw when their ammunition ran low. When the smoke cleared, the British held the field but had suffered a staggering 40 percent casualty rate.

Meanwhile the Americans launched a two-pronged attack toward Canada in an attempt to eliminate Canada as a base for British operations. The rebel leaders hoped the appearance of an American army would spark an uprising of French Canadians against their British overlords. One prong of the attack proceeded up the river-lakes route in good order and defeated the small British garrison at Montreal. The second column was to proceed to Maine and up the Kennebec and Chaudière rivers, then down the Saint Lawrence to Quebec. Unfortunately, poor planning caused by faulty intelligence doomed the expedition. Travel took twice as long as expected, only 30 percent of those who set out from New England actually arrived at Quebec, and those who did arrive were in deplorable physical condition. Somehow the Americans mustered an assault on the city but were repulsed. The remnants of this pitiful force remained before the city throughout the winter of

1775–76. Their hopes of sparking a revolt were dashed, however, as the French Canadians were uninterested in the entire affair.

Washington was having much greater success before Boston. In the spring of 1776, he was able to mount the heavy artillery pieces seized at Ticonderoga on Dorchester Heights and directly threaten the city and its British garrison. Sir William Howe, the British commander, saw discretion as the better part of valor, evacuated the city, and retired to Halifax to regroup and refit.

To this point the almost accidental rebellion had been a surprising success. The rebels had won a series of offensive victories. Even their setback in Canada had been on an ambitious offensive expedition. The militia had defeated the Redcoats; the British troops had been thrown out of the rebellious colonies along with the authority of the Crown. The first phase of the Revolutionary War was over. Now would come phase two of the Revolution. Independence had been declared and the rebels would have to defend their gains against an opponent who would no longer be caught unaware and unprepared.

During the first phase the advantage had been with the rebels. Only one sizable contingent of British soldiers was in the colonies (at Boston). In spite of the make-shift nature of the American army, it outnumbered the Redcoats and was nearly as well armed. Now, however, Washington faced a fundamentally different situation. He had to defend a long coastline that was highly vulnerable to attack along its entire length. Washington's victorious militia suddenly looked totally inadequate for the job. British sea power could transport troops to any of a hundred invasion points. Even with prior knowledge of a British landing point, ships could probably reach that point faster than could Washington's troops marching overland.

Clearly, the initiative had passed to the British and they seemed to have all the advantages, but the British also faced many disadvantages. First, they would be fighting at the end of a very long line of communications, supply, and reinforcement. Second, political sentiment at home was anything but united; the American colonists had many sympathizers in the home islands. Third, fighting such a war would be an expensive proposition, one not easily supported. Finally, the American colonies were but one colonial responsibility of the British. The Union Jack flew around the world and with it went responsibilities for support and protection.

Just as the British strategy had to be offensive at this point, so did Washington's strategy have to be defensive. Above all he had to prevent the destruction of his fledgling army. Decisive defeat of the Continental Army meant the death of the Revolution, for it was the only means of resistance. Washington knew he had no reasonable hope of victory against the British army in open battle. At best he could wage defensive battles, judiciously withdraw after inflicting casualties, and wait to fight again another day. With some good fortune (and poor British tactics) Washington might be able to fall upon isolated portions of the British force and inflict small defeats. Washington's objective had to be to buy time, raise the cost of the war to the British, and hope they would tire of the whole affair. The other

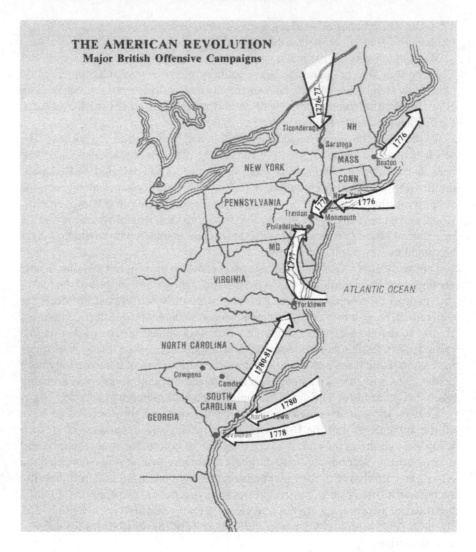

THE AMERICAN REVOLUTION
Major British Offensive Campaigns

American hope was for foreign help from France, Britain's traditional enemy and colonial rival.

The British strategy for 1776 was complex. First, a two-pronged attack was planned to isolate New England from the rest of the colonies. Sir Guy Carlton was to drive the remaining American forces from Canada, pursue them down the river-lakes route, and then turn into New England. Sir William Howe was to land at New York (clearly an important target and excellent base of operations), drive up the river-lakes route, and also turn into New England. Meanwhile Henry Clinton would lead an amphibious expedition to the southern colonies where he had been assured that thousands of Tories would rise up and subdue the rebels.

Few things went as planned for the British. Clinton's effort in the south failed because of a lack of military coordination. Tory uprisings failed to materialize as the rebel militia kept those loyal to the Crown in check. In Canada, Carlton had considerable success as he chased the Americans all the way back to Ticonderoga, but he failed to press his advantage after a naval victory at Valcour Island. Carlton's lethargy and the approach of cold weather put the victorious British forces into winter quarters.

Howe's position at New York was particularly interesting. He came to wage peace as well as war. He was empowered to negotiate with the rebels and offer concessions. The rebels, still flushed with the success of their revolution, were in no mood for serious negotiations. Howe was left with no choice but to wage war. To do this he brought 32,000 professional troops and a considerable naval armada to New York. Washington was able to assemble about 20,000 men, most of them ill trained.

New York would have been difficult to defend in any case, but the task was nearly impossible since the enemy possessed control of the waterways that surrounded and divided the city. Washington was determined to offer a significant defense for political and moral reasons, if no other. If the Continental Army would not defend such an important place, what would it defend? What good is an army that defends nothing? These questions would plague Americans if Washington let New York fall without a struggle. Refusal to try to defend such an important target could be as dangerous to the Revolution as a decisive military defeat. Thus Washington had to offer resistance and then retreat to fight another day.

New York was lost, but not without a series of bloody battles. At Brooklyn Heights, Washington barely escaped total defeat. After additional defeats at Fort Washington and White Plains, Washington retreated across New Jersey as his army slowly disintegrated around him. Finally, Washington crossed into Pennsylvania as winter fell. Howe assumed the campaign was over, that Washington was beaten, and that the rebel army had disintegrated. He ordered his deputy, Cornwallis, to post detachments across New Jersey and then go into winter quarters. Howe would wait comfortably for spring. If the Americans had not sued for peace by then, he could easily reestablish British sovereignty in good weather without having to worry about the defunct American army.

Howe's lack of aggressiveness provided the breathing room Washington needed. When Howe had Cornwallis post detachments across New Jersey, he provided Washington the opportunity to achieve the victory the Americans desperately needed. Had Howe pressed his advantage and chased the remnants of Washington's army or pressed on to the rebel capital at Philadelphia, the American Revolution might have come to a quick and inglorious end. But winter campaigning was no easy task in that era, particularly in the primitive conditions in the colonies.

By going into winter quarters, Howe was following the rather leisurely tradition of eighteenth-century European warfare. He seemed unaware that this would be a very different kind of war. Moreover, Howe failed to realize that the American Revolution was an ideological conflict with its attendant passions.

From Howe's perspective organized resistance seemed at an end. Any disorganized or passive resistance remaining was a job for the diplomats and politicians to resolve, with the help of the army. This task could be left safely for spring when the government in London would have its policy toward the rebels worked out, and when the weather would make implementation of that policy easier. Howe's lack of appreciation for the kind of war in which he was involved led him to miss his best opportunity to put down the Revolution.

Washington was in such a desperate situation that he was forced to seize aggressively every opportunity presented. He had been badly defeated in New York and driven in headlong retreat across New Jersey. His army was disintegrating around him as terms of enlistment expired or as the dispirited simply quit. Enlistments lagged as few were willing to risk their lives for a cause that appeared doomed. Washington needed a victory to boost the morale of his little army, keep the men in camp, and boost enlistments.

Washington could now concentrate his force against one of Cornwallis's isolated detachments rather than facing the bulk of the British army. The object of his attack would be the Hessian garrison at Trenton. On Christmas night Washington took a force of more than 2,000 Continental Army regulars across the Delaware River (the incident portrayed in the famous painting of Washington crossing the Delaware) and surprised the defenders, who, after a brief fight, surrendered. Washington then retreated back across the river. His success so shocked the British that they withdrew their detachments back from the river to Princeton. Washington was then able to recross the river into New Jersey. Cornwallis reacted by coming south from New York, gathering up British forces as he proceeded toward Washington's New Jersey encampment. On the night of 2 January 1777, Washington slipped around Cornwallis's camp and struck Princeton. Cornwallis turned about to advance on Princeton and upon his approach Washington led his army toward Morristown in the rugged New Jersey highlands. From there Washington could easily fend off British attacks and threaten the British line of communication to their posts in southern New Jersey. Howe, recognizing the threat to his posts, withdrew them to the area immediately around New York to await spring.

In strictly military terms, the American victories at Trenton and Princeton were anything but decisive. They were hit-and-run raids, the classic method of war of a much weaker adversary hoping for little more than to wear down a stronger opponent. In the broader sense, however, these two small victories were nearly as decisive as the later victories at Saratoga and Yorktown. Washington had served notice to the British that subduing the Revolution would be no easy task and could not be accomplished quickly. More important, these victories served notice to the American people that victory was possible in spite of seemingly insurmountable difficulties. Victory was an elixir for the Continental Army that cured many of the ills brought on by the defeats in New York. As both armies licked their wounds in winter quarters, 1777 promised to be a decisive year.

The British campaign plans for 1777 were a picture of confusion. Howe, previously entrusted with the capture of New York, was again about to set off to capture a city, Philadelphia. It was unclear what Howe intended to do with Philadelphia once it was in his hands. He may have believed that the Tories would flock to him and that the seizure of the capital would dispirit the revolutionaries. He may have believed that Washington would have to defend the capital, presenting an opportunity to inflict a decisive defeat on the Americans. Perhaps Howe still believed the rebel army was of little consequence.

While Howe planned to seize Philadelphia, Gentleman Johnny Burgoyne had a London-approved plan to split the rebel colonies by moving down the river-lakes route from Canada to isolate the New England colonies. Under this plan Howe's forces in New York were to aid Burgoyne by moving north on the river-lakes route. Unfortunately for Burgoyne, Howe was on his way to Philadelphia.

Howe's campaign against Philadelphia took advantage of British sea power. His troops embarked from New York, put to sea, sailed up the Chesapeake Bay, and landed at Head of Elk in Maryland. This route was, as Washington said, a strange choice. Overland from New York the distance was less than 100 miles. By sea the journey took 33 days and the expedition landed at a point still 50 miles distant from the rebel capital. As Howe advanced on Philadelphia, Washington rallied an army of more than 15,000 for the city's defense.

Philadelphia was not an important target in military terms. Certainly it was the seat of government, but a government can move and it eventually did. Washington needed to defend the city for the same reason he needed to defend New York. Thus Washington felt compelled to lead his army to Chadd's Ford on Brandywine Creek south of the capital. There on 11 September 1777, Washington was outmaneuvered, badly defeated, and nearly surrounded and destroyed. Again, as at New York, the head-on confrontation with major professional forces led to defeat and near disaster.

Washington retreated to Warwick while Howe advanced and seized Philadelphia. Washington resorted to the tactics of the previous winter as he led his army to Germantown, a community just north of Philadelphia, where he fell upon an isolated British garrison. Washington was narrowly defeated, but Germantown was an impressive performance for a recently defeated army that many had written off. Following the battle at Germantown on 4 October 1777, Washington again led his army into winter quarters, this time at Valley Forge, Pennsylvania. While the Americans froze and starved at Valley Forge, Howe wintered comfortably, enjoying the pleasures of the former rebel capital.

Two weeks after the battle of Germantown, the Americans achieved a victory whose impact went far beyond the confines of the battlefield. At Saratoga, New York, Burgoyne surrendered his entire command to an American militia army led by General Gates and Maj. Gen. Benedict Arnold. Burgoyne's surrender was positive proof that the British were not invincible and that the American rebels could win. The news of the American victory convinced the French to enter the conflict on the side of the Americans, and this intervention ultimately made victory possible for the rebels.

Burgoyne had begun his trek down the river-lakes route from Fort Saint John, north of Lake Champlain, with 8,000 professional troops. He had easily moved to Fort Ticonderoga, which quickly fell after a token defense. Rather than taking the water route down Lake George to Fort George, Burgoyne pursued the Americans retreating from Ticonderoga on an overland route. His men hacked a 23-mile road through rough, heavily wooded country, overcoming numerous obstacles placed in his path by the retreating rebels. The head of Burgoyne's column took three weeks to complete the trip, and his heavy artillery lagged well behind the column. Gentleman Johnny knew little about the countryside, and he had taken far too few horses, oxen, mules, and carts to supply his army at the end of an overextended supply line. His base of operations was 185 miles away in Canada. His choices were to retreat to Canada and admit defeat or to press on to Albany, where he could obtain succor.

Burgoyne tried to relieve his problem by sending a Hessian foraging party to Bennington, Vermont, where it was reported that a large number of horses were available. At Bennington the Hessians were virtually destroyed by a militia group commanded by Gen. John Stark, and a British relief column arrived just in time to also be destroyed by the Americans. Burgoyne lost 900 of his best professional troops and achieved nothing except to bolster the confidence of the American militia army, which was swelling rapidly with the arrival of more militia units.

Burgoyne finally decided to move to Albany even though he knew this would be an extremely difficult task. The Americans had cut his supply lines and blocked the path to Albany by entrenching themselves along Bemis Heights. Gentleman Johnny's hungry and demoralized troops would have to attack a fortified position. On 19 September Burgoyne's attack was repulsed with heavy losses, and his situation was nearly hopeless. Surrounded by an army that outnumbered his own almost two to one, Burgoyne had nearly exhausted his food supplies. He made one last desperate effort to break the American lines on 7 October and again was repulsed. Burgoyne retired to Saratoga to consider his limited options, and ten days later surrendered his entire command. Included in the booty were 7 generals, 300 officers, and 5,600 soldiers. A considerable array of artillery plus a large stock of powder and shell also fell into American hands.

Four months after the American victory at Saratoga, France and the American government signed a military alliance. The surprising American performance at Germantown and the smashing victory at Saratoga provided the grease for the diplomatic wheels. The alliance was the pivotal political act of the war.

With the French in the war, Washington's situation was entirely different. In the past Washington had to concentrate on keeping his army intact and avoiding serious defeat. His military objective had been to survive to buy time, to tire the British, and to raise the cost of the war to a level unacceptable in London. Now the French fleet could challenge and perhaps defeat British sea power, so that the British would not have the luxury of unlimited mobility. With a strong ally Washington could afford to take risks; he had more than his own meager resources to fall back on should those risks lead to defeat. Perhaps more important in the long run, the British were seriously distracted because the war had suddenly broadened in scope, and other British possessions were threatened. Finally, the hard work of Washington and his officers, particularly such foreign advisers as Baron Friedrich Wilhelm von Steuben, produced a trained core of Continental soldiers capable of acquitting themselves well on the open battlefield. Not only could Washington take risks, he had the tools to make success possible.

The British, meanwhile, were thrown into great turmoil by the new situation. Clearly, they would have to retrench their forces in the American colonies and be prepared to embark to other destinations that might be threatened by the French. The first move was to relieve the lethargic Howe, who was still in Philadelphia, and replace him with Gen. Sir Henry Clinton. After some confusion Clinton was ordered to evacuate Philadelphia and consolidate his forces in New York, where they could easily embark for other ports.

Washington struck hard at Clinton as he retreated toward New York. Deploying nearly half of his total force of Continentals against the bulk of the British army in the colonies, Washington attacked Clinton at Monmouth, New Jersey, on 27 June 1778. The two armies fought for several hours and the Continentals acquitted themselves with distinction. Clinton, however, was able to escape to New York. The British, rebuffed at every turn, virtually abandoned their hope of recapturing the northern colonies. They remained in New York, closely guarded by Washington and his Continentals.

The British still hoped to take advantage of assumed Tory sentiment in the southern colonies. To this end London dispatched Lord Cornwallis in 1780. The British won a great victory at Charleston, moved into the interior, and established a line of posts reminiscent of those in New Jersey during 1776. Meanwhile the American Congress appointed Gen. Horatio Gates, the self-proclaimed "Hero of Saratoga," as commander of the southern armies over the objection of Washington. The British soundly defeated Gates at Camden, South Carolina, on 16 August 1780. Although the Continental contingent fought well, militia units broke and fled from the field along with a panicked Gates, giving the British one of their most complete victories of the war. Only about 700 of the nearly 4,000 Americans involved escaped. Following that debacle Congress bowed to Washington's wishes and appointed Nathanael Greene as army commander in the southern colonies.

Greene fought a classic guerrilla-style war against the frustrated Cornwallis. Greene led Cornwallis on a hard chase and turned and fought only when British forces were spread out, tired, and ill supplied. Along with Francis Marion, another brilliant tactician of partisan warfare, Greene harassed the British, wore them out, and occasionally defeated them. Finally, the frustrated Cornwallis broke away from the chase and retreated to the coast at Wilmington, North Carolina, to refit. Greene quickly headed back into South Carolina to attack the posts that Cornwallis had earlier established and now could not defend.

Disgusted that he could not destroy the rebel forces once and for all, Cornwallis headed north into Virginia hoping for better luck. He had little success there, again chasing elusive American forces about the countryside. Clinton, who was still in New York watching Cornwallis's campaign with great concern and a degree of disbelief, ordered Cornwallis to the coast at Yorktown to meet with the British fleet for refitting and possible embarkation. The fleet Cornwallis found at Yorktown, however, was not British.

Cornwallis's retreat into Yorktown presented Washington with a rare opportunity for a complete and politically decisive victory. Cooperating with the French, Washington quickly made his plan; speed was necessitated by knowledge that the French fleet in the area around Yorktown would soon be returning to the Caribbean at the end of hurricane season. Washington had to mass a superior army to besiege the British from the land side. Washington and the French army commander, Comte de Rochambeau, marched a mixed army from New York (where they left forces demonstrating for Clinton's benefit) to Yorktown. The movement took just over a month, a considerable feat in eighteenth-century conditions. Washington managed to mass 16,000 troops while the French fleet of 30 ships blocked a British escape by sea. Siege operations began under direction of the French engineers. Less than a month after the siege began, Cornwallis surrendered his entire army of 8,000 to Washington on 19 October 1781 after his plan to escape across the James River was foiled by a storm that destroyed his boats.

After five years of hard fighting, the British had suffered two crushing defeats at Saratoga and Yorktown and were worse off than they had been in 1776. They still held New York and Charleston (both closely watched by the Americans), which would be good bases for further operations. Also the British faced a global war against powerful enemies, and an American army that could stand and fight in the best European tradition. Worse yet, political opinion at home was badly divided.

For all practical purposes, the war was over for the Americans. Although some fighting still occurred in the south and west and the British had to be guarded at New York and Charleston, the real war was over and the Americans were victorious.

The Revolutionary War reintroduced ideological conflict, a significant step in the democratization of war. One unfortunate result was that, led by the memories of Breed's Hill and Saratoga, for the next 160 years many Americans believed that they did not need a professional army of any size and could instead rely on militia units. Although this had disastrous consequences in 1812, in the Civil War, and in the early battles of two world wars, the legend of the minutemen died hard. Only after 1945, when America entered the world of international power politics as the leader of the Western democracies, would the United States establish a sizable professional military force.

Who fared better? It is clear that in 1776 the British squandered their best opportunity to end the rebellion. Had Howe pressed his advantage after defeating Washington in New York, the war might have ended quickly. The American Revolution was, in effect, a civil war. As clearly demonstrated later in the American Civil War, rebel momentum and morale grow if they are not quickly checked.

Time was clearly on the side of the Americans. Not only did American confidence grow, but British will declined at home. The Americans were aided and abetted by poor British planning (Howe going to Philadelphia instead of aiding Burgoyne in 1777) and military blunders (Burgoyne's disaster and Cornwallis's entrapment at Yorktown).

Although ultimately successful, Washington's strategy can also be questioned. Washington played for time well and did an excellent job of keeping his fledgling army together. He recognized the proper time to take the offensive and to take the necessary risks to achieve decisive victory. However, one can question his campaign in New York in 1776 and Philadelphia in 1777. On one hand, the failure to at least attempt a defense of those two vital points might have been a serious blow to American morale. On the other hand, Washington's army was nearly trapped and destroyed both in New York and at Brandywine Creek. The question remains, however, were the benefits worth the possible consequences?

Overall, Washington, the amateur soldier, must receive higher marks than his professional British opponents. Although the British had a clear political objective, none of the British commanders seemed to have a clear conception of how to reach that objective. Washington, however, seemed to have had a much clearer picture of how to achieve his objective. His immediate objectives accurately changed with the circumstances, and each one was geared toward the ultimate goal. The proof of this contention is in the outcome.

Better State of the Peace

The American Revolution achieved its political objective of independence with the signing of the Treaty of Paris in 1783, ending eight years of formal hostilities between Great Britain and her former colonies. Just as de facto independence was achieved in 1775 when the British evacuated Boston, so was permanent and legal independence a fact when the British sailed out of New York, Charleston, and Savannah in 1783.

Unlike most wars, and certainly the major conflicts of the twentieth century, American objectives were not truly won on the battlefield. The yardstick that suggests that the defeat of the enemy's armed forces is prefatory to imposing those peace terms by which the victor defines the better state of the peace simply did not apply in the American Revolution. The British army was certainly not destroyed as a fighting force: in fact, it won most of the battles, and with the exception of some inept generalship, it acquitted itself well. If a goal in war is to overcome hostile ability, the Revolution was at best a draw.

The reasons the war could be won in the absence of decisive military success can be boiled down to two factors. The first was the nature of warfare as practiced in the eighteenth century. Unlike modern wars, it was fought by limited means (basically because of technological limitations) and with relatively small forces (largely because of budgetary limits on raising and supporting large forces). The size and kinds of armies available might be capable of decisive action in the comparatively confined space of continental Europe, where the capture of a critical city would lead to peace. The American colonies, however, were not well suited to this kind of war; the territorial expanses were simply too great for control by 30,000 British troops, and there were no geographical points critical to the revolutionary cause. Washington could not have eluded the British had the war been fought in Belgium, but the British could not corner him in America.

The second factor that made the war militarily inconclusive was the contrast in objectives and military strategies adopted by the combatants. The European tradition called for the open, frontal confrontation of standard, stand-up fighting forces, but this was not what the British faced. The Continental Army was simply neither large enough nor good enough to take on the British in that manner. Given the military balance, a strategy of attrition, featuring generous doses of what we now call unconventional warfare, was the only available means. It was, moreover, a style of warfare that better suited the rugged, heavily wooded American topography.

This style of warfare both confused and frustrated the British, who never did devise an effective means for dealing with an enemy who hid behind trees in ambush and simply melted away into the vast countryside when confronted by a superior force.

If the rebels failed to overcome British hostile ability, they did succeed in overcoming that element of British hostile will defined as the willingness to continue to bear the costs of fighting. British cost-tolerance was, in the end, exceeded, and that was what proved conclusive. The lesson was, or should have been, instructive for future generations. The British were forced to fight an unfamiliar kind of war on unfamiliar and hostile territory. The war was fought far from home, straining supply capacities and raising economic costs (in more contemporary terms, the British had a very long logistics tail). Moreover, the war dragged on and on with apparently inconclusive results as casualties continued to mount. The longer it lasted, the stronger the protests became, and ultimately those who had opposed the war all along gained the upper hand. Finally, a lack of support forced the British to pack up and go home, undefeated militarily but with their will to continue shattered. So constructed, the parallels with Southeast Asia are stark and painful.

The question of overcoming hostile British will (defined as acceptance of American policy preferences-independence) is more difficult to assess. At one level, the British clearly did accede to the American objective by signing the peace treaty that formally created the new independent state. At the same time, the British resented deeply what had happened to them and showed disregard for American sovereignty, as evidenced by their cavalier treatment of former British sailors (impressment) who had been granted asylum by the U.S. government. That arrogance had a great deal to do with leading the two nations to their second conflict, the War of 1812. Many observers maintain that it was not until that conflict was resolved that British hostile will toward the American state finally ended.

3

CIVIL WAR

The American Civil War represents the greatest American national trauma. It was the first conflict in the American experience that clearly and unambiguously met the dual criteria of total war. The issue of union or disunion of the country was all-encompassing and could be resolved only on the field of battle, and the war became a match between totally mobilized societies. The result was the bloodiest war in American history. When the last gunpowder haze rose and Robert E. Lee and Joseph E. Johnston surrendered the remnants of their armies, more than 600,000 had died and nearly another 500,000 had been wounded. And it had all been done at our own hand.

In many ways the Civil War was warfare in transition, a junction between the classical conflicts of the eighteenth century and the massive carnage of the twentieth century. Tactically, battles were organized and fought along the lines of the Napoleonic campaigns by armies led by officers who had learned to fight that way at West Point, and both sides (especially Lee's) were obsessed with the "decisive battle" concept integral to Napoleon's success.

At the time no one really appreciated how the face of war was changing. Partly this was the case because it was a war fought and led by men who were largely inexperienced in combat when it began. The soldiers available were largely untrained militia (in the American tradition) who could neither drill nor, in many cases, fire a gun. The war's first major battle, First Manassas (Bull Run), showed both sides how poorly prepared they were for the war and how difficult the task would be. After First Manassas both sides mobilized their societies, and in the following spring when the real fighting began, war machines of unprecedented size were ready to grind against one another and produce equally unprecedented carnage.

Another reason most contemporary observers did not recognize that the Civil War presaged the new face of war was that the armies that fought never resembled the highly drilled and disciplined troops of Europe. This led to the conclusion that they and the war were aberrational. One foreign military observer, the Prussian Helmuth von Moltke, typified the war as "two armed mobs chasing each other around the country, from which nothing can be learned." Missed were such harbingers of the future as the elaborate entrenchments around Petersburg, Virginia, that foreshadowed the awful trench warfare in France a half century later.

The Civil War was our bloodiest conflict, and it is also the war we best remember and most romanticize. Certainly a large part of our obsession is deserved because the Civil War, in terms of its effect on American society, stands with the American Revolution as one of the two most important events in our history. It was a major event in the American experience in both a political and a military sense. Only the Revolution rivals this great conflict as a political event; no war before or since comes close to matching it in terms of American blood expended. It is also the great American tragedy; Americans turned the weapons of war on their fellow Americans. Why and how it all happened occupies the pages that follow.

Issues and Events

It has been difficult to achieve anything resembling consensus about the causes of the war. Slavery, the imposition of the Northern industrial system on the South, and states' rights have all been argued as the basic issue. No attempt is made here to add to that debate or resolve the question of whether the war was inevitable based on which root cause one picks. Rather, our perspective is that there is truth in each of the ways of looking at the issues and that each contributed to the final result. Moreover, the issues leading to the war can usefully be organized as a clash between what had evolved as two distinct cultures that manifested themselves in progressively diverging political, economic, and social systems. In this view issues like slavery or states' rights are significant symptoms of the deeper incompatibility between two distinctly regional cultures and the values they represented. The American culture of 1860 was in fact two very different cultures. In the long run, those differences had to be resolved before there could be a truly *United* States of America. Whether these differences could have been reconciled differently than they were is one of history's moot points.

Northern industrialization was at the heart of the divergence between the sections. By the eve of the Civil War, Northern society was undergoing the pervasive change that attended the industrial revolution, but there was no parallel transformation in the South, which remained an agrarian society. When both sections had been agricultural, the differences between the free-holding farm pattern of the North and the slave-based plantation organization of Southern agriculture were not critical. When Northern society moved from an agrarian to an industrial base, the differences between the sections became more pronounced and vexing.

The issue of labor was at the heart of the friction. Although the majority of white Southerners owned no slaves, the plantation system that sustained the Southern economy depended upon chattel slavery. Cotton production was the core of this system, and slave labor was appealing to the culture of cotton for several reasons.

First, growing cotton was a labor-intensive enterprise, but it did not require highly trained workers or great efficiency. Unpaid labor in the form of slaves kept costs low enough to turn a profit.

Second, the planters perpetually suffered from a "cash-flow" problem. Cotton did not produce a steady flow of income; rather, it produced revenue in spurts when crops were sold. In addition there were frequently substantial lag times between harvest and payment; the unpaid slaves never knew the difference.

There were costs that made slave labor unattractive in the industrializing North. Slaves, because they were not paid, had no incentive to work harder than necessary to avoid punishment. Hence, slaves were not efficient, which was intolerable in an industrial setting. Industry, after all, has efficient production as its ultimate measure of success.

This is, of course, a very pragmatic way to look at the slavery question, but it is closer to how the average Yankee viewed the issue than can be found in abolitionist literature. The abolitionists were noisy but few in number and marginal in political clout (although many Southerners overestimated their influence). To average Northerners slavery was at worst an unfortunate institution that they could not actively support. To most Northerners who thought at all about such matters, slavery was not so much evil at it was inappropriate.

The two cultures were diverging in other ways as well. Southern plantation society had become highly stratified and stagnant. The planters were an elite who dominated the South politically, economically, and socially. In that position the planters were natural conservatives, seeking to preserve a position from which they benefited. Beneath them were merchants and artisans who benefited from the planters' largesse. Beneath the merchants and artisans, was the larger population of poor whites, who toiled on rented land at subsistence agriculture, but who, at least, had the slaves to look down on.

At the same time, the existence of the slave system effectively precluded change and development because slavery was custom-made for the cotton plantation system but was of dubious economic viability otherwise. The plantation needed slaves, and the institution of slavery needed the plantation. The consequences of breaking the circle and freeing the slaves were something few Southerners were willing to face.

The North was evolving very differently. Unfettered by slaves and nurtured by waves of European immigrants and foreign investment, the North was on its way to becoming a modern industrial state. With industrialization came the emergence of an urban working class, a merchant class, and industrial entrepreneurs, groups largely absent in the plantation South. The result was great social fluidity and social and economic leavening. The North was becoming a society of workers and shopkeepers; the South remained a society of aristocrats and farmers.

Coexistence became more difficult as the differences magnified. Northerners argued for high tariffs to protect their new industries. Southerners resisted, preferring European goods at lower prices. Northerners pushed for legislation to require Southern cotton to be sent to Europe on more expensive American (which meant New England) ships. Southerners resisted because such shipping was more expensive and made their cotton less competitive on international markets. Dual cultures were increasingly coming into conflict in practical ways; something had to give.

The issue that broke the camel's back and that, combined with the election of Abraham Lincoln, provided the proximate events leading to secession and the war was the extension of slavery to the territories. It was an issue that had been brewing for some time. The Missouri Compromise of 1820 had defused it for a short time, but it returned in the protracted fight over admitting slave-holding Texas to the Union. The Great Compromise of 1850 attempted to settle the problem, but the compromise was followed rapidly by such unsettling events as the Dred Scott decision, Bloody Kansas, and John Brown's raid at Harper's Ferry.

On the pragmatic, political level, if slavery were allowed in a territory, that territory would ultimately enter the Union as a slave state and the converse was also true. In turn, a new slave-holding state would elect proslavery representatives to Congress who would generally support Southern positions, just as free states would elect antislavery representatives of the opposite bent. In a system where there was a rough balance between slave-holding and nonslave-holding representatives (especially in the Senate), additions on either side would tip the balance.

The question of the extension of slavery also created problems at the deeper level of competing cultures. Extension was particularly vital to the South, because cotton cultivation rapidly depleted the soil. If the cotton and hence the plantation system were to prosper, it had to be able to move from depleted soil westward to fertile soil. Hemmed in, the cotton culture would die; thus, the absence of extension amounted to slow strangulation. In the North the extension of slavery was opposed because slavery was an anachronism that had no place in the kind of society that Northerners wanted and expected to build in the new lands. In this light the extension issue emerges as a lightning rod for the entire clash of cultures, and it boiled down to a zero-sum game: if one side was to win, the other had to lose.

This irresolvable, irresistible conflict came to a head with the election of Abraham Lincoln in 1860. Lincoln was not politically an abolitionist, but his candidacy had been supported by the more radical abolitionists. Lincoln's stated position (he found slavery personally offensive but protected by the Constitution) was lost in guilt by association in the minds of many Southerners.

When the calls for secession came, the issue of states' rights was the rallying cry. In the South particularly, there was great sentiment for a weak central government and primary investment of political authority in the states. This position, of course, was more than abstract and academic. The South had its "peculiar institution" and social system to protect. The closer to home political authority lay, the more compatible public policy would be with maintaining that system. In the North the development of a modern industrial state required a comparatively strong central government that could adopt national policies conducive to continued growth (protective tariffs are a good example). The South generally did not benefit from these policies and sometimes suffered from them. At the heart of Southern opposition, however, was the lingering fear that a strong central government might adopt legislation directly attacking Southern institutions. The election of a president believed to be actively sympathetic to abolitionism produced a greater strain than could be borne.

If the critical political issue was the supremacy of the rights of the states and central governance was equated with tyranny, then the new, alternative government had to reflect those beliefs. The Confederate States of America chose a confederal format, a state-dominated political system that conformed nicely to philosophical predispositions but which featured a central government with limited authority to make wartime decisions, a political problem that dogged the Davis administration and the Confederate military command throughout the war. Ironically, the very principles for which Southerners fought hampered their ability to fight.

Once the South Carolina legislature voted unanimously to dissolve the union between itself and the federal government on 20 December 1860 (an action quickly duplicated by six additional states and later by four others), the question was how the rest of the country should react to secession. The answer was not as clear then as it may appear today. There was, for instance, considerable disagreement, mostly in the North, about whether states had the right to secede, and both those who said they did and those who said they did not based their arguments on the Constitution.

The argument boiled down to the states' rights versus strong central government debate. Those advocating the legality of secession were in fact arguing states' rights, and those who maintained that the states could not secede were arguing the supremacy of the Union over its constituent parts. The latter belief formed the basis for Lincoln's famous statement that "a house divided against itself cannot stand," which clearly reflected how the president-elect felt about the matter. When South Carolina seceded, however, James J. Buchanan was a lame duck president, and he reacted officially to the secession by ignoring it. When Lincoln was sworn in as the nation's sixteenth president, some action would be necessary.

Political Objective

Lincoln determined that the Northern political objective was to reestablish the Union, by force if necessary. As a statement of purpose, this was disarmingly simple, but there were powerful politico-military problems confronting its realization.

This first and most obvious problem was that the South did not intend to return to the Union voluntarily. The North would have to fight for reunion and that led directly to Lincoln's second problem. The objective was not popular in the North. As a political objective, in other words, reunification lacked the moral power and persuasiveness to galvanize Northern public opinion. A more morally lofty objective was necessary to gather and sustain support.

Lincoln realized his problems, and part of his answer was the Emancipation Proclamation, which added the end of slavery to reunion as the political objective. This goal was announced in September 1862 and took effect in January 1863, but emancipation had to overcome a problem of timing: throughout 1862 the Union suffered a succession of defeats in the Eastern theater. To change the objective in the midst of calamity would have appeared an act of desperation that could backfire and diminish rather than increase support. What Lincoln needed was a military victory to precede his announcement. He got his wish when Lee forayed into Maryland and was stopped by George B. McClellan at Antietam (Sharpsburg). The battle itself was a draw, but it forced Lee to retreat back into Virginia and thus looked enough like a win to serve the purpose.

The effects of adding the abolition of slavery to the political purpose of the war were mixed. In the South the reaction was mortification, reinforcing the citizenry's worst fears about Lincoln and increasing their will to resist. In the North the result was a sort of backhanded success.

The basic negative was that freeing the slaves was not an overwhelmingly popular objective to most Northerners, who were about as racist in their attitudes toward blacks as were Southerners. With the exception of the abolitionists, most Northerners shared Southern beliefs in the inherent inferiority of the Negro, and even if they found slavery repulsive in the abstract, many did not think the destruction of the institution worth dying for.

Making emancipation a major objective did, however, add moral weight to the Union cause in at least two important ways. First, opposition to the war became tantamount to being proslavery, a position that relatively few in the North held or at least were willing to admit. Second, this moral elevation of the objective effectively ended any possibility that the Confederacy would gain recognition by the European powers, notably Britain and France, which had been primary prewar markets for Southern cotton. No European state could politically align itself with human chattel slavery. The proclamation thus ended the possibility of foreign support for the Confederacy.

If the Union political objective was to restore the Union and free the slaves, the Confederate objective was just the opposite: to maintain its independence by whatever means necessary and to avoid the emancipation of the slaves. As long as the South maintained control over its territory and had a functioning government, military forces, and a loyal population, the Union could not achieve its political objective.

Unlike the situation in the North, the Southern political objective was overwhelmingly popular and sustained citizen support for the war effort until nearly the end. The Southern cause was to defend their homeland and their society from a foreign enemy who could accomplish his purpose only through physical invasion, subjugation, and occupation.

The popularity of defending home and loved ones from an invading force added greatly to Southern political support for the war, and the Confederate political objective was never seriously challenged from within. Support for the objective and its translation into military activity was one of the great advantages the South had (other primary ones being fighting on the defensive and on familiar ground and having generally superior military leaders), and this advantage was particularly obvious in contrast to the marked ambivalence about the objective in the North. In turn these political objectives translated into military objectives and strategies for waging the conflict. As we shall see, the political objectives sometimes became blurred or distorted in the process, but political concerns and considerations were never far from the field of combat.

Military Objectives and Strategy

For the Union, that part of its political objective involving restoring the Union was simple and straightforward. It required that the Rebel government be disbanded. But how does one destroy a rival government? Clearly, the armed force that defends and supports the government must be overcome, neutralized, or destroyed. Thus, the Union military objective was also unambiguous: overcome the Confederate military so that the Rebel government could be disbanded.

An offensive strategy was certainly required, but it needed to be a strategy for a very rapid and decisive offensive. If the fighting lasted for a lengthy period, the suffering and destruction might be such that full union could be impossible to achieve for generations. Quick victory was also an imperative because of the political situation within the Union. A drawn-out struggle would breed war weariness and undermine the Union war effort. Additionally, the Union victory had to be decisive, because there could be no compromise with the supremacy of the Constitution and the illegality of voluntary secession. If either of these principles were not maintained, the concept of a United States would be in constant danger from recalcitrant states.

The Union's situation was not favorable for achieving either a quick or a decisive victory. The regular army was pitifully small and equipped to fight frontier Indians. Many of the army's most capable officers resigned to serve their home states in opposition to the Union. Burdened with these difficulties, Union forces had to take the offensive, but how?

Winfield Scott, veteran of the War of 1812, hero of the Mexican War, and general in chief of the United States army, had a plan that would exploit the Union's crushing superiority in manpower, resources, and industrial power and attack the Confederacy's weaknesses in those same areas. Scott envisioned a tight naval blockade of the Confederacy's long coastline and seizure of the Mississippi River to cripple the South's economy. The blockade, combined with the limited industrial capabilities of the Southern states, would deny the Confederate army the wherewithal to wage war effectively. The Confederate army would slowly deteriorate as would the entire Rebel economic situation. The consequences of rebellion would be brought home to average Southerners in terms of empty stomachs and pocketbooks.

While the South deteriorated under the pressure of the blockade, the Union army would expand to the proportions required. Northern factories would provide the finest equipment and most sophisticated weapons. Time would be available for proper training and the selection of capable officers. Finally, using river lines of approach, particularly from the west, the Union army would crush the demoralized and ill-equipped Rebels. Scott's "Anaconda Plan" aimed at a militarily efficient victory, but it would not be a speedy victory.

Lincoln, a man with no military experience, needed a quick victory. He wanted a cordon offense, that is, simultaneous offensive pressure around the periphery of the Confederacy. Such an offensive would make maximum use of vastly superior Union resources and present the Rebels with the impossible situation of trying to be militarily strong everywhere with inferior resources. The result, he believed, would be rapid Confederate disintegration. If such a plan could be executed, the Union victory would be both quick and decisive.

The actual differences between Scott and Lincoln had more to do with time than concept. However, Scott realized that the Union army and navy were simply not capable of such a massive undertaking and had to be greatly expanded. Proper equipment had to be provided and capable officers found. Finally, raw recruits had to be trained and disciplined if the army was not to be a mob. Scott realized that time was needed, but Lincoln demanded immediate action.

What eventually evolved was a strategy similar in concept to Scott's plan but compressed in time to suit an impatient Lincoln. Rather than a fully coordinated cordon offensive, the Union effort was, in the beginning, two separate wars on two different fronts. West of the Appalachian Mountains, Union generals (the most successful being an obscure man by the name of Ulysses Simpson Grant) struggled to capture the length of the Mississippi River. The plan was to cut the Confederacy in two from north to south. From this base of operations, the forces in the west could then attack to the east, particularly toward the vital rail centers at Chattanooga and Atlanta.

In the Eastern theater, Lincoln and many others in the government believed that Washington was seriously threatened with a Confederate attack because the capital was, in effect, on the front lines. Such an attack would not only cause panic and destruction, but would damage Union credibility with foreign governments and perhaps add fuel to the fire fanned by the Copperhead movement. The fear that a Confederate army would march on Washington caused Lincoln to insist that significant forces guard the city at all times. As a result, inordinate attention was paid to Lt. Gen. Thomas J. "Stonewall" Jackson's Shenandoah Valley campaign in 1862, which was only a Confederate diversionary movement. The demand for troops to protect Washington also frustrated Maj. Gen. George B. McClellan as Lincoln withheld troops that McClellan had designated for the Peninsular campaign against Richmond.

The Confederate capital beguiled the Union leaders. It, too, was on the front lines, barely 100 miles from Washington. Union planners envisioned a drive toward

Richmond, which, they believed, would be fiercely defended. The decisive battle that would destroy the Confederate army would be fought in front of the city, and the war would quickly be over. Such a plan had much to offer as long as the purpose was to draw the Rebel army into decisive battle. However, as time wore on, Richmond itself became the objective.

The basic idea of a decisive battle fought for the Rebel capital had considerable merit when considered in light of Union political objectives. Lincoln's desire that the war be short and decisive was wholly appropriate and a campaign toward Richmond offered the opportunity for a quick and decisive victory. Execution of the plan rather than the plan itself was the problem. In this respect, some of the blame must be laid at Lincoln's feet for forcing the action before the Union army was fully prepared. The most serious deficiency was in senior leadership. The army suffered defeat after defeat as George McClellan, Gen. John Pope, Ambrose Burnside, and Gen. Joseph Hooker successively tried to lead the blue-clad troops to Richmond. Had Lincoln taken the time to ensure that the army was well trained and well officered, the traumatic defeats at Manassas, Fredericksburg, and Chancellorsville, to name but a few, might have been avoided. The instant action Lincoln desired would have been delayed by some months, but in the long run the course of the war might have been significantly shortened.

The Confederate objective was also simple and straightforward—to defend itself from "foreign" invasion and thus protect the sovereignty of the Rebel states. Such objectives clearly dictated a defensive strategy making it possible for the Confederates to fight, in Clausewitzian terms, the strongest form of war. Despite the military advantages peculiar to a defensive strategy and the moral advantage of fighting to defend hearth and home, the Confederacy was in a disadvantageous position.

Economically the Rebel states were the poor cousins of the Union, particularly in terms of those heavy industries important to a war effort. Northern factories produced over 90 percent of the nation's firearms and railroad equipment. Perhaps most important, the North had a comprehensive rail system, while the South had a series of independent railroads built primarily to get plantation products to port cities. Only one trunk rail line connected the far-flung eastern and western Rebel states.

Manpower and political organization were also areas of Rebel weakness. Many figures have been used to estimate Northern and Southern manpower ratios. Some authorities count the slave population and others do not. Still others treat slaves as less than a full person available for combat, but count them an advantage because of the work they accomplished on the home front. All things considered, the best estimate of relative combat potential seems about five to two. Added to the South's manpower problem was its fragmented command of available manpower. The Confederacy was built on the concept of states' rights. The Rebel government never achieved the required centralized control over Confederate assets and never achieved an effective centralized command structure for its military forces.

The Confederate problem was to defend a vast territory despite the disadvantages of an inferior economic base and lesser manpower. One solution to the dilemma was to obtain foreign allies much in the manner of the rebels during the American Revolution. Cotton, needed by the factories in Europe, offered an economic bargaining chip. However, Europeans would not back a sure loser on the battlefield. Thus it was incumbent upon Southern armies to demonstrate their viability.

Robert E. Lee, first as the military adviser to Confederate President Jefferson Davis and later as commander of the Army of Northern Virginia, settled upon an offensive-defensive strategy. Although strategically on the defensive, the Rebel armies would often be tactically on the offensive. By taking the offensive, Lee hoped to dictate the time and place of battle. With tactical victories, particularly successful forays into Union territory, he hoped he could set the stage for intervention by sympathetic foreign governments.

As a result, after fending off Union attacks toward Richmond, Lee invaded the Union, first Maryland and later Pennsylvania. Both expeditions ended in disastrous losses for both sides, but losses that the Confederacy could afford less considering its manpower disadvantage. Meanwhile, in the West, the Union captured the length of the Mississippi River and began a methodical campaign to seize Chattanooga and Atlanta.

Lee has also been criticized for a Richmond or Virginia fixation. Lee was a Virginian who resigned from a senior post in the U.S. Army to offer his services to his state, not to the Confederacy. It must be said in Lee's defense, however, that the abandonment of the Confederate capital without a spirited defense would have been looked upon unfavorably by possible foreign allies and certainly would have demoralized the home front. Virginia was the most important and prosperous of the Confederate states, and had it not been defended, its return to the Union would have been a political and military disaster of the first magnitude.

Lee's offensive-defensive strategy led to terrible and irreplaceable losses during his forays into the North. Thus the battlefield execution of these plans left much to be desired, but the basic strategy was probably correct considering the circumstances. Lee had to make some attempt to control the pace and place of the action or risk being overwhelmed and outmaneuvered by vastly superior resources.

Political Considerations

The line between purely military and purely political considerations was vague and shifting. The result was inevitably some level of tension between political and military leadership on both sides and a certain amount of what one might call the "low politics" of war (marked by petty bickering, political posturing, and the like). The Union side probably had the more severe problem, partly because President Lincoln was not himself a military man and did not fully comprehend the military mind (a problem so severe that he had to appoint Gen. Henry Wager Halleck as his chief of staff to translate messages to and from his field commanders).

At the same time, President Lincoln had extraordinary difficulty finding military leadership willing or able to carry out the types of operations that would achieve the political objective, especially in the Virginia theater. The Union record was abysmal until Grant was transferred from the West (where, with little publicity, he had been doing quite well). Tactically, Grant made his share of mistakes; but he understood the strategic objective of the war and with the considerable assistance of William T. Sherman and Philip H. Sheridan, he was able to translate the political objective into a successful military objective where others had failed.

In the Southern case, President Davis had graduated from West Point, seen service in the Mexican War, and been secretary of war in the James Buchanan administration. As a result, he considered himself well qualified to direct the military effort and did so personally. He acted as commander of all Confederate forces until the war's waning months when that title was given to Lee.

Despite these kinds of political diversions, the war was fought with more important political considerations in mind. Domestically, a major concern throughout the war was influencing public opinion in the North. Since the war was not particularly popular and there was a sizable peace movement in the North, a prime Southern purpose in following a strategy of attrition was to drag out the war to the point that Northern cost-tolerance, always a fragile commodity, would be exceeded and the Union would simply quit the contest (a strategy closely paralleling that of Washington during the Revolution and subsequently used against the United States by North Vietnam). In other words, the South did not need to win the war to achieve its independence; rather, it needed to avoid losing only long enough for Northern public opinion to turn decisively against the war. Had it not been for the succession of Southern military reverses beginning at Gettysburg and Vicksburg and culminating with the fall of Atlanta, the strategy might well have succeeded.

Even if the political objective had not suggested a war of attrition aimed at undermining the Union's willingness to persevere, the South's physical circumstances made such an approach the most reasonable way to fight. Fighting on the defensive meant the North, which had to attack and destroy the Confederate armies to win, would be fighting away from home in hostile, unfamiliar territory, which was bound to create military and morale problems. At the same time, fighting on both the strategic and tactical defensive was likely to eventuate in low casualties for the manpower-poor South, especially given the emphasis on frontal assaults at the tactical level.

There were, of course, variant opinions about how to achieve the political objective, most notably Lee's concept of the offensive-defensive. Beyond its sheerly military aspects as already described, this strategy sought to have a political impact as well, attacking Northern morale by demonstrating Northern vulnerability to attack. The purpose of invading the North was, of course, not conquest, an objective clearly beyond the Confederacy's political aim as well as its military ability. Rather, part of the purpose was demonstration: in the 1862 invasion of Maryland to show foreign governments the Confederacy was a military force worthy of recognition; in the 1863 Pennsylvania campaign to put a major federal city (Philadelphia, Baltimore, or even Washington) in danger and hence to stir antiwar sentiment.

If the South's purpose was to exceed Union cost-tolerance, Lincoln's problem was how to avoid that fate. As suggested already, he was hampered early in the contest by the absence of a compelling political objective around which to unite his population and the inability to identify competent commanders. The ideal solution would have been a quick and decisive victory that would nominally test popular will, and it was this hope that gave rise to cries of "On to Richmond." Once thwarted, Lincoln's major need was for victories that would show progress toward the desired end, but for the first two years of combat, the only successes were in the Western theater, whereas the Army of the Potomac faced a seemingly invincible force in Lee's Army of Northern Virginia.

The differing levels of support for the war in the North and South and the delicacy of Lincoln's problem in maintaining public willingness to continue are well illustrated by the two sides' approach to conscription. In the South where the war was very popular, a universal conscription system was quickly adopted and effectively enforced, with the result that approximately 80 percent of eligible males either volunteered or were drafted into the service. By contrast the Lincoln government was reluctant to institute any kind of draft early in the war for fear of antiwar backlash. Instead it relied on appeals to governors to raise volunteer militia units to meet manpower needs. This system had serious military disadvantages. Usually the units were organized by local politicians who had no military experience, but who were elected as commanders. At the same time, since these troops arrived as units, they could not be integrated into existing veteran units, meaning the Union was constantly fighting with inexperienced units and veteran units were perpetually undermanned. When a conscription system was finally introduced, political necessity (the draft's unpopularity) required that it be easy to avoid. Thus the system featured multiple sources of exemption, and a draftee could meet his commitment by hiring someone to take his place. Only about 6 percent of Union forces in the war were conscripts, yet even this limited form of draft resulted in numerous riots.

International politics was also a concern. Recognition of the Confederacy by the European powers would create legitimacy for the Confederate government as the representative of a sovereign nation-state, meaning that, in international legal terms, reunion could only be achieved by aggression across an international border. The U.S. government's legal justification for the war rested on refusal to recognize the right to secede, meaning that its military action was legally no more than restoration of order within territory still part of the nation. Recognition of the South by third parties would have brought that rationalization into question and strengthened the case for the Northern peace movement.

Materially, the Confederacy needed recognition to ensure continuing trading relations with Britain and France, the traditional consumers of Southern cotton.

Both countries were major processors of cotton, but more important, they were potential suppliers of armaments Southern industry could not produce in adequate supply. Recognizing this weakness of the Southern economy, the Union blockaded Southern ports, both to deny the Confederacy access to outside supply and to ensure that European ships would be subject to seizure in the event they tried to trade with the South.

There were, at the outset, some temptations for the British and French to offer recognition. Both countries were heavily dependent on Southern cotton for their textile industries and the plantation system provided a good market for European goods, especially luxury items. Freedom from protective tariffs erected for the benefit of Northern industry and from having to ship cotton to market on American ships would mean lower cotton prices. Thus trade could be expected to expand. Moreover, there was growing recognition that an expanding United States would become a power to be reckoned with sometime in the future. Fragmenting that developing giant into two smaller and weaker states had its own independent appeal.

The Confederate leadership tried to push the British and French governments to grant recognition, and a lively competition between Union and Confederate diplomats in European capitals ensued. The Southern strategy for forcing positive decisions revolved around using "King Cotton" as a weapon, but the strategy proved disastrous. To create pressure in London and Paris, the Confederacy decided to withhold the 1861 cotton crop from the market, letting it pile up on Southern wharves until diplomatic recognition occurred. Although getting that crop past the Union "paper blockade" would have been relatively easy in 1861 (thus gaining needed foreign capital to buy weapons), by 1862 the blockade was real and cotton's commercial potential was greatly reduced. In effect the South squandered a year's crop and the profits it could have brought. As the blockade tightened, Britain turned to and nurtured cotton production in its colonies (especially Egypt and India) so that by war's end, dependency on Southern cotton had largely evaporated. At the same time, the Union expanded its Midwestern grain trade with Europe. King Corn replaced King Cotton.

The Confederacy also had to establish its political viability before European nations would recognize the young government. To do so required demonstrating the ability to resist reunion, which translated into appearing to be a military winner. It is generally conceded that demonstrating that capacity was a major reason that Lee decided to extend his unsuccessful 1862 campaign into the North. If Confederate armies could successfully forge their way into Union territory, it was reasoned, their prowess would be established and European qualms would be overcome. When McClellan stopped the Confederate advance at Antietam and forced the Army of Northern Virginia to retreat back across the Potomac, that hope began to fade, as did the chances of European recognition of the Confederacy. Lincoln's declaration of the Emancipation Proclamation provided the final coffin nails for those hopes.

Military Technology and Technique

Although often called the first modern war because of the nature of the implements used, in truth the war served as bloody transition from the limited wars of the eighteenth century to the mechanized wars of the twentieth century. The contrasts between the old and the new were particularly stark. Steam power, particularly the railroads, was one of the ingredients critical to victory, but reliance on muscle power remained pervasive. Impersonal and unseen military staff work was critically important to the successful operation of mass armies and yet so were the personal leadership and bravery of frontline commanders. The "indirect approach" exemplified by deft maneuvering of troops was common and yet so were old-fashioned and bloody frontal assaults. The Civil War was warfare in transition.

The most influential technological development was the railroad. Railroads made possible the vast expansion in the scope of war. Unlike the individual small battlefields of the past, the Civil War featured separate and far-flung theaters of war, each populated by mass armies transported and supplied by rail. The addition of the telegraph meant that not only could mass armies fight across vast areas, but they could also be centrally controlled in a common coordinated effort. Railroads offered considerable advantages in mobility and speed, but with these advantages came considerable "baggage." Because of their importance to both sides, strategy began to revolve around rail lines. Armies in the field became tied to rail lifelines and thus had to protect those lifelines at all costs. Often, offensive maneuvers were aimed at seizing or cutting vital enemy rail links while defensive maneuvers were often aimed at protecting those same lifelines. Thus such relatively insignificant (at the time) settlements as Chattanooga, Atlanta, and Petersburg became vitally important because they were major railroad junctions. Although railroads may have been the most important technological advance, the industrial revolution did not overlook improvements in the tools of war themselves.

The standard infantry weapon remained the single-shot, muzzle loaded rifle. However, the Civil War weapon had a rifled barrel giving it much greater accuracy over a much longer range than smoothbore weapons used previously. The most common rifle had a .58-caliber bore and was fired using a percussion cap. Although it had a 1,500-yard range, it was most effective and quite accurate at 500 yards, a tenfold increase over Revolutionary War smoothbore weapons.

Both breech- and magazine-loaded weapons had come into use before the Civil War. The Sharps and early Winchester rifles and the Hall and Spencer carbines were all in private use. They were also used to a limited extent by the contending Blue and Gray armies. However, neither government chose to make them a standard weapon because of the extensive retooling time and expense required to convert government production facilities.

The standard artillery pieces of the Civil War were the 6-pounder bronze gun and the 12-pounder howitzer. Although really effective only at close range (less than half a mile), these smoothbore guns remained effective anti-infantry weapons using grapeshot. Rifled artillery pieces were also used but in far fewer numbers. Although they were much more accurate at longer ranges, they could not be sighted accurately against distant targets. More important, the rifled shell had a relatively small explosive charge, an important factor when the enemy is well dug in.

The Civil War saw numerous other advances in the technology of war. The first rapid-fire weapons (based on Richard Jordan Gatling's concept of revolving barrels) saw limited service during this conflict, as did a primitive submarine. Neither had a significant impact upon the war's prosecution or outcome.

However, the use of armored ships not only had considerable impact but also foreshadowed the all-steel fleets that would become standard by the end of the century. Armor more than proved its worth in combat and made unprotected wooden-hulled vessels obsolete for close combat. The design of the Northern ironclad "monitors" with their revolving gun turrets was the first attempt at a design that, in modified form, would become standard in the age of the great battleships.

The use of rifled infantry weapons with their highly accurate fire at long range meant that the linear infantry tactics of the eighteenth century and the Napoleonic era would have to change. Infantry was now under constant and accurate fire while still hundreds of yards distant from the intended goal. The close-order formations of the previous age were suicidal and quickly disappeared. Officers still attempted to align the looser formations to a certain degree in both offensive and defensive modes. On the offensive, a reasonably straight wave of attackers would increase the shock effect when the attackers struck the defender's lines and allow the defender little opportunity to reinforce points of breakthrough. On the defensive, aligned troops could increase the effectiveness of volley fire.

Infantry, whether attacking or defending, went to ground to avoid the accurate long-range fire of opponents. On the defensive, breastworks became the order of the day. On the offensive, "attack by rushes" eventually became a common practice. Infantry would charge forward and then fall to the ground after a short rush. They would regain their feet for another short rush and then again seek cover. The objective was, of course, to reduce exposure to hostile fire.

In the Civil War firepower dominated the battlefield. Casualty rates increased dramatically, with instances of 80 percent casualties in a given unit during a single engagement. Today such a casualty rate would be shocking. The impact during the Civil War was even greater considering the primitive medical treatment available. A serious wound in the trunk of the body was likely to be fatal. Serious wounds in the limbs usually resulted in amputation. To make matters worse, the causes of infection had not been discovered. Surgeons typically did not clean their instruments before or between operations, and the result was added suffering and death. The army of survivors maimed by the surgeon's knife represented an embittering postwar legacy that hindered the process of reunification.

All of these factors led to the inescapable conclusion that "modern war," as practiced during this period, depended upon a strong economic base. Mass armies required massive amounts of weapons, munitions, and other supplies. (Munitions were required in previously undreamed of quantities because of the size of armies and because longer-range weapons were fired more often both on the offensive and on the defensive.) Such massive amounts of weapons and munitions could not be provided by cottage industry. Moreover, the armor of ships and the heavy equipment required to operate railroads could be provided only by an industrialized economy.

Because of these factors, the Confederacy was in an almost untenable position. The South had almost no heavy industry (the exception being the Tredegar Iron Works in Richmond), and it did not have a first-class rail system. The South trailed even in such mundane requirements as the production of uniforms. Although blessed with an abundance of the raw material (cotton), Southern uniforms were handmade. In the North, the Howe sewing machine and the McKay shoe-stitcher manufactured uniforms in great quantity.

Military Conduct

Tradition has it that one Edmund Ruffin, a Confederate firebrand from Virginia, fired the first shell at Fort Sumter in Charleston Harbor and thus the first shot in the Civil War. The time was 4:30 A.M. on Friday, 12 April 1861. Federal authority was physically challenged, and shot and shell were used against Union soldiers; there was no turning back. The Confederate commissioners sent to Washington by Jefferson Davis to negotiate a settlement short of war had been rebuffed, and they departed from the Union capital on the day before the guns fired upon the beleaguered fortress in Charleston Harbor.

Lincoln moved rapidly to prepare the Union for war. On 15 April he declared that an insurrection existed and called out 75,000 militiamen from the various Northern states. On the 19th he declared a naval blockade of the Confederacy. In May, Winfield Scott proposed his Anaconda Plan, and federal troops began massing in the Washington area. By July an impatient Lincoln was more than ready for his army to move south. "On to Richmond," was the cry. The Confederate legislature was due to meet in Richmond on 20 July, and Union patriots wanted to overrun the new Rebel capital before the meeting took place.

General McDowell and his Union Army departed their base on the Potomac on 16 July and proceeded toward Manassas, Virginia, to face the first obstacle on the road to Richmond, General P. G. T. Beauregard and his Rebel army. McDowell outnumbered Beauregard 35,000 to 20,000, but west of Manassas in the Shenandoah Valley, General Joseph E. Johnston had 12,000 Confederates ready. Leaving a small covering force to demonstrate and deceive the local Union commander, Johnston's forces boarded trains on the Manassas Gap Railroad and arrived at Manassas in time to tip the scales in favor of the Confederates. Among those who came with Johnston was one Thomas J. Jackson, who during this battle earned the nickname "Stonewall" for his fortitude under heavy fire as he rallied retreating Confederate units.

The battle was fought on 12 July 1861. Initially, McDowell's forces were successful and the Rebel forces fell back (it was Jackson's famous action of standing fast "like a stone wall" that helped to stop the retreat). Regrouping, Beauregard and Johnston counterattacked, and the Union forces began to fall back. Green Union troops turned an orderly retreat into a disorganized rout as they fled toward Washington amid bag, baggage, and the many spectators from the capital that had come to see the expected great victory. Had the Southern forces been able to mount an organized pursuit, it is entirely possible they could have swept into the Union capital.

McDowell's rout highlighted the need for training, discipline, and better leadership. First Manassas (there would be another battle in this same area) also indicated how costly the war would be. In this brief battle, the combined casualties numbered nearly 5,000 and both sides ended where they started. McDowell's forces licked their wounds (physical and mental) on the Potomac while the victorious Rebels remained in northern Virginia.

Few major battles were fought in the Eastern theater until the spring of 1862, as Lincoln and his generals argued over what course of action they should follow. In February 1862, however, the focus of the struggle shifted to the Western theater. There, Union Brig. Gen. Ulysses S. Grant moved boldly to seize Confederate Forts Henry and Donelson on the Cumberland and Tennessee rivers and thus began clearing the upper reaches of the Mississippi River and its tributaries. Grant's victories were the opening salvos in the campaign that would seal the Confederacy's doom.

Grant moved south from Fort Donelson as part of a three-pronged Union offensive. Grant moved down the Tennessee River toward the northern borders of Mississippi and Alabama, and Brig. Gen. Don Carlos Buell and his Army of the Ohio moved south from Kentucky to Nashville (vacated by Confederate Gen. Albert Sidney Johnston after the fall of Fort Donelson). Buell was to continue his southwestward movement and join forces with Grant at Pittsburg Landing on the Tennessee River. Meanwhile Gen. John Pope proceeded down the Mississippi River from Cairo, Illinois, clearing the river of Rebels as far south as Memphis by June.

Confederate General Johnston faced a serious situation. Rebel forces were scattered throughout the western portion of the Confederacy and the Union forces were concentrated in a well-coordinated offensive that threatened to divide the Confederacy from north to south. Johnston's first move was thus to begin concentrating his forces, bringing General Beauregard, erstwhile hero at First Manassas, from Mississippi and Gen. Braxton Bragg from Alabama. Johnston concentrated about 40,000 men at Corinth, Mississippi, just south of the gathering Union formations encamped at Pittsburg Landing and around nearby Shiloh Church.

Johnston attacked early on the morning of 6 April and surprised Grant's ill-prepared forces. Only Grant's personal efforts on the battlefield finally established a defensive position that held the Rebels as the day drew to a close. The next

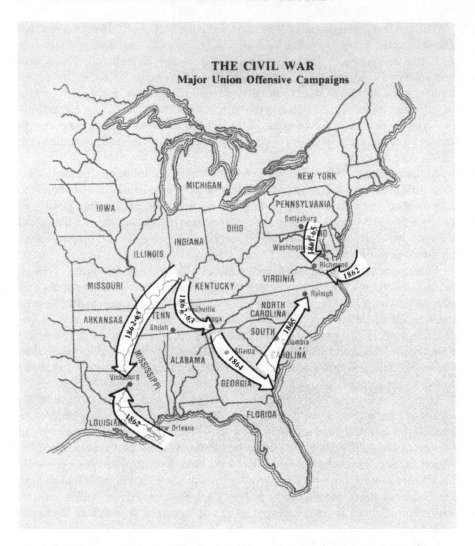

THE CIVIL WAR
Major Union Offensive Campaigns

morning Grant counterattacked. After hard and bloody fighting, the Rebels were forced from the field in a disorganized retreat. Shiloh was a terrible defeat for the Southerners, who suffered 11,000 casualties, including Gen. A.S. Johnston, but Grant's victory was none too sweet. He had nearly been badly defeated, and had lost about 14,000 men.

For the Confederates in the West, the situation now approached desperation. In addition to the defeats in Kentucky, Mississippi, and Tennessee, Adm David Farragut forced his way up the Mississippi River in April. In May he sailed upriver and occupied both New Orleans and Baton Rouge. The Union admiral also had the audacity to sail farther up the river and bombard Vicksburg, the last major Rebel river stronghold. But Vicksburg was not New Orleans (which fell without a shot). Vicksburg would require all the talent and power that Grant could muster.

While the situation in the West became ominous for the Rebels during the first six months of 1862, events in the Eastern theater were totally different. As the idle armies sat facing each other after First Manassas, George B. McClellan, the new Union commander, devised a bold plan to outflank the Rebel forces of Joseph E. Johnston that blocked the route from Washington to Richmond. McClellan planned to sail down the Potomac to Chesapeake Bay and land at Fort Monroe on the Virginia coast. From there he would quickly march up the peninsula between the York and James Rivers and take Richmond "from the rear" before Johnston could react.

Thanks to Confederate sympathizers in the Washington area, Johnston was soon aware of McClellan's plan. While McClellan waited until 17 March to move, Johnston withdrew his army from Centreville and marched toward the same area. Thus the real purpose of McClellan's giant flanking maneuver—surprise—was lost.

By early April, McClellan had landed 50,000 troops at Fort Monroe and had begun to move up the peninsula. A skillful defense by Confederate Maj. Gen. John B. Magruder delayed the advance and Union forces did not come into position near Richmond until 24 May, having advanced at an average rate of only two miles per day.

Johnston was outnumbered almost two to one (although faulty Union intelligence convinced McClellan that Union forces were badly outnumbered) and preparations were under way to remove the government from Richmond. Near Centreville were General McDowell and 40,000 more Union troops, who sat guarding the route to Washington. These forces were to move overland and join McClellan to add the final crushing weight to Union forces. Johnston and General Lee, then military adviser to Confederate President Davis, realized a successful Union linkage would seal the fate of Richmond. To prevent this situation they planned a large-scale diversionary action in the Shenandoah Valley by the brilliant and reclusive Gen. Stonewall Jackson.

The Shenandoah Valley was a natural invasion route from western Virginia into Maryland that threatened both Washington and Baltimore. For defensive purposes, the Union had stationed Gen. Nathaniel Prentiss Banks with 23,000 men near the head of the valley. Jackson's mission was to prevent any of Banks's troops from joining McDowell and to prevent McDowell from joining McClellan at Richmond. Jackson began his campaign in the valley with fewer than 4,000 effectives.

Jackson's campaign was one of brilliant rapid maneuver, in which he isolated elements of the superior Union forces in the valley and defeated them. At one point, Jackson had totally routed Banks and chased the Union forces from the valley and across the Potomac in complete disarray. Lincoln, fearing for the safety of the capital, ordered Gen. John C. Fremont into the valley from the west and McDowell into the valley from the east (and away from Richmond) in an effort to trap Jackson. Jackson escaped the trap and defeated the forces of Fremont and McDowell in separate battles.

Jackson was entirely successful in preventing McDowell from joining McClellan. After his final battle in the valley, Jackson added insult to Union injury by slipping away undetected and joining Confederate forces in front of Richmond. Lee was now in command of the forces around Richmond, and he massed his army to attack an isolated portion of McClellan's army on the north side of the Chickahominy River. Although the attack was badly handled, the Union forces were defeated and began a skillful withdrawal. Finally, McClellan yielded to fears that he was badly outnumbered and ordered a general withdrawal to Harrison's Landing, a base of operations on the peninsula. Desperately trying to turn the retreat into a rout and a decisive Rebel victory, Lee attacked the retreating Union Army again and again in a series of engagements known as the Seven Days' Battle. But the Union troops would not be routed and Lee's cherished decisive victory escaped his grasp.

Thus by July 1862, the status of the contending armies depended upon the theater considered. In the West the Confederate situation was rapidly deteriorating. Only Vicksburg remained as a major bastion on the Mississippi River, and the vital rail junction at Chattanooga would soon be threatened. Major action for the remainder of the year would continue in the Eastern theater, where the South had been more successful.

In mid-July, while McClellan and his army huddled in Harrison's Landing, Jackson was sent north toward Gordonsville to deal with the reorganized forces that had earlier opposed him in the Shenandoah Valley. The Union commander, Maj. Gen. John Pope, had collected 47,000 troops. This formidable force posed a serious threat to Lee's flank, particularly when combined with McClellan's force of 90,000 at Harrison's Landing.

On 6 August Jackson attacked Pope at Cedar Mountain but achieved only a tactical draw. Later, however, Jackson withdrew and moved his forces around the Bull Run Mountains to a position near Manassas, behind and directly across Pope's communication and supply line. Pope reversed course, advanced on Manassas, and on 29 August attacked Jackson. The result was another disaster for Union forces and Pope withdrew into Washington. Second Manassas was a defeat every bit as bitter as the first battle on that bloody ground. In general the Union cause in the Eastern theater was in shambles.

Lee now seized the opportunity provided by federal disarray and took the offensive. On 7 September Lee and his army crossed the Potomac River near Leesburg and plunged into Maryland. Although strategically on the defensive, Lee had several objectives for this tactical offensive. First, he still sought a decisive Napoleonic-style victory over the Union Army that might spell permanent success for the Rebel cause. Second, the Confederates desperately needed to get both sides' armies out of northern Virginia because they had stripped the area of food and forage. Third, an invasion of Union territory might cause Maryland to secede (Maryland's status had always been questionable) and might also impress the British and French governments enough to bring recognition and badly needed help to the Confederacy.

After considerable maneuvering, Lee and McClellan (who was still in command of the Army of the Potomac in spite of losing Lincoln's confidence) met near Sharpsburg on Antietam Creek. McClellan had an opportunity to inflict a decisive defeat on Lee because he had come into possession of Lee's campaign plans, but squandered the opportunity by delaying the movement of his forces. As Lincoln would later comment, McClellan was afflicted by "the slows." Finally, on 17 September McClellan attacked. Although McClellan outnumbered his Confederate foe nearly two to one, the battle was a draw. However, in the face of a continuing influx of Union reinforcements, Lee began withdrawing to Virginia on 18 September. Antietam was the bloodiest *single* day of the war, with more than 22,000 casualties.

Lee's foray into the North had disastrous consequences for the Confederate cause. He did not transfer the fighting out of northern Virginia for long, he did not win his decisive victory, he suffered a large number of casualties, Maryland did not secede, and foreign recognition was now little more than a forlorn hope. On a personal level, however, Lee's reputation as a commander was enhanced as was the reputation of the Rebel army. Lee and his army had carried the war to the enemy and had beaten back a Union army nearly twice the size of the Confederate force.

McClellan, with a typical lack of aggressiveness, failed to follow up on his "victory" as Lee slipped back into Virginia. In fact McClellan did little but rest and resupply his army. Lee made use of the respite to do the same. Finally, on 7 November Lincoln had had enough of inaction and relieved McClellan of command in favor of one of McClellan's lieutenants, Ambrose Burnside. Burnside presented, and Lincoln approved, a complex plan to move on Fredericksburg, Virginia, and to use this important road and rail junction as a base of operations against Richmond. The plan required an assault across the Rappahannock River to seize both Fredericksburg and Marye's Heights just beyond the town.

By the time Burnside had finally forced a crossing on 12 December, Lee was well entrenched on Marye's Heights. On 13 December Burnside ordered a disastrous frontal assault on the heights. At the base of the heights, the Union troops had to attack a sunken road bordered by a stout stone wall. As the Union troops advanced, long-range rifle fire from behind the wall and artillery fire from the heights shattered their ranks. The fire was so withering that no Union soldier ever got to within 25 yards of the wall. It was not for want of effort. Union soldiers surged forward toward the wall 14 times, and the field in front of the wall became a killing ground littered with 6,000 blue-clad casualties. The battle in front of the stone wall at Marye's Heights was, perhaps, the worst example of outmoded eighteenth-century tactics applied in a more deadly era.

On the following day a truce was arranged to tend the wounded and bury the dead. That night Burnside withdrew his entire army back across the Rappahannock River. The war in the Eastern theater was over for 1862. It ended as it began, with Union blundering, Union defeat, and Union retreat. In the West, however, the Union continued to fare well in the last half of 1862.

Following his narrow victory at Shiloh, Grant proceeded west to begin his campaign against Vicksburg. Meanwhile Buell advanced east toward Chattanooga. The defeated Confederate forces under Gen. Braxton Bragg regrouped and, by a circuitous route, marched north through Chattanooga, across Tennessee, and into Kentucky. Buell was obliged to follow and finally forced a fight at Perryville, Kentucky. The result was a draw, but Bragg withdrew south to Chattanooga. Buell retraced his steps back to Nashville, where he was relieved by Gen. W. S. Rosecrans. Bragg and Rosecrans met at Murfreesboro, Tennessee, in a savage battle in which Bragg was forced to retreat south to Tullahoma, Tennessee. Both armies, exhausted after lengthy marching and savage fighting, spent the next six months resting and refitting.

As 1863 opened in the Eastern theater, Lee's army remained in winter quarters on Marye's Heights and Lincoln had relieved Burnside in favor of Gen. Joseph "Fighting Joe" Hooker (whose nickname was more self-designated than earned). Hooker had an audacious plan to defeat Lee decisively. Hooker's plan called for small demonstrations in front of Lee's position at Fredericksburg to "fix" the Confederate army while he moved the bulk of his forces up and across the Rappahannock and Rapidan rivers to descend on Lee's rear. Trapped between this giant "right hook" and the two corps that remained in front, the Army of Northern Virginia would be destroyed. Hooker began his maneuver on 27 April 1863 and by 30 April had consolidated his forces at Chancellorsville, a crossroads in Lee's rear just nine miles from Fredericksburg. And then Hooker stopped!

Hooker's sudden timidity gave Lee a chance to react with a plan even more audacious than Hooker's. First, he divided his forces, leaving only about 10,000 men on Marye's Heights, and moved his remaining 43,000 troops toward Chancellorsville. He split his forces again by sending Jackson and 26,000 men on a flanking march across Hooker's front, while he faced Hooker with only 17,000 men. Jackson was in position by 6:00 P.M. on 2 May and attacked the Union right flank, which quickly broke and retreated toward the center of the Union position. Only darkness and exhaustion halted Jackson's drive from collapsing the entire Union line. That night, while scouting ahead, Jackson was mortally wounded by one of his men, a loss from which the Confederacy never recovered. Jackson was, perhaps, the most talented commander in either army.

The following day Lee continued to attack and Hooker continued to withdraw slowly. Meanwhile Gen. John Sedgwick and his two Union corps attacked the thin Confederate forces on Marye's Heights and drove the Rebels back toward Lee's main forces. On 4 May Lee left Maj. Gen. J. E. B. Stuart (Jackson's successor) with 25,000 men to watch Hooker and turned the bulk of his army against Sedgwick. Hard fighting finally forced Sedgwick back across the Rappahannock, and Hooker joined him on 6 May.

Although shaken by the loss of Jackson, Lee's total confidence in his army led to his second foray into the North. Lee moved westward into the Shenandoah Valley and north across the Potomac, through Maryland and into Pennsylvania. Hooker shadowed this thrust, remaining between Lee and Washington. En route Hooker was relieved by Gen. George Gordon Meade.

The two armies met at Gettysburg more by accident than by design on 1 July 1863. Minor skirmishes occurred during the first day as both armies hurried to concentrate their scattered forces. On the second day Lee attacked and a vicious battle raged, but the Union line held in hand-to-hand fighting. On 3 July Lee made his famous assault on the Union positions along Cemetery Ridge. After a lengthy Confederate artillery barrage, 12,000 infantrymen under Gen. George Pickett advanced toward the distant federal lines. Pickett's charge collapsed under the weight of Union shot, shell, and bayonet, and the next evening Lee and his battered army withdrew during a driving rainstorm. The Rebels had suffered an estimated 28,000 casualties, which proved nearly impossible to replace. The tide had finally turned in the Eastern theater, and the following day disaster befell the Rebels again.

Vicksburg, Mississippi, was a natural fortress blessed with a commanding view of the great river from high atop sheer bluffs. It remained the last major impediment to opening the Mississippi to Union use and was the final link between the eastern and western portions of the Confederacy. After Grant's victory at Shiloh, Vicksburg had to be the next major objective.

Grant's task was not easy. Assault directly from the river was out of the question because of the high bluffs, and interlaced rivers, bayous, and lakes limited the dry ground upon which armies could maneuver. Finally, scattered throughout Mississippi were 35,000 Confederate troops under the command of Lt. Gen. John C. Pemberton, who was bent on keeping Vicksburg in Southern hands.

Grant needed to get his troops on solid ground on the Vicksburg side of the river. Attempts to establish a base of operations north of the city had failed. Solid ground was available south and west of the city, but the problem was how to move his army downriver past the guns of the fortress. After several schemes failed, Grant marched his army down the western riverbank through muddy swampland and forced a crossing south of the fortress, aided by Union gunboats and transports that had run past the city.

On 30 April 1863 Union troops began crossing to the eastern shore at Bruinsburg and embarked on one of the most brilliant, daring campaigns of the war. Grant, remembered today as the relentless "butcher" for his later northern Virginia campaigns, wielded a rapier rather than a cleaver at Vicksburg. In the 18 days that followed his crossing, Grant moved his army more than 200 miles, fought five major battles (all of which he won), and bottled up Pemberton's army in Vicksburg. After twice unsuccessfully attempting to force his way into Vicksburg, he lay siege to the city. More than a month passed before the tired and hungry Rebels finally surrendered on 4 July 1863.

Disaster had struck the Confederates in both Pennsylvania and Mississippi. Meanwhile, General Rosecrans began moving his army, idle since the battle at Murfreesboro the previous January, toward Chattanooga. Rosecrans cleverly maneuvered against Bragg and forced his withdrawal from the city without a fight on

7 September. The Confederates reinforced Bragg with Gen. James Longstreet's corps, which was transferred rapidly to the area by rail from northern Virginia. After some clumsy maneuvering by both sides, Bragg narrowly defeated Rosecrans at Chickamauga Creek just southeast of Chattanooga and forced the Federals back into the city.

Now it was the Union's turn to rush in reinforcements. After securing an adequate supply line into Chattanooga, Grant assaulted Confederate positions overlooking the city on Lookout Mountain on 24 November and successfully assaulted the main Confederate forces on Missionary Ridge the next day. The broken Rebel army retreated southward into Georgia. The vital gateway into the heart of the deep South was open. In the long run, the fall of Chattanooga with its opening of the deep South was as decisive as Gettysburg or Vicksburg. The tide had turned, and the collapse of the Confederacy was only a matter of time.

Grant's reward was to be named general in chief of all Union armies. In the Eastern theater, he ordered Meade to attack Lee's army continuously, using the overwhelming resources of the Union Army and allowing Lee no time to recuperate. Additionally, he ordered Maj. Gen. Franz Sigel (and later General Sheridan) to attack in the Shenandoah Valley and to destroy Confederate war resources. Meade engaged Lee in a continuous series of bloody battles from early May through the end of June 1864. Beginning near Chancellorsville, both armies sidestepped to the south and east, fighting the major battles of the Wilderness, Spotsylvania Court House, North Anna, and Cold Harbor. Finally, they arrived at Petersburg, another vital rail center that serviced Richmond. Petersburg had been heavily fortified and there the federal campaign bogged down until March 1865.

In the west Grant placed General Sherman in command and directed him to destroy the forces of Joseph E. Johnston (who now commanded the forces in northern Georgia that had retreated from Chattanooga) and to destroy Confederate war resources to the maximum extent possible. Sherman set out from Chattanooga toward Atlanta. Johnston fought a skillful delaying campaign hoping to hold off Sherman until the Union presidential elections in the fall of 1864. If a major Union victory could be avoided, growing antiwar sentiment might sweep Lincoln out of office. Jefferson Davis, seeking a victory rather than a skillful retreat, replaced Johnston with Gen. John B. Hood in front of Atlanta. Hood attacked Sherman unsuccessfully three times in late July and was forced to withdraw from Atlanta on 2 September. Hood retreated and then set out northward into Tennessee and reached Nashville where his army was destroyed by Gen. George Henry Thomas on 15 and 16 December 1864.

On 15 November 1864 Sherman's 62,000 men departed for Savannah. Leaving the railroads in Atlanta in ruins and the city's public buildings and warehouses in flames, the Union Army systematically burned and plundered its way across Georgia on a 60-mile front. On 21 December Sherman entered Savannah and, after resting and replenishing his supplies from Union ships, turned north into the Carolinas. On 17 February he entered Columbia, South Carolina, and then pressed on toward North Carolina. Joseph E. Johnston was restored to command, but he had only ragtag forces and Sherman brushed him aside first at Averysboro and then at Bentonville.

Meanwhile Grant finally broke through Lee's defensive positions at Petersburg on 1 April 1865 at the battle of Five Forks. Lee retreated west and Grant pursued, sending General Sheridan's mounted forces to cut the Confederate commander's line of retreat. Within a week, all hope for the Rebel army was lost and Lee surrendered on 9 April 1865 at Appomattox Court House. On 26 April Johnston surrendered his forces near Raleigh, North Carolina. The war was effectively over.

Better State of the Peace

The fighting ended after nearly four years of bloody conflict that produced more than a million dead and wounded. The hostilities formally ended when Gen. Kirby Smith surrendered Confederate forces west of the Mississippi River on 26 May 1865. The North had won the war on the battlefield and was hence able to impose its political objectives of reunion and emancipation of the slaves.

In the terms we developed earlier, the Union had overcome two of the three aspects of Southern hostile will and ability. Lee surrendered his army because he believed that it no longer had the ability to continue to contest successfully. By the time the Army of Northern Virginia reached Appomattox, a combination of Grant's hammer blows and desertion had reduced Lee's force to a shadow (about 15,000 effectives) of its former size. Lee faced the Army of the Potomac that had more than 100,000 effectives. Given those odds and the physical and material condition of his forces, Lee concluded further resistance could result only in needless, futile carnage. At the entreaties of such leaders as Lee and Davis, the soldiers went home in peace, rather than continuing the fight through guerrilla warfare (the outcome that Lincoln feared most). When the last Rebel army surrendered, Confederate hostile ability ceased.

By the end, Southern hostile will measured by willingness to continue the war (morale) had eroded badly as well. Until the winter of 1864–65 public support had never wavered, but the effects of bringing the war directly to the Southern people were beginning to take their toll. The chief architects were Sherman, Sigel, and Sheridan through their scorched-earth campaigns. Sherman's unopposed march through Georgia and the Carolinas and Sigel's and Sheridan's rampage through the Shenandoah Valley reduced the Confederacy's chief sources of food to ashes. The result was large-scale hunger and starvation in the Confederate armies and civilian population.

In these circumstances desertion became a major Southern problem for the first time in the war (it was a problem for the North throughout the conflict). Many Confederate soldiers, hearing of the plight of family and loved ones, simply laid down their weapons and went home to try to alleviate the suffering. This manpower drain contributed to the depletion of troop strength and the diminished ability to resist.

The question that remained was how or whether the Union would overcome hostile will defined as resistance to the policies (political objectives) the South had opposed. As is usually the case, overcoming hostile ability on the field was a necessary precursor to imposing political will. The U.S. government then needed to bring about compliance with and acceptance of politics that had led to secession in the first place, and that had provided fuel to continued resistance for four bloody years.

The question was fundamental because the issues that led to war were fundamental. At issue, after all, was whether two distinct ways of life (societies) could continue to coexist in America, and the answer that made war inevitable was that they could not. Since the solution of separation into two sovereign nation-states had failed in combat, ultimately the peace could be won only by creating a national societal structure in which the regions would become compatible. That meant the transformation of Southern society from its plantation basis into something that resembled the rest of the country.

The direct political objectives of the war were symptoms of this basic issue. The goal of reunion had been accomplished militarily. Politically (partially through the counsel of Confederate leaders who beseeched their population to accept reunification), it ceased to be an issue shortly after the fighting ended. Emancipation, on the other hand, remained contentious because of its social effects. If, as argued earlier, slave labor was the key to the Southern economy and hence social system, then the destruction of this form of labor was the key to changing the Southern way of life. Military victory made the political emancipation of the slaves a reality. The problem was to gain the acceptance of the white population of the former Confederacy for the new status of blacks. The solution required transformation of the racial attitudes that had justified holding blacks in bondage and adjustment to the consequences of a new economy in which everyone was a wage earner. For the peace truly to be won, the citizens of the South not only had to accept the implications of emancipation, they had to embrace them as right.

For those who would direct the reintegration of a slaveless South into the Union, there were two broad options. One approach was conciliatory, a peace settlement wherein the Southern states would be readmitted into the Union with a minimum of fanfare or conditions, wherein efforts would be made to lessen the ravages of war, and wherein physical coercion would be minimized. This was the solution that Lincoln desired, but an assassin's bullet ended his ability to pursue it five days after Lee surrendered.

The alternative was a punitive peace, and with President Lincoln dead, the forces who felt the South should be punished for its attempted secession prevailed. Led by a group aptly called the Irreconcilables (the Radical Republicans), the Union imposed punitive policies on a prostrate South with little attention to convincing the former Rebels that these policies were proper. The South was not asked to accept reunion or abolition of slavery; instead, these policies were dictated by the

Irreconcilables, enforced by an army of occupation, and administered by legions of "carpetbagging" Northern politicians.

How much the punitive peace contributed to a residue of hostile will toward accepting Union political objectives is, of course, conjectural. What is certain is that hostile will, particularly toward the emancipated slaves, remained, and it is at least arguable that some of it is still with us today in the vestiges of racism that continue as part of our social fabric. Certainly "reconstruction" and the harshness by which it was imposed contributed to nurturing hostile will in such visible ways as formation of the Ku Klux Klan under the leadership of the brilliant Confederate cavalry general Nathan Bedford Forrest. The North, itself largely indifferent and even hostile to the freed slaves, did little to ameliorate the conditions of the Negro beyond the formal grant of freedom. White Southerners were punished for their misdeeds, but little was done to rebuild Southern society on the Northern model or to create prosperity in which all could find substance.

Could reconciliation have occurred in another way? Was there an alternative better state of the peace that would have removed hostile will more quickly and with less resistance? Removed as we are by more than a century from the suffering of the war that gave rise to the spirit of vindictiveness, it is easier for us to see that a reconciliatory peace might have healed the national wounds far faster than was the case. Had Lincoln lived, reconstruction might have been different, or if the war had been shorter and less bloody, there might have been less cause for bitterness. For better or worse, the model of a punitive peace was imposed and would be repeated after the next major conflict in which the United States would participate, World War I.

4

WORLD WAR I

They called it the Great War. For many it was the "war fought to end all wars" and, for the United States, it became a war "to make the world safe for democracy." Battle after battle, campaign after campaign, and year after year passed inconclusively amid unprecedented carnage, suffering, and destruction. It was the largest, bloodiest war in human history to that time. Literally millions were mobilized to fight it, and millions died in the no man's land between the opposing trenches that scarred the Western Front in virtually unbroken lines from the Alps to the North Sea. According to demographers, France would need 66 years merely to recoup the young men who died during the war. The war's bloodiest battle claimed 650,000 lives and when it was over, the lines had scarcely moved. That outcome symbolized the futility of the fighting generally and helped create an enormous cynicism in those ordered to fight and die for no apparent reason or effect.

World War I changed the face of Europe and the face of war. The decline of Europe as the center of Western civilization began during this time and would be completed 20 years later in the second world conflagration. The once dominant countries of Europe bled themselves dry of manpower and treasure and thereby lost the physical wherewithal to control international politics after 1945. Militarily, the Great War carried the logic of total war previewed in the American Civil War forward toward its grisly fulfillment in World War II.

Although the United States was eventually drawn into the awful fray, it was not really an American war. Certainly the issues that gave birth to the war were, at most, peripheral to American concerns. Moreover, the United States entered the fighting at an extremely late date. Our contribution to pushing an exhausted Germany over the brink to defeat, while psychologically important, was minor compared to the effort of the other combatants.

Political scientists, historians, and others have struggled ever since to understand how and why this great human tragedy happened, and there are nearly as many explanations as there are explainers. Because our primary concern is with America at war, it is neither necessary nor fruitful to add to the mountainous literature on what ignited "the guns of August." Rather, we will look briefly at some of the common themes that run through that literature to show some of the flavor of the times that made it all possible.

As the vital center of the international system, Europe had been at relative peace with itself for the century following the Napoleonic Wars. Certainly there had been conflicts. Modern Germany and Italy had been forged on the anvil of war, and Russia had been restrained in the Crimea, but these were relatively short and isolated breakdowns in the structure of international peace. The major themes of European politics had instead been internal, adjusting both politically and economically to the impact of the industrial revolution, coping with nationalism, and witnessing the birth of the German Empire and Italy. This process itself was wrenching and consuming of energy and effort. Such foreign adventurism that occurred centered around colonialism, and the subjugation of much of Africa and Asia, where colonial territory was relatively abundant and clashes between aspiring colonialists were infrequent and comparatively mild.

This tranquility between states began to break down around the turn of the twentieth century, and one of the major themes underlying the war emerged. This theme was a more contentious struggle for influence. One reason for the growing contentiousness was that the process of colonizing Africa and Asia was largely complete by the 1890s. After northern Africa fell under European control, there was essentially no place left where a European power could gain influence or control without challenging other powers. Closely related to this struggle for influence, the map of Europe was beginning to redraw itself. Two of the major empires of Europe, the Austro-Hungarian and the Ottoman, were disintegrating under the dual weights of atrophy and resurgent nationalism that had their roots earlier in the nineteenth century. As these empires crumbled, the other powers scrambled for influence in the newly emerging states. The competition focused particularly in the volatile Balkan States, and that area provided the spark that started the war when the Archduke Franz Ferdinand and his wife were assassinated by Bosnian Serb extremists at Sarajevo.

A second theme relates to the unwillingness of the major powers to prevent war once its possibility loomed on the horizon. Some Europeans actually relished the prospect, believing that war would be beneficial and that it would re-create a spirit of discipline in a generation that had not known war and that had grown soft and decadent as a result. While this purifying, martial view may not have been dominant, it was present. Despite the sentiment of those people and others, the issues that underlaid the road to war were clearly inadequate to justify what followed. Yet no one acted decisively to keep it from occurring. When it began, young men rallied willingly to the banner and marched off to war. An embittered generation of widows and veterans, as well as countless analysts, would later ask why.

A large part of the answer was that no one understood the kind of war it would be. There were two visions of modern war available, and the Europeans chose to believe in the wrong one. One model was the quick, decisive, highly mobile warfare of the Franco-Prussian War. The other was the protracted and bloody American Civil War. For reasons of ethnocentrism that suggested the inherent superiority of the highly disciplined European soldier, they rejected the model of warfare based on Helmuth von Moltke's image of the "two armed mobs chasing one another across the countryside" and instead believed the 1866 and 1870 European models more appropriate. Moreover, Germany and France (the two major Western Front antagonists) believed that each could win quickly. In the process, both overestimated their own capabilities and underestimated those of their adversaries.

As the first troops left Berlin in the summer of 1914, the Kaiser promised them that they would be home amidst glory before the first leaves fell from the trees. Instead, the war quickly stalemated, the trenches were dug, and four years were spent in futile frontal assaults against heavy entrenchment, a tactic long since obsolete but all the generals could think to do. To make such tactics all the more futile, machine guns, barbed wire, and poison gas had been added to the defensive advantage. Had the leaders and people on either side possessed a premonition of these horrors, the war might have been prevented. We will, of course, never know.

A final theme that runs through the web of causation is the mediocrity of the political leadership when the war began. The institutional arrangements by which the European powers had moderated their conflicts through much of the nineteenth century, the so-called Concert of Europe, had fallen into disrepair after the Franco-Prussian War (some would argue earlier than that). Diplomatic relations had become personalized around such leaders as the German Kaiser and the Russian Tsar. As the Serbian crisis eventuated in the mobilizations and countermobilizations that greased the slide toward war, what was needed was the leadership, statecraft, and diplomacy of a Metternich, Talleyrand, Castlereagh, or Bismarck, but none was available. The war happened partly because the leaders could not figure out how to avoid it.

Americans watched these events from the sidelines. Although the United States had evolved after the Civil War into the world's largest industrial power (Germany was second), it seemed to be only peripherally involved in the European-centered international political system. Separated from Europe by a broad ocean, the issues and problems that led to war did not appear greatly to affect American interests, nor did the Europeans have much of a sense that a totally immobilized United States could make any differences in the quick and decisive war they anticipated. Only when the war had dragged on for some time did Europeans peer across the Atlantic and ponder the contribution Americans might make to breaking the stalemate.

The initial American response to the European war was remarkably similar to our early attitudes toward the wars of the French Revolution and Napoleonic Empire a century before. That response was to declare American neutrality, a posture we maintained officially through the election of 1916 (President Woodrow Wilson campaigned vigorously and won largely on the promise of continued American noninvolvement). Also reflecting our reaction a century earlier, we adopted the policy of trading with both sides. As time went by, our trade with the Western Allies increased while trade with the Central Powers declined, but up until the eve of American entrance, the president was still calling for a negotiated settlement based on "peace without victory."

Formal American entrance into the conflict did not come until April 1917, when the hostilities were well into their third year. Other than the psychological lift it may have given the Allies, the declaration of war at that point had little practical effect on the fighting. As had been the case before, the United States entered the war totally unprepared to fight. We had essentially no standing armed forces, and it would take over a year to recruit and train the American Expeditionary Force that took part in the final push to Allied victory. America raised a 1.75 million-man army, but it was not declared fit for nor did it begin to engage in large-scale combat until the summer of 1918, about six months before the armistice.

Issues and Events

When the war began in Europe, the initial American response was that it was none of our business. Reflecting well-entrenched attitudes and beliefs, most Americans agreed that we should remain aloof from the intramural European struggle and that sentiment prevailed during the early years of combat. This urge to neutrality in part reflected traditional American preferences that dated back to the formation of the republic. Both George Washington in his Farewell Address and Thomas Jefferson in his First Inaugural had echoed the theme that American interests were best served by remaining separate from and uninvolved in the tainted power politics of the old world. "Friendly relations with all but entangling alliances with none" had been U.S. foreign policy for a century. At the same time, America contained sizable populations of both English and German extraction. It was therefore difficult to ascertain conclusively on which side popular sentiment lay at the outset. Neutrality was a convenient way to skirt the issue, while simultaneously providing the rationale for trading with both sides.

The American attitude was based on a myth (and it can only be described as such) that American destiny can somehow be fulfilled in isolation from the affairs of Europe. This belief was pervasive and did not really disappear until after World War II. Unfortunately, it was a myth based on a historical accident that never had much to sustain it for at least two reasons.

First, the myth was nurtured during an atypical time in European history, a period from the end of the War of 1812 until the First World War. This was a period when there were few major European upheavals. The idea took hold, however, that since the United States had remained above European power politics for nearly 100 years, this was a normal and preferable condition that was also a matter of American choice. Certainly, aloofness made a great deal of sense during the period of state-building that dominated nineteenth-century American history. Such a period lends itself to turning inward and that is what we did. What was missed, however, was that the same processes were occurring in Europe with the same effects. We did not perceive a need for Europe, and the feeling was mutual. However, at each end of the period, the United States became involved in the major European struggles that did occur. The War of 1812 was really only an extension of the Napoleonic Wars, and we eventually became involved in World War I. When the affairs of Europe have affected us and required our participation, we have not been able to avoid the call to arms. The myth could be nurtured only because Europe did not need us for a century.

The second reason was largely economic. As the United States moved to become a major industrial and commercial power during the nineteenth century, our prosperity increasingly required extensive trade with the world. The world's significant markets, of course, were in Europe and access to those markets was vital. As well, the immigrant waves that provided the manpower for industrial expansion came from Europe, and a great deal of the developmental capital that financed industrial growth came from private European banks. Without those sources, our pattern of development would have been quite different, but it has always been a curious aspect of the American worldview that economics and politics can be separated. The two in fact are cut from the same cloth.

The mythology of and desire for political aloofness while retaining commercial ties made neutrality appear attractive. Despite the illusion that the war was none of our concern, there were in fact economic and political issues that would eventually impel the United States into the war.

The economic issue had two essential aspects, one of which was the already mentioned desire to maintain normal commercial patterns with the belligerents. Such a posture was understandable from a sheerly economic viewpoint, but it was untenable politically. The reason, of course, was that both the Central Powers and the Triple Entente wanted trade with them to the exclusion of trade with the other. Although there was some U.S. attempt to be evenhanded in the flow of trade early in the war, the pattern gradually shifted to a much closer relationship with the Western Allies, especially Great Britain. Germany eventually found this pattern intolerable, which helped create the proximate events leading to the American declaration of war. In many ways, the situation was a replay of the problems that had drawn the United States into the War of 1812, except that Germany and not Great Britain would be the enemy.

The other side of the economic coin arose from the prospect of a German victory that looked increasingly probable as German troops previously committed to the Eastern Front began to move to France in 1917 after Russia withdrew from the war. America's prosperity required trade with Europe. The security of that trade, in turn, rested on guaranteed access to European markets and safe, reliable means to get American goods to those markets. A German victory threatened to alter both of those conditions.

A German victory over France would, in all likelihood, ensure that continental Europe would be dominated by Imperial Germany. Since Germany was the chief industrial rival to the United States in many important areas of trade, such an outcome offered the reasonable prospect that Germany would exclude or sharply restrict American access to continental markets.

The defeat of Great Britain would also allow German naval domination in the Atlantic Ocean, and thus posed a threat to open and secure access to the sea-lanes between North America and Europe (as well as presenting a possible future menace to the U.S. homeland). Gradually during the nineteenth century, the United States and Great Britain had reached a condominium ensuring the freedom of the high seas for commercial purposes. By the end of the century, the informal arrangement was that Great Britain enforced that policy in the North Atlantic and the United States enforced it in the Caribbean Sea.

The policy was to the clear advantage of both countries. Both were commercial, mercantile states, and Great Britain had the additional requirement for secure access via the oceans to her far-flung colonial empire. The commonality of interest between the United States and Germany, however, was not so obvious; commercial competition could easily spill over into naval competition for control of the trading routes.

There was an underlying political issue that Americans sought to avoid but could not. After the fall of the Tsar in early 1917 (and especially after the Bolshevik Revolution resulted in the removal of the new Soviet Union from the war), the contest did, after all, pit the world's major democracies, Britain and France, against the world's major autocracies, Imperial Germany, the Austro-Hungarian Empire, and the Ottoman Empire. In the long run, a democratic United States could scarcely avoid greater sympathy for those who shared our political form over those whose political philosophy was diametrically opposed to our own.

Each of these underlying economic and political factors manifested itself in proximate events that made neutrality progressively less tenable. The economic issue came to a head over German attempts to interrupt the lucrative flow of American war materiel to the Western Allies and specifically focused on the question of unrestricted submarine warfare. The political issue gradually emerged in the depiction of the war as a moral crusade between democracy and autocracy.

The issue of German submarine attacks on Allied shipping, especially on ships carrying American passengers (for example, the Lusitania), became the most volatile issue between the United States and Germany. It was an issue, however, not entirely lacking in irony. That irony involves both the restrictions that were supposed to be placed on submarines in war and the reasons the German navy failed to stress the submarine more in her naval competition with Great Britain.

Anticipating the introduction of this new weapon system to naval arsenals at the turn of the century, the participants at the Hague Convention had attempted to devise and include in the rules of war permissible and impermissible uses of the submarine. The provisions that came into force included requirements that virtually ruled out effective employment of the submarine; before attacking any vessel, the submarine had first to surface, announce its intention to attack, and be prepared to take aboard any and all survivors after an attack.

These were, of course, totally unrealistic requirements that, if adhered to, would destroy the usefulness of the submarine as a naval weapon. One advantage the submarine possesses is the element of surprise, which surfacing takes away. When on the surface, submarines are especially vulnerable to attack, since they carry no effective surface armament. Moreover, they are too small to take aboard more than a few survivors after an attack. Thus the only way to use them effectively was in direct violation of the Hague Convention, and that is precisely what the Germans did. When they did so, they were loudly condemned in the United States for barbaric activity in violation of the laws of war. The result for Germany was thus a classic catch-22. They could use submarines legally but ineffectively, or they could use them effectively but illegally. It was clearly a no-win situation. To make the irony deeper, the Germans were prepared to back down to American objections on the eve of our entrance into the war, in effect offering to suspend as a matter of policy (it had already been suspended in fact) unrestricted submarine warfare in return for continued American nonbelligerency.

The irony is made even greater because the U-boat was the most effective weapon (actually about the only one) Germany possessed for the naval competition with the British. Because of the influence of an American naval strategist, however, they entered the war with too few of them for decisive effect.

In 1890 the American strategist Alfred Thayer Mahan had published his seminal *The Influence of Sea Power upon History*, which argued the critical importance of control of the seas in warfare, and the book was widely read in Europe as well as America. A central tenet of Mahan's analysis was the need for a large navy with heavy capital ships as the crucial element in naval control and the consequent deprecation of naval strategies emphasizing what he called "commerce raiders," lone marauding vessels whose purpose was disrupting commercial trade routes through the capture or sinking of individual ships. The submarine, of course, perfectly fit Mahan's definition of a commerce raider, and one of Mahan's most ardent students was Kaiser Wilhelm. As a result, German naval development concentrated on the construction of heavy capital ships (in the extreme, the dreadnoughts). Germany was never able during the course of the war to get its fleet of these large ships out of the North Sea. At the same time, the Germans neglected the construction of submarines, which were able to escape the British blockade and which were quite effective against Allied shipping until the convoy became common practice.

Politically, the road to war was paved by the gradual conversion of American popular opinion away from neutrality and disdain for the entire war to a black-and-white depiction of the valiant democracies fighting desperately against evil autocracies. British propagandists were particularly influential in this effort, whereby the issues were simplified into terms of good and evil, so that, when the declaration of war occurred, the full support of the American people could be rallied behind the Allies and support for the "Huns" was equated with treason. The German submarine campaign, which enraged the American public, and devious actions like the Zimmermann note to Mexico, which proposed that Mexico declare war on the United States, simply added fuel to a growing anti-German fire.

Political Objective

World War I was, of course, an allied operation and as is usually the case within coalitions, each of the allies had its own distinct political objective and its own vision of the better state of the peace. When the war broke out, it is fair to say that none of the original combatants had particularly clear objectives beyond a generalized belief in the need to honor alliance agreements that committed various states to one another. This lack of clarity was not entirely surprising because none of them had any clear idea what the war would be like.

In some ways similar to the way the colonial side's objectives were formed in the American Revolution, political objectives flowed from the military state of affairs rather than the other way around. The major influences of the battlefield were to produce tremendous frustration, bitterness, and hatred. The result was an increasing impulse toward vindictiveness. Total warfare produced total political objectives. One side or another had to be defeated completely and the loser would be forced to pay. No one had talked this way as the troops were rallied in the summer of 1914, but after futile years in the trenches, revenge became a common desire.

These feelings were held with varying levels of intensity by the individual Allies, depending on the amount of suffering that they had endured. Among the Western Allies, the feeling was definitely strongest in France. Most of the war was fought in France; hence French territory bore the deepest physical scars of war, and French blood had flowed freely.

In this context, French objectives came to dominate Allied political aims. In its simplest form, the objective was to create a structure of the postwar peace wherein the war could not be repeated. Stated as an aim to make the Great War "the war to end all wars," there was substantial agreement among the Allies. There was disagreement, however, about what kind of structure would best ensure a peaceful world in the future. Determining what was necessary to guarantee the peace was partly a matter of determining who was responsible for the war in the first place.

In the French view, Germany bore special and unique guilt for the war and for French suffering. From the French premise, which was debatable (as we shall see in the final section), it followed that the structure of the peace must include a Germany incapable of instigating another war, and that meant a disarmed German state and a pastoral German society. Everyone wanted peace. France wanted peace and revenge. The disagreement over whether both should be objectives would dog the Versailles peace talks at the end of the war.

American wartime political objectives are what most concern us here. The figure of President Woodrow Wilson was of overarching importance in framing American objectives and engineering the process of change from neutrality to belligerency. As a result, one cannot fully understand American objectives without some insight into the character of the American leader.

At least three characteristics of Wilson are relevant in understanding his view of the war. The first is that Wilson was an academic and an intellectual. He had gained early fame within the academic community by writing *Constitutional Government* (a study of political democracies first published in 1885, and one of the most respected works in the field we now call comparative politics). He later served as the president of Princeton University. Wilson's intellectual background predisposed him to democracies, which, the prevailing view in political science argued, were inherently superior systems. Second, Wilson was a deeply religious man and a lay Presbyterian minister. This aspect of his background predisposed him to see matters in moral terms and would assist him in framing the war's objectives in the terms of a moral crusade. Third and finally, Wilson was a Southerner. He had been born and reared in Virginia during Reconstruction and had witnessed the embitterment and suffering that the punitive peace created in a defeated Southern people. As a student of and participant in the aftermath of a particularly bloody, total war, Wilson sought consistently to avoid seeing the same mistakes made in Europe.

American objectives toward the war necessarily changed from the period of neutrality to that of military participation. When the United States was neutral, Wilson's hope was to act as a peacemaker. The operative phrase forming that objective was "peace without victory." This approach flowed naturally from Wilson's boyhood experiences and consequent conviction that a punitive peace settlement imposed by a victor would unnecessarily slow healing of the war's wounds. This theme, although later abandoned by the United States, had great appeal within Germany after the American entrance into the war. Those who led the movement that overthrew the Kaiser and then sued for peace cited Wilson's statement as hope for a reasonable negotiated peace.

Fighting a war not to win, as peace without victory implied, was not the kind of cry that would rally the country to the banner, and it had to give way once the United States decided to join the hostilities. Military victory replaced military stalemate as the objective and, in a manner reflecting both Wilson's religiosity and the prevalent American worldview, had to be phrased in appropriately moral, lofty tones. Aided by the efforts of those British propagandists who had painted the issues in black and white, a rallying cry emerged to form the basis for the crusade. American purpose became an effort "to make the world safe for democracy." The moral tone is well captured in Wilson's war declaration to the Congress: "The day has come when America is privileged to spend her blood and her might for the principles that gave her birth and happiness. God helping her, she can do no other." So armed, the United States began to prepare for its entrance into the Great War.

Military Objectives and Strategy

Each belligerent evolved specific political objectives as the war went on. Each objective required the defeat of enemy armed forces and in some cases their destruction. Events quickly showed, however, that victory would not be quick and clear-cut. Military victory could only be achieved by enemy exhaustion, and thus the war wore on endlessly. In a sense, the war itself became the objective.

No one dreamed the conflict would be so long and senseless. The general consensus at the time it began was that no country could wage a long war because of the incredible costs in blood and treasure. Modern wars, so the experts said, would be short and sharp, their brevity ensured by the perceived dominance of the offense over the defense. The side which could muster and mobilize huge modern armies first and put them on the offensive would have an overwhelming advantage. Attackers would smash into and crush ill-prepared defenders who would be unable to swiftly maneuver their own massive forces. In France, the cult of the offensive reached its zenith in the teachings of Ferdinand Foch at the Ecole Supérieure de Guerre (i.e., French War College). According to Foch, the essence of war was to attack and any improvement in firepower ultimately benefited the attacker. To attack successfully, morale was critically important. No battle was lost, Foch reasoned, until the soldiers believed it was lost. With the proper *élan*, the French soldier was irresistible in the attack. And thus was born the rigid French doctrine of *offensive à outrance*—the offensive to the extreme. The birth of the French offensive doctrine brought death to a generation of Frenchmen.

Of the major combatants, the Germans faced the most difficult military problem. Situated in central Europe, Germany faced potential enemies on opposite fronts. In the east, massive Russian armies threatened to overrun East Prussia and crush the Germans. In the west, the French waited to avenge the humiliation of 1870. The only solution was to take advantage of the interior lines afforded by their central position, to mobilize more rapidly and efficiently, to eliminate one or the other of the opponents quickly, and then to concentrate on the remaining enemy.

The original plan to accomplish this complex task was devised years before the war and modified several times. Its original author, Chief of the German General Staff Count Alfred von Schlieffen, assumed that the great masses of the Russian army could not be fully mobilized for at least six weeks after the commencement of hostilities. Thus the Germans must concentrate their forces in the west and knock out France with one quick crushing blow. The blow would be a giant "right hook" of German armies marching through the low countries into France along the English Channel coast. The invasion would wheel inward, envelop Paris, roll up the French armies, and crush them back against the German border. Schlieffen expected the decisive battle to occur east of Paris within the six weeks' "grace" period while Russia mobilized. With the demise of France, the Kaiser's troops could be transferred to the east to dispose of the Russians. It was a bold plan and it was nearly successful.

When the war began, the Germans mobilized quickly in accordance with their elaborate plans and the offensive got under way. It moved through the low countries at a steady pace, pausing only to reduce the fortifications at Liège with giant siege mortars. On into France the Germans marched as French and British forces fell back, all going according to the master plan. Unexpectedly, however, the German armies on the right flank began wheeling toward the German border before they enveloped the French capital city. Thus the German right flank was exposed to an attack from the garrison of Paris. Additionally, because of a breakdown in communications, a gap developed between two of the wheeling German armies, and this mistake presented an opportunity for the Allies. The result was a successful counterattack and the retreat of the German armies away from Paris.

Both sides dug in. Both quickly began a series of attempts to outflank the enemy's position—the so-called Race to the Sea—that resulted in the extension of defensive positions on both sides. These positions eventually stretched from the English Channel to Switzerland. The seeds of stalemate were sown. There were no longer any flanks to turn and both sides continued to improve their already formidable defensive positions. Thus the trench war that dominated the Western Front began within weeks of the outbreak of hostilities. The next four years saw a series of massive offensives, each yielding thousands of casualties but precious little progress. The war became one of physical attrition, and the loser was determined by exhaustion, not military skill.

Schlieffen's plan was boldly conceived, and its failure can be traced to many causes. First, the younger von Moltke, the chief of staff in 1914 and nephew of the victor of 1866 and 1871, weakened the right hook by shifting some troops to the east to meet a surprisingly rapid Russian thrust into East Prussia. Second, German communication and supply lines on the right flank became extremely long because of the rapid German advance. Third, German soldiers were exhausted by their long march while the Allies fell back on shorter and shorter supply lines. Fourth, the French high command did not collapse under pressure as it had done in 1870, and as the Germans suspected it might in 1914. Finally and most important, the plan assumed victory at every juncture, and that everything would go well and smoothly. There was little room for error or successful enemy counteractions. In other words, the plan was arrogant. It paid scant attention to the capabilities of Allied armies or their commanders.

By 2 April 1917, when the United States entered the war, it seemed that the Germans and their allies might finally win. They controlled a rich portion of northern France, as well as most of Belgium and Holland. At sea, German submarines threatened to starve Britain out of the war. Moreover, revolt was brewing in Russia. If the Russians dropped out of the war, the Germans could shift large formations to the Western Front. The question was whether or not fresh American troops could get to the front in time. The American Army was minuscule, with an authorized strength of only 100,000. Men had to be inducted, trained, equipped, and sent across the Atlantic. Despite the enormity of the task, more than one million American soldiers landed in France by the summer of 1918.

The United States entered the war late and pursued no innovative strategy of its own. Essentially, American troops were fresh blood with which to continue the war in the established manner. American numbers tipped the balance scale toward victory for the Allies as the exhausted Germans could not compensate for the fresh and numerous American soldiers.

The commander of the American Expeditionary Force, Gen. John J. "Black Jack"

Pershing, found his principal strategic battles to be with other Allied commanders who wished to use American soldiers as piecemeal replacements in their battered formations. Pershing resisted with President Wilson's support. Wilson, however, eventually bowed to Allied pressure and intervened on their behalf. Pershing recognized the authority of the civilian commander in chief and placed American units, rather than individuals, at the disposal of Allied commanders. However, after tactical training in France was complete and the American Army was fully ready to take to the field, Pershing managed to get back the units that had been amalgamated into the Allied armies.

Pershing has been praised as a leader, organizer, and manager. He has been criticized as a less than innovative strategist and technician. The praise is certainly justified as the task of building, equipping, and transporting a massive army overseas in a short period of time was formidable. The criticism also may be justified. Pershing offered little in the way of new strategic thought to the Allies and produced no novel tactical ideas. However, one must remember Pershing's circumstances. The war was four years old and the die was cast in terms of a strategy to exhaust and defeat the enemy. There was little room and no time for new strategic visions. Tactically, it is no wonder that American tactics mimicked those of our European allies. The rigors of trench warfare were new to the Americans (Petersburg being long forgotten) and, upon their arrival in Europe, the American units were often trained for combat by European veterans.

Political Considerations

World War I was one of America's most popular wars. Public support for American intervention was high when Woodrow Wilson made his war declaration and that support never wavered. Given that neutrality had enjoyed overwhelming support only a year before, this was a remarkable turnaround. Only years after the war would the United States develop some degree of cynicism over having plucked European "chestnuts from the coals." Yet, once the fighting was over, the American urge to withdraw, to return to "normalcy," was overwhelming. It was a curious domestic situation, at least partially the result of factors in the American culture and, more specifically, the ways Americans go to war.

One reason that Americans rallied to arms and then recoiled quickly from their efforts is traceable to the American missionary zeal. The war was advertised as a moral crusade, and, as such, it appealed to that part of our national character that flows from a feeling of America as a special, morally superior place. Our goal was to "save" Europe from itself, and when that end was accomplished, we packed up and returned to the more normal state of affairs. When it became clear in the 1920s that Europe had not been cleansed as we had hoped, we turned inward once again, seeking to isolate ourselves from the vagaries of European power politics, but at the time support for the war was unquestioned.

Another reason for the high level of support was that the American part of the war was so short. As stated earlier, the United States was a formal belligerent for a year and a half, but engagement in combat was limited to the war's final six months. We began the war entirely unprepared to fight (a common theme in American military history) and set about mobilization in what our Allies found an irritatingly languid manner. The legislation to create a means to induct armed forces had passed at President Wilson's request in 1916, but our standing military was no more than a shadow in April 1917, when war was declared. Building the apparatus to induct, equip, and train a force that could be placed responsibly into combat required time. Starting from scratch with mechanisms and personnel inadequate to a task of such proportions further lengthened the process.

The effect was to romanticize the war. It was a time of parades, of soldiers coming home for leave from the training camps (most of which had had to be built for the occasion) in impressive uniforms. There was little mud and no blood and dying. The American Expeditionary Force did not arrive in Europe *en masse* until June 1918, and General Pershing was reluctant to throw it into combat even then. Instead, he maintained that the force required a period of orientation and further preparation for fighting in a new and strange environment.

The leisurely pace of American preparation and commitment created political tensions within the alliance, especially between the Allied military command and the American command. Reflecting the nature of the war and where it was being fought, the Allied general staff was dominated by France, and the supreme commander was Marshal Ferdinand Foch. The Allied command had a very different viewpoint of the American role than did its American counterpart.

From the vantage point of the Allied command, the chief contribution of the Americans should have been to provide fresh manpower to British, French, and other forces badly depleted and exhausted by the long years in the trenches. With France particularly near the brink of physical exhaustion and collapse, fresh bodies were the requirement against Central Power forces that appeared increasingly menacing after the fall of Russia. To that end, the Allies argued long and loud, if with little effect, for the Americans to speed the process of building their army and for getting that army into the field.

Pershing resisted these efforts successfully. He viewed these requests, probably correctly, as no more than an attempt to make cannon fodder of the American forces in the futile fighting, a usage that would have been unpopular at home. Moreover, he, like most observers, was less than overwhelmed with the tactical brilliance that had thus far marked the war effort on both sides. As a result, he insisted that Americans not be integrated into existing units and that, instead, there be created a distinctly American place on the lines. Foch resisted, but Pershing insisted, refusing to commit the Americans at less than the unit level until his end was achieved. Ultimately, the Allies' need for the American troops exceeded their feelings about how they should be used, and the Americans were given their own front for the final offensive.

There were, moreover, important political differences between the leaders of the principal Allied partners that simmered beneath the surface but which would become apparent when it was time to settle accounts in the peace negotiations. The major disagreements were between American President Woodrow Wilson and French Premier Georges Clemenceau. The heart of these disagreements centered on why the war was being fought and what its outcome should be.

As we have said before, the First World War was a total war. However, the war was more total for some than for others. For France and Britain, it was a total effort being fought for national survival, and both countries totally mobilized to fight it. For the United States, national survival was never a problem, and although we raised the largest army in our history to fight it, it was not a war of total mobilization for us. As an example of the contrast, the United States did not develop an armaments industry to support the effort; instead, the United States relied largely on arms purchased from the British and the French.

The different levels of desperation with which the war's outcome was viewed translated into discordant views about what would constitute a satisfactory peace. While the fighting continued, these divergences were hidden behind the veil of common effort. When the guns were stilled, the discord came to the fore. The United States was disadvantaged by the combination of a lack of experience at coalition decision making (the only previous war in which we had allies was the American Revolution) and the intensity of French claims based in greater experience, sacrifice, and proximity.

The basis of disagreement was over the question of a punitive or a reconciliatory peace. Reflecting on his background as a Southerner and an academic, Wilson wanted a reconciliatory peace that would feature self-determination for all countries (or at least those in Europe). He wanted a postwar world founded around the Fourteen Points, which declared these ideals and featured the League of Nations as its centerpiece to guarantee the peace. The Wilsonian vision was lofty and idealistic; Clemenceau viewed it as naive in the real world of European politics. Rather than seeking reconciliation, Clemenceau sought punishment of Germany and the reduction of the German state to impotency. During the war Clemenceau humored Wilson because he needed American help. When it came time to restructure the world, the gloves came off and the disagreements were laid on the table. The results of those interactions are discussed in the final section of this chapter.

Military Technology and Technique

World War I has often been characterized as history's most senseless and poorly contested war. Given the objectives of the belligerents, particularly as the war progressed, the unending stalemate on the Western Front, and the incredible number of casualties caused by endless frontal assaults against enemy lines, it is difficult to argue against this unfavorable judgment. At the same time, however, World War I was a watershed in the evolution of modern warfare because of the technological innovations first demonstrated during the conflict. Two technological developments were of primary importance, either having a direct impact on the conduct of the war itself or foreshadowing the nature of warfare in the future.

The development that had the most immediate impact was the widespread use of rapid-fire weapons. Key to this development was the perfection of smokeless powder that did not obscure the field of fire or foul weapons to the extent previously common. Although rapid-fire weapons were used experimentally in the American Civil War (the Gatling gun, for example), it was not until 1882 that Sir Hiram Maxim designed the first machine gun widely adopted by military forces. By World War I, reliable designs permitted rates of fire from 200 to 400 rounds per minute. The weight of such a machine gun was approximately 100 pounds including its mount, which made it a defensive rather than an offensive weapon. Weapons carried by individuals also improved greatly by World War I. Modern military rifles could fire up to 20 rounds per minute and their effective range was limited primarily by the vision of the soldier.

Field artillery followed the rapid-fire trend of smaller weapons. Without question, the finest field artillery piece in the world in 1914 was the 1897 French model 75 mm. It could fire 6 to 10 rounds per minute with great accuracy. Other excellent field pieces were the U.S. model 1902 field gun and the Austrian 88 mm developed and produced by the Skoda Works. In the years leading up to World War I, heavy artillery became much heavier. Large-caliber guns were common, including huge siege mortars. For example, the Germans possessed 420 mm mortars that delivered a one-ton projectile.

The development of rapid-fire weapons and heavy artillery pieces significantly affected the way in which World War I was fought. The most obvious effect was in the number of casualties. The human toll of the war dwarfed all previous experience. The second major effect was to give the advantage to the defense, a phenomenon which thoroughly surprised most military planners. Their failure to cope with superior defense added to the human carnage. Against rapid-fire weapons and heavy artillery, the techniques of previous wars led only to failure and casualties of unprecedented proportions. Finally, rapid-fire weapons used prodigious amounts of munitions. As a result, industrial capacity on the home front became of paramount importance in determining success or failure on the battlefield.

Although the development of rapid-fire weapons had a significant impact on the war itself, a revolution in transportation would portend the nature of wars to come, even if its impact on World War I was less than decisive. The transportation revolution was caused by the invention and application of the internal combustion engine, which led to the development of not only cars and trucks but also tanks, submarines, and heavier-than-air aircraft. These weapons would eventually change the face of warfare.

Although mules and horses continued in common use, all belligerents used trucks extensively during the war to overcome the inherent weaknesses of railroads. Railroads had revolutionized military transportation, but they also brought with them some unwanted baggage. First, they were relatively inflexible, since a great deal of preparation and construction was required to establish a rail line, particularly one that would be used heavily. Second, because they were both important and inflexible, they tended to dominate strategy. During the Civil War, campaigns revolved around rail lines as opponents sought to protect their own and cut those of the enemy. Trucks, on the other hand, provided flexibility. They required no ties and rails. Troops and materials could be hauled rapidly from railheads to far-flung battlefields. Large-scale battles could now be fought wherever there were roads. Railroads did not, however, lose their military importance. In World War I (as well as in later wars) they remained critical to military success because of the quantity of men and materiel they could carry.

The tank was first introduced by the British in 1916. Its purpose was to break through enemy trench lines and clear the way for infantry to advance. Some visionaries thought that tanks could not only break through enemy lines but could also range far to the enemy rear and capture command centers, disrupt communications and supply lines, and spread panic. The World War I vintage tank did, however, have severe limitations. Although impervious to small arms fire, their slow speed, especially when used as infantry support weapons, made them tempting artillery targets. Near the end of the war, Allied planners (the Germans paid scant attention to the use of tanks) began to recognize better uses for tanks. In the last offensive thrusts of the war, the Allies massed tanks for attack rather than using them in small concentrations. In the interwar years, the relationship between tanks and infantry began to reverse, and infantry would be used for the support of armor. With this change of tactics and improved tank design, armored warfare would come of age in World War II.

The internal combustion engine was also important to the development of submarines. An undersea craft that could attack enemy warships and merchantmen had long been a dream of naval designers. The internal combustion engine provided a compact means of surface propulsion and a means to recharge the batteries used for submerged operations. Designs improved rapidly in the years preceding the war, and submarine warfare became an important part of overall German strategy as the Germans sought to starve Britain out of the war by cutting her sea-lanes.

With the development of the airplane, war entered a third dimension. Initially, the airplane was intended only for observation of enemy movements and artillery spotting. By the end of the war aerial photography of enemy trench lines was an indispensable tool of military planners. Of course, such a valuable tool had to be denied to the enemy, and as a result aerial combat began shortly after the war commenced. By the war's end, airplanes were conducting extensive air-to-air combat with sophisticated machine guns that fired directly through the propeller by the use of an ingenious interrupter gear. Airplanes also strafed enemy troops at the front and conducted bombing raids in rear areas. Finally, both sides made tentative attempts at strategic bombing. However, aircraft were not an overly important weapon in the Great War, primarily because aircraft and engine design were still primitive arts. But aerial visionaries of the war, such as America's William "Billy" Mitchell and

Britain's Hugh Trenchard, saw possibilities for a much greater role for the airplane in the future. Between the two world wars they pressed for better designs and for the development of air power doctrine that they believed would make air power a decisive factor in modern warfare.

The development of weapon systems based on the internal combustion engine represented a watershed in the evolution of modern warfare, and yet these weapons had only a limited impact on the techniques of warfare in World War I. Military tactics remained rooted in the past. The tactical problem on the Western Front, once the trench lines were firmly in place, was to achieve a breakthrough in the enemy's linear defenses and then exploit that breakthrough to bring a degree of mobility back to the war. The trench lines for both sides had no tactical flanks, and thus the only attack possible was a frontal assault.

Generals on both sides used several techniques in an attempt to achieve a decisive breakthrough. The most common method was to conduct a massive artillery barrage (some lasted for weeks) calculated to obliterate the enemy trenches. Once a gap was created, the infantry was supposed to pour through and exploit the advantage. Unfortunately, such a tactic sacrificed all surprise. Defending troops simply withdrew to deep dugouts during the barrage, waited for it to end, and emerged ready to fight. In addition, the barrage was often counterproductive for the attackers because it turned the no-man's land between the opposing trenches into a cratered, muddy moonscape through which the attackers wallowed at a snail's pace. With no surprise and slow movement, reinforcements could fill any gap in the defender's line. At that, even if an enemy trench was seized, no real breakthrough was achieved because the trench "line" generally consisted of three trenches separated by some distance. Thus the exhausted attackers, having captured the initial objective, still faced fresh troops and fresh defensive works just ahead. It is no wonder the front lines moved so little during the war.

Another innovative method was gas warfare. When first used on both the Eastern and Western fronts, it had considerable success. However, its effects often depended on the weather (especially which way the wind was blowing), and the introduction of protective equipment made gas a progressively less effective weapon.

The most innovative tactics (aside from the use of tanks discussed earlier) to achieve breakthrough were developed by the German Gen. Oskar von Hutier. He employed a very short barrage combined with infiltration by specially trained assault troops who avoided strong points. Regular infantry, who reduced the by-passed strong points, followed the assault troops. The Germans used these tactics extensively during their final offensive in 1918 and had considerable success in achieving deep breakthroughs quickly. However, they could not move artillery and supplies forward fast enough to sustain the attacks. Hutier's tactics foreshadowed the blitzkrieg tactics of World War II.

For the most part, the tactics of World War I resembled the worst displayed in the American Civil War. Time after time masses of men lunged across open ground to assault well-entrenched defenders and were slaughtered at an incredible rate.

It seemed the generals had learned nothing from experience. World War I tactics were not a tribute to human wisdom.

Military Conduct

The Allies were exhausted after the hard fighting of 1917, and a revolt of the soldiers in the French army had further shaken their leaders. In these circumstances, they were content to sit on the defensive and await the arrival of the fresh American troops. The same was not true for the Germans. With the collapse of the Russians, the Germans moved massive numbers of troops to the west and, so reinforced, hoped to strike a decisive blow before the Americans turned the tide.

Gen. Erich Ludendorff planned a series of German offensives all along the Allied lines in 1918. In addition to the new troops from the Eastern Front, he planned to incorporate the new tactics developed by Hutier. The first German attack came in March and was aimed at the British at the Somme River. The attack achieved astonishing success, advancing 40 miles in eight days, before it slowed. In early April the Germans struck the British again, this time at the Lys River, and again achieved some success. This was followed in late May by an attack on the French at the Chemin des Dames, an important ridge line northeast of the French capital. Once again the Germans were astonishingly successful by the standards of trench warfare. By the end of May they again threatened Paris from newly won positions on the Marne River at Château-Thierry.

At Château-Thierry the Americans were first sent into heavy combat. When the Germans attempted to cross the Marne, they were thrown back by the newly arrived American troops. Ludendorff attempted to renew his stalled offensive during the summer months, but the effort failed. The Germans had spent themselves and the Americans were now pouring into the lines. In all, ten American divisions took part in the summer operations; their presence was crucial to Allied success in stopping the German drive.

Allied counterattacks began as early as mid-July and were aimed at retaking the vast salients created by the surprising successes of the earlier German offensives. On 8 August the Allies massed several hundred tanks for an attack on the salient around Amiens. German resistance collapsed in front of the armored assault and the Allies penetrated nearly 10 miles into German-held territory. Ludendorff seemed to realize that the game was up as he referred to 8 August as the "Black Day" of the German army.

The final salient to be erased was at Saint-Mihiel and the task was entrusted exclusively to the Americans. Pershing massed half a million troops and support equipment for a difficult fight. However, the Germans offered little resistance as they evacuated their men and equipment from the exposed salient. With the salients erased and the Americans bloodied, it was time for the final grand offensive that would end the war.

The grand offensive was just that, grand. Troops all along the front were expected

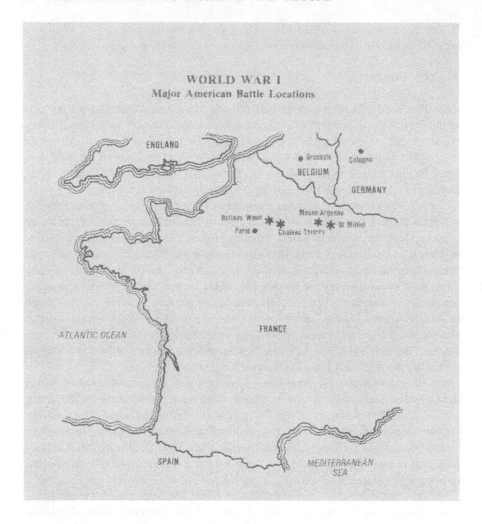

WORLD WAR I
Major American Battle Locations

to advance and place intolerable pressure on the German line. The American sector in the offensive was surrounded by the Meuse River and the difficult Argonne Forest. The fighting was bitter and extended, particularly in the Argonne, but by 10 October the Germans had been driven from the forest. By 5 November the Americans forced a crossing of the Meuse. The grand offensive rolled on, not just in the American sector but all along the front as the exhausted German defense finally disintegrated.

Germany would not be the master of Europe, at least not for two decades. But neither would the French or British, both of whom had suffered grievously. Nor would the Soviets reign supreme, as they sought to consolidate their internal position after the revolution. Austria-Hungary simply disappeared as a political unit. Europe would have to wait 22 years for a master to proclaim itself.

The United States entered the war late, but American participation was vital to the eventual Allied success. Fresh American troops stopped the final German offensive in 1918, and they broke the back of German resistance thereafter. American troops and their leaders acquitted themselves well on the battlefield, and the lessons learned about massive mobilization, training, and deployment would prove valuable in the future. The cost, in human terms, was staggering. During approximately six months of combat operations, 50,000 Americans died in battle, while 75,000 more died from nonbattle-related causes. But the cost in American lives paled in comparison to the losses sustained by the other Allies. Nearly 1.5 million Frenchmen died in combat, as did nearly 1 million British and nearly 2 million Russians. The small price America paid in its short war affected the amount of influence the United States had in the peace treaty negotiations that followed the war. The United States also was spared the incredible casualties suffered by civilian populations during the war. Some estimates put the civilian death toll at over 15 million. Perhaps the price would have seemed a bargain if this war had been the war that ended all wars.

Better State of the Peace

The armistice that took effect on 11 November 1918 ended the formal hostilities that had raged for more than four years. When the victorious Allies gathered at Versailles, the palace of the Bourbon kings located outside Paris, their purpose was to re-create the peace that had preceded the outbreak of the Great War, to make a better state of the peace.

In essence, the Allies had three tasks facing them in their collective quest to ensure that the conflict had indeed been the war to end all wars. These were dealing with the enemy, restructuring the peace, and redrawing the political map. The first task was how to deal with the vanquished Central Powers. German hostile ability had been finally overcome through exhaustion of German resources in its summer 1918 offensive and the Allied counteroffensive. The coup that overthrew the Kaiser and resulted in a suit for peace by the new German government demonstrated that German cost-tolerance had been exceeded as well. What remained to be decided was how to integrate Germany back into the international system. This was a question of fashioning a peace that Germany could or would accept and embrace for reordering a peaceful world. Two alternative visions were laid on the negotiating table.

As suggested earlier, the two chief protagonists in this debate were Wilson and Clemenceau. Harking back to his own childhood experience and his earlier call for a peace without victory, Wilson preferred a reconciliatory settlement toward Germany that would reintegrate the German state into the international system with a minimum of recrimination and punishment. He believed this to be the best way to overcome those vestiges of German hostile will that might resist acceptance of the policies governing the peace. Clemenceau, representing an embittered France, had other ideas. Given the suffering and privation France endured (admixed with some long-smoldering resentment arising from the settlement of the Franco–Prussian War 48 years earlier), France insisted upon a punitive peace. French insistence prevailed and Germany was punished for the war.

In essence, France had two preferred outcomes to the peace talks. The first was to recover the enormous physical and economic costs of the war (the French national treasury as well as those of virtually all the major combatants had been drained). The second objective was to destroy what France viewed as the cause of the war, which was an expansionist, militaristic Germany. The solution to the first problem was to make Germany pay for France's war expenses; the answer to the second was to transform Germany into a pastoral, permanently weakened state that could pose no future military threat to France.

The key to achieving those goals within the framework of the peace settlement was the infamous Article 231 of the Versailles Accords, the war guilt clause. That article, which the new German government was forced to accept, placed sole and complete blame for starting the war on Germany, and it served as the necessary justification and underpinning for the other punitive parts of the settlement. With German cupidity and responsibility formally established, France could justify exacting retribution, including a severe schedule of reparations aimed at compensating France for its wartime costs (a provision, one might add, in which a number of the victorious allies, notably Great Britain, rapaciously joined). Germany would pay for the war, although the effects of those payments would prove ruinous to the German economy. At the same time, provisions were included that substantially disarmed the German state and reduced its territory to ensure against a militarily resurgent Germany.

A prostrate Germany had no choice but to accept these humiliating conditions, but the leaders who did so would later be vilified as traitors for their actions. The conditions were so severe and humbling that the German people could not possibly embrace these policies and hence have their hostile will (resistance to policies) permanently overcome.

There were several factors that made acceptance impossible. The first was the war guilt clause. That provision was, at best, questionable. As one tries to unravel who was to blame for the war in the first place, Germany emerges as but one candidate among several. There was too much guilt and stupidity to lay on a single doorstep.

Another factor that ensured the survival of German hostile will was the size of the reparations payments. These virtually ensured that Germany would not recover fully from the war economically (as some observers such as Lord John Maynard Keynes prophetically but futilely pointed out at the time). The result would be enormous inflation in Germany during the 1920s with which the democratic Weimar regime could not deal effectively and economic devastation when the Great Depression took hold. At the same time, the demilitarization of Germany left that country at the mercy of the rest of the international system and perpetuated German enmity. Suddenly, Germany was back where it started before unification, at the military mercy of the system. The seeds for the emergence of a Hitler-like figure could not have been more skillfully sown had that been the purpose of the peace treaty.

Although he objected to these aspects of the settlement, Wilson finally acceded, because his attention was focused on the second task which confronted the conferees, the establishment of a mechanism to ensure the future peace. The essential question was what had gone wrong with the Concert of Europe that had allowed the war to occur.

When it had operated properly, the Concert had served as a collective security arrangement: all the major states were members, and each agreed that a threat to or breach of the peace threatened their own interests and must be resisted. With that principle established, a potential aggressor knew that the community of states, and hence overwhelming power, opposed its action, making it futile and thus deterring aggression. The Concert appeared to work for a time, but ultimately the system failed. What had gone wrong?

The answer Wilson and others (rightly or wrongly) devised was that the Concert system (actually a series of irregularly scheduled meetings of the major countries called when the need arose) was inadequately institutionalized. What the system lacked, they reasoned, was a formal mechanism, an institution that would always be available whenever crises arose and which consequently could act in a timely and authoritative manner. That mechanism was to be the League of Nations, which would serve as guarantor of the peace. Unfortunately, that noble institution never had a chance to perform its role.

A collective security arrangement requires two conditions for effective operation. The first is that the mechanism must have at its disposal obviously overwhelming power to deter transgressors. Meeting that standard requires that all major powers be represented in the arrangement. This condition was never met. The United States never joined the League (to Wilson's bitter and debilitating disappointment), the Soviet Union was not permitted to join the organization until the 1930s, and it was during that decade that the countries that formed the Axis in World War II resigned. The second condition for success is the willingness of the major powers to enforce the peace, and that translates into accepting the justice of the peace system that one is upholding. For reasons embedded in the third task facing the negotiators, this condition was also unmet.

The third task was redrawing the political map of the world. The operative principle, as part of Wilson's Fourteen Points, was to be self-determination, the right of all nations freely to determine their status as states. The principle in truth applied only to Europe (nationalists in several Asian colonial states would discover this truth to their dismay). Although the principle was applied with reasonable effectiveness in central Europe, there remained another part of the agenda, which consisted of territorial rewards for the victors and penalties for the losers.

The Austro-Hungarian and Ottoman empires collapsed and disappeared as a result of the war. The European sectors of those empires were allowed to engage in self-determination, resulting in such new states in eastern and central Europe as Czechoslovakia, while the Ottoman Middle East was placed largely under the trusteeship of Britain and France. The German Empire was dismantled and apportioned among the victorious allies, and Germany's prewar boundaries were reduced through transfer of territory to France (the return of Alsace-Lorraine) and through the recreation of a Polish state. Some of these territorial adjustments placed German populations under non-Germanic rule (for example, the Sudetenland), but Germany had no choice but to accept its dismemberment—at least for the time being.

At the same time, some members of the victorious coalition, notably Italy and Japan, had territorial claims that were not honored by the dominant members of the alliance (Britain, France, and the United States). Italy had claims, for instance, in the Balkan region and Japan expected to receive the bulk of the former German dependencies in the Pacific, but most of these were placed under American stewardship instead. Germany simmered under the loss of empire and territory it considered rightfully a part of Germany, and Japan and Italy resented the rejection of "rightful" rewards for their contributions to the war effort. None of the three could be expected enthusiastically to endorse or enforce the territorial status quo, as participation in the League of Nations collective security system required. Ultimately, of course, none of them did.

All of this is to say that the better state of the peace was doomed from the beginning. The victors' policies were punitive and viewed as unjust by the vanquished and even by members of the victorious coalition. From an American vantage point, the irony was redoubled because, when Wilson returned with the peace treaty in hand, it was rejected after an acrimonious national debate by a resentful Senate. The official argument that led to the failure to ratify was the commitment to the League, which would tie the United States irrevocably to the affairs of Europe. Such a commitment, which Wilson felt was the linchpin in an enduring peace, was unacceptable to those Americans who continued to believe in the myth of American aloofness from European affairs. Underlying the failure, however, was a tactical blunder on Wilson's part; he had failed to take any senators to the Versailles conference, and hence they had no stake in the outcome (a mistake no subsequent president has made).

The outcome of the settlement of World War I is generally considered the classic case of winning the war and losing the peace. The war had been won militarily; German and other Central Powers' hostile ability and willingness to continue (cost-tolerance) had been broken on the battlefield. But that was not enough. Adversary hostile will, defined as resistance to the victor's policies, not only was not overcome, but the terms of the peace were almost guaranteed to increase that hostile will. Resistance to accepting these policies was, of course, most strongly felt in a humiliated Germany. Given the circumstances and conditions imposed upon the German state, it is hard to envision how the German people could have reacted otherwise. For policies to be embraced, they must be accepted as just, and the peace terms could hardly have been looked upon that way in Germany. In the end, shortsightedness and vindictiveness ruled the day. In the long run, the peace was lost. The manifestation of that loss was the need to fight World War II to resolve the differences that had been dealt with so abysmally in the peace ending the war to end all wars.

5

WORLD WAR II

The Great War had been the largest and bloodiest conflict to date in human history, but it was in many ways only a preview of what would follow. Approximately 20 years later World War II erupted and became the largest military event in history. It was a conflict that was total in all senses of that term and a *world war* in which virtually every corner of the globe served as a theater of action at one time or another. Although records are inadequate, the best guesses are that about 80 million people were in military service at a given time. Of those, between 15 and 20 million were killed and probably about the same number of noncombatants perished. There were around 10 million combatant casualties and almost that many additional civilian casualties in the Soviet Union alone. Even more than the First World War, World War II was a war between whole societies, a war of factories. The entire resources of the major combatants were dedicated to the war's conduct, and whole populations were mobilized for one aspect or another of the effort.

America's role in World War II was unique in at least two ways. World War II was almost two distinct conflicts: one in Europe, where the Western Allies (including the Soviet Union) faced Germany and her European allies, and the other in the Pacific, where Imperial Japan was the major antagonist. The U.S. position was unique in that only the United States had major responsibilities in both theaters. In Europe the Allied effort was dominated by the triumvirate of the United States, the Soviet Union, and Great Britain, but American presence was of vital importance: the British depended heavily on the United States for the materiel and manpower necessary to open the Western Front, and the Soviets relied to some degree on American lend-lease. In the Pacific the war was essentially a conflict between the Americans and the Japanese. Certainly others were part of the Pacific campaign: the British on the peripheries (e.g., Burma) and Chiang Kai-shek's Kuomintang Chinese (who occupied a million-man Japanese army that could not be used elsewhere). The task of defeating the military might of the Japanese Empire, however, was clearly an American task. As we shall see later in the chapter, this unique position created some friction within the government and with our Allies over which enemy should receive the greatest attention and even some interservice rivalries about resource allocations (the war in Europe was basically an Army enterprise, whereas the Navy dominated the Pacific war).

The other unique aspect from an American perspective is that the United States was the only Allied power that emerged from the war stronger than it entered. When the United States entered the fray following the attack on Pearl Harbor, the vast (and, thanks to the lingering effects of the Great Depression, underutilized) American industrial base was turned into the "arsenal for democracy." The conversion and the stimulation it provided the economy ended the depression and allowed the United States to emerge in 1945 as the unquestioned economic colossus of the world. The other Allies were "winners" in the sense of being on the prevailing side, but all the other Allies were wounded seriously by the effort. Britain's expenditure in blood and national treasure accelerated its gradual decline from great power status, a circumstance with which British governments continue to grapple today. The other major ally, the Soviet Union, arguably emerged more politically unified because of the enormity of effort necessitated by the Great Patriotic War (as the war was officially known in that country), but its land was scourged by the Nazi invasion that left two-thirds of its industry destroyed, countless towns, villages, and buildings reduced to rubble, and nearly 20 million citizens dead.

The United States avoided those disasters. After the surprise attack at Pearl Harbor, American continental soil was never seriously attacked during the war, so that there was no physical reconstruction to deal with after the war's end. Our materiel contribution to the war had been enormous (the war cost the United States more than $500 billion in the dollars of the day), but our 300,000 casualties were comparatively light; the war did not bleed us dry in the literal sense of that phrase. Moreover, the war effort revitalized an American industrial plant gone flabby during the hard years of the depression. American industry was more productive at war's end than at the beginning.

The major effect of World War II was to critically alter the power map of the world. In the broadest sense, the roughly 150 years of European history from the onslaught of the French Revolution through World War II was a contest between France and Germany to dominate the continent and hence to dominate the international system. Ironically, World War II ensured that neither of them would. France had been defeated, humiliated, and occupied, and even though it rode to "victory" on the coattails of the victorious Allies, France clearly emerged from the war diminished in spirit and power. For Germany the outcome was even more disastrous. Its armed forces were decimated, it was occupied by its former enemies, it bore the unique moral stigma of Nazi excesses, and it was once again physically divided. Division was the cruelest blow of all, both because it returned the German people to the weakened status of a divided state and because the shadow of the Nazi past raised serious questions of when, if ever, the international system would allow a German resurgence.

The other actors were not in materially better shape. Great Britain was a member of the victorious coalition and hence technically a winner, but the British economy lay in ruins. The war would force Britain through the agony of gradual reduction from a global to a regional power. The British Empire, like that of its principal rival, France, would wither under nationalist demands for independence, setting in motion a whole new series of dynamics that are yet unfolding. Japan, like Germany, was defeated and occupied, and its reemergence would require massive assistance and nearly two decades to accomplish. The other major combatant, China, simply resumed the civil war that had raged between the Communists and Kuomintang in the 1930s.

The war ended with only two states possessing significant power, the United States and the Soviet Union. Of the two, the United States was clearly the more powerful. Although allied against the Nazi and Japanese menaces, the Soviets and the Americans moved into the postwar power vacuum with very different worldviews and motives that almost guaranteed a clash as they sought to reorder the power map.

Issues and Events

There are, of course, various ways to look at the question of what caused World War II, and numerous explanations have been put forward. In essence, however, there were two interactive underlying issues: Franco–German competition for dominance of the European continent that went back over a century, and the failure of the peacemakers at Versailles to create a structure for the interwar peace that could be embraced and supported by the participants in the Great War.

Competition between modern France and Germany is historical and long-standing, and treating it in its breadth and richness would only divert us from present concerns. In the modern era, however, one can usefully date it back to the Napoleonic campaigns, when Prussia, the precursor to and leader in the unification of the German state, was soundly defeated by the French *levée en masse*. From that humiliation arose the Prussian determination to unify Germany and to produce a strong, militarized state that would no longer be forced to suffer such indignities. The process of unification took nearly a half-century and was climaxed by Prussia's easy victory over Louis Napoleon in the Franco–Prussian War of 1870. A critical outcome of that conflict was the reannexation of Alsace-Lorraine to Germany, which in turn was an important element in forming the static alliance systems that contested World War I.

The Great War, round three in the competition, left Germany in essence back where it had started: politically dismembered, territorially reduced, and economically and militarily debilitated and vulnerable. Germany was once again reduced to being the "weak sister" of Europe, and the history of the German states had taught that this condition was intolerable. Moreover, the economic system that the settlement erected was saddled by reparations that made economic recovery virtually impossible. It was also burdened with a democratic political system that was both alien to the German political tradition and, by virtue of signing the Versailles peace treaty, held responsible for German humiliation by sizable parts of the population (a vulnerability that Hitler used to form the basis of his assault on democratic elements in Weimar Germany).

Given this unacceptable outcome of the third round of Franco–German rivalry, the Versailles peace was doomed from the beginning. Its centerpiece, the League of Nations, never had a realistic chance to organize the peace effectively: too many of its members were unwilling to defend a status quo they viewed as unjust, and others excluded themselves (the United States) or were excluded (the Soviet Union, on the premise that the way to arrest the spread of the "cancer" of bolshevism was through isolation).

Proximate events of the 1930s would transform these underlying issues into the bloodiest conflagration in human history. These events, in turn, can be traced to two related sources reflecting dissatisfaction with the peace ending World War I: economic nationalism and the effects of the Great Depression, and the rise of fascist regimes that would become increasingly aggressive and expansionist in the face of tepid responses from the Western democracies.

Economic nationalism and the depression are related events. The core of economic nationalism, which was to manifest itself in unprecedented protectionism of national industrial plants, can be found in the attempts of the drained countries of Europe to recover and recuperate from the ruinous economic effects of the First World War. National coffers had been emptied, industries had been turned to the war effort and had to be reconverted, and there were widespread scars of war (especially in France) that required rebuilding, all at considerable cost. Governments were forced to foot these bills, and one of their strategies for recovery (especially of the industrial plant) was to erect high protective tariffs against goods and services from elsewhere. Combined with the artificial flow of wealth resulting from reparations, the economic base of Europe became increasingly shaky.

The Great Depression represented the final blow to the international economic system. As businesses failed, banks defaulted, and the jobless lines increased throughout the continent, commerce between states came to a virtual standstill. The result was even more protectionism and an economic maelstrom that continued to get worse. As the times worsened, so did the political situation. In this climate, the rise of regimes that promised an end to the economic chaos, even if through escapism and adventurism, became progressively stronger. In no place was the cry louder and more inexorable than in Germany.

Fascism was not, of course, entirely a phenomenon of the 1930s. Mussolini came to power in Italy in the early 1920s, and the Japanese imperial monarchy predated the Great War. The factors that made fascism different in the 1930s and that led to war were the coming to power of fascism in its most virulent form through the

National Socialist (Nazi) party in Germany and the progressively expansionistic form that the various fascist regimes exhibited as the decade progressed.

The rise of Nazism was the key factor, because only a resurgent Germany had the potential to mount a major threat to the peace; Japan and Italy could also engage in mischief, but their reaches were limited. Germany, particularly when expanded to something resembling its pre-1919 borders, could pose a threat to the whole of Europe, as had been the case in the Great War.

Hitler's Germany and its Führer have been and continue to be the source of enormous, if macabre, fascination, and there is little we can add to the voluminous literature that surrounds the Nazi era. A few points can, however, be made that are germane to our general theme.

The first is that the terms of Versailles virtually guaranteed that something like Nazism would emerge in interwar Germany. Although the monstrous directions that Hitler's policies took were not preordained by the Paris peacemakers, the combination of humiliation and degradation that Germany suffered, the artificial nature of the political system imposed on Germany, and a German political culture that associated authoritarian rule with prosperity (many Germans even today consider the Kaiser's Second Reich the golden age of German history) certainly made a militaristic, authoritarian movement seem quite appealing.

Second and significantly, the appeal of Nazism was inadequate to gain power in the relatively affluent 1920s but was inexorable in the depression-plagued 1930s. When Hitler began to organize his political movement and made his first clumsy attempt to seize power (the "Beer Hall Putsch"), he was ridiculed, rejected, and thrown in jail. A little less than a decade later, with Germany in the depths of the depression, his simplistic analyses of Germany's woes and his grandiose solutions met a more responsive audience. Granting that many of the power brokers who helped him come to power viewed him as a comic figure they could manipulate and control, nonetheless Hitler had enough popular appeal to be elected chancellor.

The pattern of unchecked aggression provided the proximate events on the road to war. Each of the three major powers in what became the Axis participated in this process, and in each case timid responses (when there was any response at all) not only did not deter future actions but almost gave them tacit approval.

Some observers maintain that World War II really began in 1931. In that year Japan made its first major expansionist move, invading the Chinese province of Manchuria (the industrial heart of the country). Through fighting of enormous ferocity, punctuated by numerous atrocities against the civilian population, the Japanese succeeded in establishing their domain and installing a puppet ruler on the throne of the country they called Manchukuo. The West stood idly by. The strongest condemnation came from the United States, which promulgated the so-called Stimson Doctrine of Non-Recognition. This doctrine stated that it was American policy not to recognize governments that came to power by force (a tenet that has been selectively applied ever since). The practical effect was that the United States did not recognize the government of Manchukuo but continued to deal with Japan virtually on a business-as-usual basis. So encouraged, the Japanese Sphere of Co-Equality continued to expand through the decade. By 1941 the proximity of that empire and American interests meant that something had to give.

The Fascist government of Benito Mussolini also got into the act, albeit in a more modest way. The major adventurism in which Italy engaged was the 1935 campaign against Ethiopia. Using the Italian colony of Eritrea as his base of operation, Mussolini unleashed his mechanized army against the pitifully underarmed Ethiopian tribesmen (some of them actually confronted tanks and other armored vehicles with spears). The world was shocked but not enough to act.

When the attack began, the Ethiopian Emperor Haile Selassie, the "Lion of Judah," went to the League of Nations and appealed to that organization to invoke the collective security provisions of the covenant and to come to Ethiopia's defense. After long debate, the League voted *voluntary* sanctions against the Italian regime and omitted petroleum, oil, and lubricants (on which Italy was particularly dependent) from the list of proscribed materials. Of the major powers, only the Soviet Union (which had been admitted to the League after Germany withdrew) argued strongly for effective, mandatory sanctions to reverse the situation. Britain and France, unwilling to risk war over that barren corner of the Horn of Africa, wavered. In the wake of the Ethiopian affair, League collective security was effectively a dead letter.

Center stage in the tragedy was, of course, reserved for Nazi Germany. Using the dual assertions of the "destiny and right" of all German peoples (broadly defined) to be ruled together and the need for *lebensraum* (living space) as his justifications, Hitler began his campaign of expansionism in 1935. Initially, the reaction of the major Western democracies was weak and ineffectual. When they finally determined to react, it was too late.

Hitler's first and riskiest action was the remilitarization of the Rhineland. Sharing a long common border with France, the Rhineland had been demilitarized by the Versailles accords to assuage French fears of a new German onslaught. Timing his move to coincide with one of the frequent crises in the French Third Republic (one coalition government had collapsed and a successor had not been organized), Hitler moved his forces into the area. It was a gamble because at that time the armed forces of France were clearly superior to his own (almost all his military advisers opposed the plan) and could have forced him to back down. Hitler, however, counted on the paralysis caused by the French political crisis to preclude effective action. He proved correct; neither France nor Britain reacted and he was able to present the world with a fait accompli. So emboldened, Hitler turned his attention to bigger things.

The list of Hitler's aggressions is familiar enough and need not be treated in detail here. Under the guise of the "Greater Germany," Hitler annexed the Sudetenland area of Czechoslovakia (which had a majority German population) and in the Anschluss, German troops occupied Austria as well. Confused and irresolute, the Western powers refused to respond forcefully, instead believing Hitler's assurances that each expansion would be the last or believing that domestic public opinion would not support a forceful response. The nadir of the process was British Prime Minister Neville Chamberlain's return from Munich and his announcement of "peace in our time."

Not everyone, of course, was deceived by Hitler's designs. In England Winston Churchill led the cry to prepare for a war he knew would come, but citizens still war weary from the Great War turned a largely deaf ear. Only when Hitler launched the blitzkrieg against Poland in September 1939 did the situation change. Because of treaty obligations with the Polish state, Great Britain and France were forced to make the declaration of war that marked the formal beginning of World War II. That declaration and the period that followed it, symbolically enough, were known as the "phony war": no fighting occurred because neither Britain nor France was mobilized to fight. Only when Hitler turned his war machine against France the following June did the British and French become directly involved.

America's role and reactions to this chain of events should be noted. Except in the Pacific, the chain of Axis advances did not directly involve American interests nor have much of an impact on the American people. Isolated by the Atlantic and Pacific oceans from the gathering war clouds, Americans could and did largely ignore these events, instead concentrating their energies and emotions on coping with the debilitating effects of the Great Depression. The rumblings in Europe were Europe's problem, and there was little sentiment for plucking European "chestnuts from the fire" yet another time.

This was, after all, the era of splendid isolationism in American foreign policy, a period when the lessons of the inextricable link between the destinies of Europe and North America were still not realized nor appreciated. During the rise of the fascist movements, there was even some support for the emerging regimes and particularly for the Nazis.

Because Americans generally opposed the idea that these events affected them, or ignored the situation altogether, there were adverse consequences as the "winds of war" approached. On one hand, those Americans like President Franklin D. Roosevelt who realized that our participation would eventually be necessary were greatly hampered in their efforts to prepare the country for war. Because of legislation enforcing American neutrality, aid for the Western Allies had to be supplied surreptitiously. The American armaments industry could only be developed slowly, and authorization for even a standby draft (i.e., preparing the mechanisms for a draft) passed only in 1940. In the American tradition, we entered the war almost totally unprepared to fight it. As a consequence, it was not until 1943 that the full brunt of American military power could be brought to bear.

On the other hand, the ostrich-like attitude that the war did not concern us affected our reactions when the conflict was finally forced upon us. Active and hot war had been going on in Europe for a year and a half before the attack on Pearl Harbor, and the forces of geopolitics, if closely watched, suggested that American involvement was, in the long run, inevitable. Moreover, many U.S. leaders believed that in all likelihood Japan would be one of our opponents. Yet, the average American did not see the war coming. The "infamy" of the Japanese attack was accentuated and American outrage was all the greater for the surprise. In our reaction to the shock, our objectives were shaped by and pursued with a moral indignation that they might not have had if we had been better prepared.

Political Objective

The American declaration of war against Japan the day after Pearl Harbor, followed by the German counterdeclaration, threw the United States into its second coalition war of the twentieth century. As had been the case in the Great War, not all who fought together shared the same political and military objectives. The Allies were united in a joint desire to defeat the Axis (especially Germany) militarily, but there were differences of opinion about what constituted that military victory. Their most serious differences were political and largely focused on the postwar map. The greatest divisions among the major Allies were between the Soviet Union and her two major English-speaking partners, but there were some items of disagreement between Great Britain and the United States.

The Americans and the British had many common objectives. Probably the clearest statement of agreed goals was articulated well before America's entrance into the hostilities, when Prime Minister Churchill and President Roosevelt met off the Canadian coast and announced the Atlantic Charter. That document, setting forth eight points, had essentially two thrusts. The first was a statement about how the postwar world should be organized. As a statement of the better state of the peace, the charter emphasized such things as abjuring territorial gains, promoting self-determination as the basis for the postwar political map, and protecting free access to trade and resources for all nations. The other thrust called for the disarmament of the aggressors and a peace that would ensure the physical security of all countries.

The first thrust was primarily political and the second military. Where the two allies disagreed was in the relative emphasis that should be placed on each, and their positions largely reflected national attitudes toward war and politics. From the American perspective, the primary purpose of the war was to rid the world of the absolute evil posed by fascism. This was a highly moralistic goal, reflecting the American tendency (so well illustrated in other conflicts) to view issues in terms of good and evil. Defining the purpose once again as a moral crusade naturally emphasized the second thrust of the Atlantic Charter rather than the underlying political purposes that made the violence necessary. Defining the better state of the peace would have to wait for the end of violence.

The British view, epitomized by Churchill, placed greater emphasis on the postwar map. Recognizing the mortal peril represented by Hitler and the consequent need to vanquish the Nazi opponent, the British view was more geopolitical, placing emphasis both on the military task at hand and on the shape of the postwar map. In Churchill's mind, the primary problem for postwar Europe would be the power vacuum created in eastern and central Europe by the defeat of Germany and how to blunt and contain Soviet aggressive, imperialistic designs on those areas. The Americans downplayed this problem, initially because Roosevelt thought he could contain Stalin's ambitions. This disagreement produced friction among the Western Allies throughout the war and became particularly evident at the time of the final offensive against Germany in 1945.

The Soviet objective was quite different and considerably more desperate. After the failure of the Soviet initiative to re-create the Triple Entente of World War I, Stalin entered into the notorious Molotov–Ribbentrop (Nazi–Soviet) Non-Aggression Treaty in 1939. The Soviet purpose, beyond the partitioning of Poland in the secret protocol, was to provide breathing space to mobilize and rebuild its military capability, which had been ravaged by the purges of the 1930s. When the Nazi onslaught (Operation Barbarossa) began on 22 June 1941, the initial Soviet objective became survival of the fatherland. This was not an easy chore as German armies spread further and further into Soviet territory. The Soviets came within a hair's breadth of losing the war (some have argued that had the invasion not been delayed for six weeks because of disturbances in the Balkans, Hitler might well have succeeded).

After the infamous Russian winter bogged down the German advance, the political objectives of the Great Patriotic War became twofold and sequential. The first objective was to ensure the territorial integrity of the Soviet homeland, and its primary imperative was the physical removal of the German army. This was, of course, an objective with which the Western Allies could scarcely disagree, although there was considerable disagreement between Stalin and his allies about the strength, location, and timing of American and British efforts to alleviate pressure on the Soviets and hence to facilitate accomplishing the task.

Disagreement was fundamental on the second Soviet objective, which was to create a physical circumstance in Europe that would preclude a repeat of Barbarossa. One aspect of this objective was to create a buffer zone between Russia and Germany to ensure that a future thrust toward Russia could be confronted in eastern Europe. A buffer zone required states in eastern Europe at least not unfriendly (preferably sympathetic) to the Soviet Union, and it was this aspect that troubled Churchill most as he contemplated the postwar European map. The second aspect was the disposition of postwar Germany. In a manner of reasoning not dissimilar to France's after World War I, Stalin wanted a permanently weakened Germany that would not be capable of again posing the menace already twice visited on the homeland during the century. This desire came to mean a permanently partitioned Germany, neither part of which would be strong enough to threaten Soviet security.

Because the Pacific theater of the war was essentially a conflict between the United States and Japan (albeit with a major theater in China and other more minor theaters elsewhere in Asia), American political objectives against Japan were neither complicated nor compromised by the problem of coalition policy-making to the degree they were in Europe. In the Pacific, Allied and American objectives were essentially the same (at least until the Soviets entered the conflict in 1945 with the apparent—now if not then—purposes of gaining a buffer zone in North Korea and a voice in the future of Japan).

The American political objective was straightforward, and was heavily influenced by the Japanese sneak attack on Pearl Harbor. The purpose was the destruction of the Japanese Empire and the abdication of the Japanese emperor, whom most Americans identified—probably erroneously—as the instigator of the Pearl Harbor raid. Roosevelt's description of that attack as a "day of infamy" set the moral tone for a crusade against the perpetrators. Nothing less than the total defeat of the enemy could create atonement; and because the emperor was, in the popular mind, the embodiment of Japan, the emperor had to go. As we shall see in a later section, this absolute requirement may well have lengthened the war in the Pacific and may even be loosely related to the later fighting in Korea.

The final consideration regarding the objective was the relative importance of attaining the political objectives of overthrowing German Nazism or Japanese imperialism: on which end should primary attention be focused? From the viewpoint of the European allies, Nazism represented the greater threat and should be dealt with first. On this point there was agreement within the alliance: Hitler should be defeated first, and then attention should be shifted to Japan. In practice military objectives and strategy only imperfectly reflected this agreement.

Military Objectives and Strategy

The military situation upon the entrance of the United States into World War II was fundamentally different from the situation faced by Americans when they entered the Great War in 1917. In World War I the United States entered on the side of viable allies and tipped the scales toward victory. On 7 December 1941 Americans had many allies, but few were in a position to shoulder a significant portion of the burden. The United States was suddenly thrust center stage into a war for which it was ill prepared.

Most of Europe had been overrun by the Nazi war machine. France, the low countries, Norway, and the Balkans were all controlled by the Germans. The Soviet Union was reeling from the lightning-war blows of the Wehrmacht and was in the painful process of trading its vast spaces for time to recover and counterattack. Britain had survived the German air assault and prevented a Nazi invasion, but its survival was still in doubt. German submarines were taking a fearsome toll on British shipping, and the British army was heavily engaged in North Africa against a superb German general bent on seizing Egypt. Besieged as they were, the Soviet Union and Great Britain were America's only significant allies in Europe.

In the Pacific and Far East, China continued to survive despite the heavy blows of the Japanese but could offer little help to the United States other than to tie down a large portion of the Japanese army. Japan was running rampant across the

Pacific. By the end of December 1941, Wake Island and Hong Kong had fallen and Japanese forces had invaded Malaya, the Philippines, and the Gilbert Islands. By mid-March 1942, Malaya, Singapore, Rabaul, and Java had all fallen; the remnants of an Allied fleet had been destroyed in the Battle of the Java Sea; and General MacArthur had been evacuated from the Philippines. The European colonial powers were besieged in Europe and could do little to stem the Japanese tide. Help was available from Australia and New Zealand, but it was clear the United States would have to shoulder the majority of the load.

If the world situation was bad, the condition of the American military was worse. Most of the Pacific fleet's firepower rested on the bottom of Pearl Harbor. The American Army was still building and was untested in combat. American air power was still a paper force. Although first-class heavy bombers were coming off the assembly lines, no U.S. fighter could match those fielded by Germany or Japan. In sum the situation at the end of 1941 was dismal.

The overall Allied strategy for the war was actually mapped out well before the United States entered the war. In the winter and spring of 1941, American-British-Canadian (ABC) staff conversations produced a generalized military strategy should the United States enter the war. The plan's first priority was to stop the enemy onslaught. The first American objective was to preserve a secure operating base in the Western Hemisphere. For the British the essential task was clearly to maintain the integrity of the British Isles and, if possible, its dominions in the Far East (particularly India). The second British priority was to maintain control of its sea lines of communication, without which all else would crumble.

The most important agreement reached during the ABC talks had to do with overall Allied priorities. Germany was considered the predominant member of the enemy camp. Thus the staffs agreed that the European theater was the decisive theater and the area for the initial concentration of effort. These priorities dictated that in the Far East the military strategy would have to be defensive at the outset. Later during the war, however, this agreement would cause considerable consternation (particularly in the U.S. Navy, which considered the naval war against Japan to be at least equally important to the war in Europe).

The offensive campaign against Germany and Italy was envisioned in stages. The first stage was to bring to bear economic pressure, including the denial of raw materials. The second stage was a sustained heavy air offensive against the German homeland. The third stage was to eliminate Italy from the war, since it was considered the most fragile of the three Axis partners. Raids and minor offensives against the enemy were envisioned at every opportunity while forces for the major offensive were built. The fourth and last stage was a major offensive against the Germans on the European continent itself.

In the European theater, the most lasting and vexing questions were where and when to invade the continent in force. The British were wary of a cross-channel invasion. The ghosts of a generation of youth lost on Flanders fields during the Great War haunted the British. They feared that they would again be bogged down in a stalemated war of attrition that they could not afford. They also feared attacking before they were finally prepared and again being thrown off the continent. The memory of Dunkirk died hard.

The Americans desired a cross-channel invasion as soon as possible and argued hard for such an undertaking as early as 1942. The situation was complicated by the need to keep the Soviet Union in the war. The Soviets badly needed a second front, and Stalin used every opportunity to press for an invasion at the earliest possible moment.

The cross-channel invasion controversy would continue throughout the war. The Americans and Soviets pressed for early invasion. The British constantly suggested such alternative (and presumably safe) invasion sites as Italy, Greece, and the Balkans.

Churchill won the first round of the controversy by posing a series of difficult questions that emphasized the enormous problems involved in mounting a cross-channel invasion in 1942. Roosevelt, however, was convinced that the Allies, particularly the Americans, must take dramatic offensive action as soon as possible. With the invasion of France put into the "too-difficult-at-present" category, he agreed to an invasion of North Africa. The object was to trap the German and Italian forces between British forces advancing from Egypt and the Anglo-American invasion forces advancing from Morocco and Algeria.

North Africa offered many advantages over a cross-channel invasion. First, and most important, the landings would not be directly opposed by seasoned German troops. Rather, the invading troops would land on shores controlled by the Vichy French. Although the French would probably oppose the landings, there was the possibility that there would be no resistance and, in any case, resistance should be far less severe than could be expected on the coast of France itself. Success in North Africa would yield several significant benefits. Victory would help to open the Mediterranean shipping lanes, facilitate the flow of supplies to the Soviet Union through the Persian Gulf, and might possibly draw German strength away from the Soviet front. Given available resources, the operation posed significant benefits at minimum risk.

The landings and the campaign as a whole were successful, and in January 1943 Roosevelt and Churchill met in Casablanca to map out the next Allied moves. Although a cross-channel invasion was discussed for the fall of 1943, Churchill won round two by convincing Roosevelt that the next step should be Operation Husky, the invasion of Sicily. Churchill argued that seizing Sicily would facilitate clearing the Mediterranean sea-lanes and would put pressure on the shaky Italians. The invasion of France was again postponed, this time until the spring of 1944. Stalin's reaction to the postponement was bitter and helped plant the seeds of disunion that would tear the Allies apart in the aftermath of the war.

The Casablanca conference is best known for two other strategic decisions. The first reaffirmed the importance of the massive bombing offensive against Germany, which was just getting under way in earnest. The air offensive had been plagued by diversions from strategic to tactical targets (support for the North African landings, for example) and controversy between British and American airmen over the advisability of the daylight, precision-bombing attacks preferred by the Americans. At Casablanca plans were laid for the combined bomber offensive in which the Americans would bomb by day using precision techniques and the British would use area-bombing techniques at night. The idea was to give German defenders and the civilian population no rest.

The other, and more famous, decision was announced at the concluding news conference, when Roosevelt stated that the Allied goal was "unconditional surrender" of the Axis powers. Ever since that news conference, the effects of that phrase have been matters of considerable debate. Critics have pointed out that such an objective probably prolonged the war because it gave the Axis powers a propaganda advantage by imbuing the enemy population with the courage of desperation. Churchill maintained that the specific retribution demanded by the Allies seemed so severe when set down on paper that unconditional surrender paled in comparison. Both Roosevelt and Churchill made significant efforts throughout the remainder of the war to explain unconditional surrender as something that the enemy population should not view with fear. Their efforts, however, had limited success.

The invasion of Sicily yielded quick success and the invasion of the Italian mainland quickly followed. By early September 1943 the Italian government surrendered, although German troops in Italy maintained strong defensive positions. Thus most of the steps in the general Allied strategy had been accomplished. The German-Italian advance had been stopped, pressure had been applied whenever possible on land and at sea, the bombing offensive against German territory was picking up momentum, and Italy had been knocked out of the war. Planning now began in earnest for the cross-channel invasion.

In early 1944 Allied air forces stepped up their strategic attacks and made their first priority the Luftwaffe itself. During the spring, the Allies seized command of the air over western Europe and began to have a serious impact on German ability to operate effectively on the ground. Control of the air was an essential ingredient in a successful invasion, and the Allies had almost total control, particularly over the channel coast.

Meanwhile the resurgent Soviet army took a terrible toll of German military power. Even though a majority of German military power was concentrated on the eastern front, the Soviet army had begun to roll toward the borders of the Reich. So great were the Soviet victories that the Anglo-American forces made plans for an "emergency" cross-channel invasion should Germany collapse and "invite" them ashore to prevent the Soviets from overrunning all of western Europe.

The successful Allied landing on the shores of Normandy and subsequent breakout from the hedgerow country sealed the fate of Nazi Germany. Unfortunately, many Germans fought on to the bitter end with fatalistic determination. Whether or not it was inspired by the prospect of unconditional surrender, the prolonged German resistance did little but cause more death, suffering, and postwar bitterness.

In the Pacific, the war against Japan progressed much faster than originally anticipated by Allied planners, thanks primarily to the incredible industrial output of the United States. The first order of business was to stop the Japanese advances. The Japanese had other ideas, including knocking the United States out of the Pacific altogether.

Thus Adm. Isoroku Yamamoto sought to destroy the remnants of the American Pacific Fleet, and the Japanese appeared to have the chance for the final decisive battle when Adm. Chester Nimitz offered battle at Midway in June 1942. Nimitz was in possession of Japanese war plans, thanks to important code-breaking operations (see the section on military conduct) and thus surprised and defeated the Japanese fleet. The Battle of Midway ended the eastward expansion of the Japanese. In the south, the Americans invaded the islands of Guadalcanal, Gavutu, and Tulagi in the Solomons. The objective was to blunt the Japanese offensive, which appeared to threaten Australia. After bitter fighting the U.S. operation was successful, and the southern expansion of the Japanese was halted. With the Japanese stopped, it was time to take the offensive.

The basic offensive strategy had four steps. The first required cutting off the Japanese home islands from sources of raw materials and thus taking advantage of a major Japanese vulnerability. In the second part of the strategy, the Allies were to keep up constant military pressure to force the Japanese to use their war reserve stockpiles. The third step was to get within range of the home islands and initiate a heavy bombing offensive to pave the way for an invasion. (Originally, Allied planners envisioned the use of Chinese air bases for the air campaign. However, Chinese setbacks in 1944 made it necessary to use island bases for the final aerial onslaught.) The fourth and last step was to invade the home islands.

Controversy arose over the best route for getting within range of the home islands. General MacArthur in the southwest Pacific argued for an advance on the Philippines from the south (his command area). The Philippines could then be used as a jumping-off point against the home islands and as the place from which to interdict the flow of raw materials from the south. To some observers, MacArthur seemed obsessed with his pledge to return and liberate the Philippines, but he had an advantage: as a former Army chief of staff, he had been the superior of those officers now running the war.

Adm. Nimitz favored a far different route. He recommended a naval assault across the central Pacific aimed at the Philippine–Formosa area to cut the Japanese Empire in two and separate the home island from their resource base. Naval strategists also leaned toward Formosa as the base of operations for the final assault on Japan. In a classic compromise, the Joint Chiefs of Staff authorized the "nondecision" that both approaches were the best method of placing intolerable pressure on the Japanese.

After the Americans seized island bases in the Marianas (part of the Nimitz plan), the aerial campaign against Japan began in earnest. The Japanese merchant marine had been virtually destroyed by naval action and the lines of supply from the south were cut. American airmen believed bombing would destroy remaining Japanese war production and directly attack the will of the people. In a turnabout, the daylight precision-bombing concept that the American airmen had fought so hard to preserve in the European theater was quickly abandoned in the Pacific. Bad weather, problems with the jet stream, long flight distances, the distribution of Japanese industry, and Japanese opposition convinced the airmen that night low-level attacks using incendiary munitions against highly flammable Japanese cities would be the most effective strategy.

Plans for the invasion progressed. The bitter experiences on Iwo Jima and Okinawa demonstrated that an invasion of the home islands would be a costly affair. However, the city-leveling bombing attack took its toll. The use of atomic bombs at Hiroshima and Nagasaki was the straw that broke the back of resistance and Japan sued for peace before the invasion occurred.

Political Considerations

As in all wars, parts of World War II gain their full meaning only when viewed within a domestic and international framework where politics affects the battlefield and vice versa. From an American viewpoint, the major political concerns were international rather than domestic because the war enjoyed broad-based internal support. At the same time, the exigencies of alliance decision making created some strains that were primarily political in character.

American participation in World War II may have been the most popular of any U.S. military excursion. Certainly it was the most strongly supported long war in U.S. history. It was the most purely total war in the American experience, and this factor may well have contributed to America's embrace of its purposes and conduct.

Although we have described other wars as total within their contexts, the Second World War more closely meets the criteria of total war. It was a war of mass mobilization wherein the entire American population was called upon to become involved in one way or another. More than 12 million Americans were brought directly under arms for its conduct, the American industrial plant was revived and converted to become the arsenal of democracy, and an unprecedented amount of American treasure was committed to its successful conclusion. Millions, including a previously inconceivable number of women and minorities, were recruited to work in factories supporting the war (many have argued that this was the crucial underpinning of later women's and civil rights movements) and everyone was encouraged to contribute in some way, if only by growing victory gardens or making do with ration coupons. Thus for Americans (other than Southerners in the Civil War) World War II demanded a depth and breadth of involvement never seen before—or since.

Fortunately, the purposes for which the war was being fought seemed adequate for the sacrifice. The enemy was an unmitigated evil: the monster Hitler and the infamous Hirohito and Mussolini. The elimination of such evils rang a responsive chord in the traditional American moral sensibility, an emotion heightened by the way America was forced into the fray. When President Roosevelt proclaimed the unconditional surrender of the various enemies as the primary military objective to ensure the eradication of fascism, he announced a moral purpose adequate to sustain American support.

If America was firmly united behind the purposes of the war, the same cannot be said of the other Allies. The most prominent of disagreements within the alliance dealt, as suggested earlier, with how to treat the Soviet Union and what the balance should be between attaining political and military objectives when these came into conflict. These problems were interactive and were the subject of considerable discussion between the Americans and the British.

Great Britain, with its greater experience in power politics generally and its longer history of dealing with Russians, was a great deal more suspicious of the Soviets than was the United States. Churchill viewed Soviet motives within the context of the expansive tendencies of the Russian Empire by whatever name. Americans, largely naive in European politics (by choice) and painfully ignorant of the Soviet Union, were more ready to embrace the Soviets and to believe in their political reasonableness. To America, all the Allies were part of the common cause, and Soviet leader Stalin was extolled by the American propaganda network as "Uncle Joe."

This difference in attitude had practical applications regarding both elements in the disagreement. Because the British believed the ultimate Soviet purpose was to extend its domain over as much of Europe as possible, they were more content to see the Soviet and Nazi armies slug it out on the Eastern Front. The Americans wished to get onto the continent, confront the Germans in classic American style, and grind the Nazi armies between themselves and the Red Army. The British demurred from this "direct" approach to the problem, preferring the more "indirect" approach of attacking at the periphery. As discussed earlier, the British approach prevailed until June 1944 while Stalin fumed that the Western Allies' real desire was to see the Nazis and the Soviets physically destroy one another so that the Western Allies could pick up the pieces (an idea put forward early in the war by a then-obscure senator from Missouri, Harry S Truman).

One result of these disagreements was the ongoing annual debate between the British and the Americans about when and where to open the second front. Certainly part of that debate represented a philosophical disagreement between Americans, whose heritage was the direct approach inherited from Grant, and the British belief in the Anaconda-like plan that made up the indirect approach. At the same time, the question of how to deal with the Soviets, which translated into the question of how fast and where to come to their aid, had something to do with the debate.

The balance between adherence to the military and political solutions was also subject to disagreement as exemplified by the final campaigns of the war in Europe. After the final German offensive in the west was broken at the Battle of the Bulge in the winter of 1944–45, Germany lay largely prostrate. The Americans and the British were approaching from the west, and the Red Army was moving in from the east.

The American command, supported by Roosevelt, wanted to concentrate on the purely military task of engaging and destroying the German army's ability to resist. Such an approach required operating over a broad front, was time-consuming, and brought relatively small amounts of territory under control. The British preferred a blitzkrieg-style movement on a narrow front, worrying about knocking out pockets of resistance later. Their motive was simple and straightforward: Churchill maintained that a lightninglike movement could capture Berlin and thus reap considerable political and military prestige that would be important in the postwar world. The Americans, however, insisted on the broad front as militarily more effective, and because of our military predominance, we prevailed.

Another political concern was how to induce the Axis powers to capitulate once the end was in sight. The goal of unconditional surrender became a problem in much the same way that the Emancipation Proclamation undoubtedly stiffened the Southern desire to resist during the Civil War. Unconditional surrender's requirements included the physical invasion and occupation of Germany that had been avoided in the Great War, and such a prospect tended to increase the Axis will to resist. Allied armed forces were gradually grinding down hostile ability, but overcoming cost-tolerance (hostile will) was a more difficult matter in the face of Allied objectives.

The primary tool used in each theater to bring about the destruction of hostile will was strategic bombardment, followed, if necessary, as in the German case, by invasion. In the case of Japan, because of the horrible prospects of having to mount history's largest and undoubtedly bloodiest amphibious assault on those islands, the United States sought help for the final invasion. Thus the United States induced the Soviet Union to declare war on Japan in 1945 to assist in the invasion (and hence to endure some of the estimated 1 million casualties that the effort would exact). The dropping of the atomic bombs on Hiroshima and Nagasaki, however, obviated the problem.

Military Technology and Technique

If defensive capabilities dominated World War I, the offense dominated World War II largely as the result of technological advances and new and inventive ways of using weapons first developed in earlier wars. Rather than a grinding stalemate, World War II saw sweeping, high-speed maneuver that revolutionized war on land, at sea, and in the air.

Mechanized armored formations fundamentally changed land warfare. Although the tank had some successes in the latter stages of World War I, its effectiveness was limited. Proponents of armored warfare between the two wars believed the tank's somewhat disappointing performance had been the result of transient problems, such as mechanical difficulties that could be resolved with better designs and materials. World War II tanks had heavier armor plating, much greater speed (up to 25 mph for even the heaviest tanks), heavier guns (up to 122 mm on the Soviet JS III), turret-mounted guns for greater firing flexibility, and far more reliable designs. In all, the World War II tank was a far more potent weapon than its lumbering predecessor.

The leading interwar theoreticians of mechanized warfare were two Englishmen. Gen. J.F.C. Fuller had been the author of "Plan 1919," the blueprint for a massive and presumably final tank offensive against the Germans. (The plan was not put into effect because Germany surrendered in 1918.) Fuller remained a leading proponent and theoretician of armored warfare after the war and long after his retirement from active service. Basil Liddell Hart, a medically retired army captain, was the second leading proponent of armored warfare. Hart was a prolific writer, and his fame, along with that of Fuller, spread around the world. However, they were prophets with little honor in their own land. (In the mid-1930s, the British army spent far more on fodder for cavalry horses than on fuel for its armored forces.) The leading practitioner of armored warfare, building on the theories of Fuller and Liddell Hart, was Heinz Guderian, who built the German panzer divisions and trained his forces in the concepts of blitzkrieg (lightning war). Meanwhile, the French, British, Americans, and Soviets fell behind in both theory and practice.

In the Second World War, as forecast by Fuller and Liddell Hart and practiced by Guderian and his disciples, the armor–infantry relationship of World War I was reversed. Now infantry supported tank operations and tanks moved at their own pace rather than at the pace of the infantry. It was the infantry's problem to keep pace with the tanks and general land force mechanization resulted. Thus modern armored warfare emphasized rapid and sustained movement. The objective was to penetrate deep to the enemy's rear, disrupt his command, control, and communication capabilities, and surround his fighting formations. Strong defensive formations were bypassed and left for reduction by the mechanized infantry formations that followed. The flanks of armored columns slashing deep into enemy areas were long and exposed. But with enough speed and movement, the flanks would take care of themselves as befuddled enemies fell back in total disarray. Mechanized battle was not only a battle of rapid movement, but it also required the use of massive numbers of tanks and supporting vehicles.

Although modern mechanized warfare depended on the development of dependable armored vehicles, another key element in "lightning" war was the development of tactical air power. When carefully coordinated with ground operations, tactical air power provided the long-range disruptive effect provided by heavy artillery in World War I. Unlike the big guns of the first war, however, tactical air power moved easily with rapidly advancing armored columns and ranged ahead of them to disrupt enemy efforts to organize an effective defense.

Not all campaigns in World War II were lightning campaigns, but nearly all displayed the combined arms operations that, if conditions were favorable, could develop into campaigns of rapid movement and deep penetration. There were, however, numerous examples of lightning war in textbook form. The German invasion of Poland was the first example, although the battle for France in 1940 remains the classic example of blitzkrieg. The largest examples were in the German offensives against the Soviets. In the summer and fall of 1941 panzer columns slashed deep to the rear of Soviet positions and trapped huge numbers of Soviet soldiers and incredible quantities of equipment. Despite setbacks in the winter of 1941–42, the Germans again achieved significant success in the summer and early fall of 1942 against the Soviets. Late in 1942, however, the Germans became seriously over-extended and got bogged down in positional "slugging matches," most notably at Stalingrad. Soviet forces linked in the German rear and trapped an entire German field army, a major turning point in the Great Patriotic War.

The foremost American practitioner of armored warfare was the colorful Gen. George S. Patton. At heart Patton was a cavalryman of the old school, but he adapted well and studied the successes and failures of others. He imitated his tutors and generally improved upon their techniques. Patton's dash across France after the Allied breakout from the hedgerows of Normandy is a superb example of armored warfare at its best. Patton's use of tactical air power ahead of his columns is an excellent example of the air-to-ground coordination required in highly mobile warfare.

The value of air power was primitively demonstrated in World War I. Air-to-ground cooperation and coordination was improved in the second war, thanks to the development of reliable radio equipment. As a result of improved coordination and the inherent speed and mobility of air power, it became critically important to successful ground operations, particularly, highly mobile ground operations. But in World War II, air power also played a much different role: to attack the ability of an enemy society to wage war.

The concept of strategic bombing flowed from the ideas of the Italian airman Giulio Douliet and the American soldier-airman Billy Mitchell. Both of these iconoclasts believed that bombing the enemy's "vital centers" could destroy the enemy's will and ability to wage war. In the United States the ideas of Mitchell, who was forced out of service, were carried to fruition by his disciples at the Air Corps Tactical School. They promulgated a doctrine of strategic bombing to destroy the enemy's industrial centers and thus destroy his ability to support warfare. The "true believers" at the school thought that strategic bombing alone could be decisive and that surface forces would become passé. The initial American plan for the air war in Europe even indicated that, if correctly executed, a strategic-bombing campaign could make an invasion of the continent unnecessary.

Two key elements were required for a successful strategic bombing campaign. The first was equipment capable of striking the enemy's vital centers. After several unsuccessful attempts to produce a heavy bomber, success was finally achieved with the development of the B-10 in 1932. However, it was the famous B-17 and later the B-24 that formed the backbone of the American bombing effort during the European campaign in the second war. In the Pacific theater, the much larger, heavier, and longer-range B-29 was the aircraft that ranged across vast distances and struck at the Japanese home islands.

The second key element was the identification of the vital centers or the key elements of the industrial web, the destruction of which would bring down the enemy. During the bombing campaign in the European theater, campaign plans were continually revised in a search for these key elements. Electric power plants, ball bearings factories, and other "vital" industrial elements were attacked with significant success, and yet the German war machine fought on. Finally, the American airmen struck at German synthetic oil production and found the key. The destruction of these facilities significantly degraded the capabilities of German ground and air forces by causing acute shortages of petroleum products essential for mechanized warfare.

The effectiveness of strategic bombing is still a subject of considerable controversy. Clearly, the idea of strategic bombing was oversold by zealous airmen before World War II. Armies and navies did not become obsolete as some had predicted. Europe had to be invaded and many bloody battles fought before the Nazis surrendered. In the Pacific, however, the destructiveness of firebombing raids on Japanese cities and the devastation caused by two atomic bombs convinced the Japanese to surrender without an invasion. The Japanese navy had been destroyed, but much of the Japanese army remained in the field undefeated. The airmen believed they had been vindicated, but the land and naval battles required to get the airmen within range of the Japanese home islands cast at least some doubt on their claims.

Naval warfare was also drastically altered by the advent of air power. The age of the battleship as queen of the fleet ended decisively at Pearl Harbor. The aircraft carrier quickly became the centerpiece of the naval battle force. Often during the Pacific campaigns, fleets fought at ranges so great that opposing ships were never in sight of one another. Aircraft ranged out from their carriers hundreds of miles to attack enemy fleets and island strongholds. Although surface engagements did occasionally occur, for the most part naval war in the Pacific was an air war. Naval guns were used to protect vulnerable aircraft carriers or to support amphibious landing and ground operations.

Other technological developments before and during World War II were less dramatic but equally important in changing the face of warfare. Of particular importance was the use of electronic warfare. Radar was an indispensable warning device against attacking aircraft, as was well demonstrated in the Battle of Britain. Counter-radar devices thus became important as did radio direction finding and guidance techniques. Radar was also important at sea both for air defense and for surface operations. Electronic eavesdropping, the gathering of intelligence by electronic message interception, became a fine art that led directly to success on the battlefield. Perhaps the best example of this was the electronic intelligence gathered prior to the Japanese assault on Midway in 1942. Not only were radio transmissions intercepted, but the Japanese code was broken, revealing Adm. Yamamoto's plan. As a result, a ragged force built around the three surviving U.S. aircraft carriers was able to ambush the Japanese invasion fleet and turn the tide in the Pacific. Equally important was the electronic intelligence gathered from the Germans. The ability to read German codes provided a singular advantage to the Allies.

Many would argue that the most significant technological development of World War II was the atomic bomb. However significant its development may have been, its primary influence lies in the postwar world, and thus is discussed in a later chapter. In World War II the development and use of atomic bombs were little more than the crowning blow that convinced the Japanese that they must surrender or face total destruction.

World War II saw the mechanization of war begun in the Civil War and advanced through World War I come to fruition. Although mechanization changed the face of war on the battlefield, the most significant change it wrought was to change the very nature of war. War, which had been a battle of masses since the time of Napoléon, became a battle of factories. The industrial base of the antagonists was increasingly important as were lines of communication, sources of raw materials, and complex logistical networks. As a result, the home front became a target of critical importance, and a target that could be struck directly because of the mechanization of war. The home front was now on the front lines.

Military Conduct

Both Germany and Japan achieved stunning successes before the United States entered the war. The Italians achieved some limited success in Ethiopia but were for the most part embarrassed by the results of their military ventures. Japan invaded China proper in 1937, and on 1 September 1939 Germany invaded Poland. By the end of December 1941, these three unlikely partners had overrun most of Europe, much of North Africa, most of European Russia, the richest portions of China, and many of the island complexes in the East, Central, and South Pacific. Their setbacks had been few in number and minor in nature. And yet, by the end of 1942, just one year after the entry of the United States into the war, the tide of conquest was reversed and the Axis powers began long and painful retreats.

Although the Allies agreed that attention would be focused on the European war first, the tide first turned in the Pacific. Adm. Yamamato, the architect of Japanese naval victories in the Pacific, attempted to lure the remainder of the American Pacific Fleet—primarily the aircraft carriers not caught in the Pearl Harbor attack—into a final climactic battle. Yamamoto planned to seize Midway Island in early June 1942 to bait the trap. When the American carriers arrived to counterattack, he would overwhelm them with an incredible assemblage of ships and planes. Unfortunately for the Japanese, the Americans were reading the most secret Japanese naval codes and knew Yamamoto's plan. As a result, the aircraft from three American carriers were able to ambush Yamamoto's carrier force north of Midway. The island base remained in American hands, but more important, U.S. aviators sank four Japanese fleet carriers and Japan lost the best of its naval aviators. The cost to the Americans

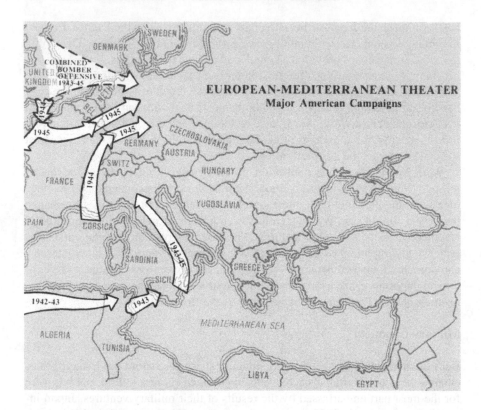

was many good aviators, one destroyer, and one carrier. The tide had turned in the Pacific only six months after the United States entered the war.

In North Africa, the seesaw war that raged back and forth across the desert finally turned for good against the German-Italian forces on 23 October 1942, when British Gen. Bernard L. Montgomery launched the second battle of El Alamein. In a well-prepared set-piece battle, Montgomery sent the Axis forces under the command of Gen. (later Field Marshal) Erwin Rommel reeling back out of Egypt and across Libya. On 8 November, with the Germans and Italians in headlong retreat, Anglo-American landings took place near Casablanca in Morocco, as well as near Oran and Algiers in Algeria. Vichy French forces controlling these areas offered some resistance but were quickly overcome, and the Allied troops began moving east to trap the Axis forces retreating in front of Montgomery. After much hard fighting and the evacuation of Rommel, nearly a quarter million German and Italian soldiers trapped in Tunisia surrendered on 12 May 1943.

The tide turned in the European war on 19 November 1942, north and south of the shattered city of Stalingrad on the Volga River. In August of that year, the German Sixth Army had reached the outskirts of Stalingrad. This advance was the high water mark of the German offensive in southern Russia. The city that bore

Stalin's name held a special fascination for Hitler, and Soviet troops defended the city with great tenacity for perhaps the same reason. The Germans abandoned the mobile, armored warfare that had given them great success and became bogged down in house-to-house fighting in which their superior armored forces offered no advantage. Meanwhile the Soviets massed forces north and south of the city to strike the exposed Axis flanks. On 19 November the Red Army attacked, and by 22 November had trapped Gen. (later Field Marshal) Friedrich von Paulus and the German Sixth Army. Despite attempts to relieve the army and to supply it by air, the shattered German army surrendered at the end of January 1943.

The Allied counterattack continued in 1943. In the Mediterranean the Allies invaded Sicily on 10 July and liberated the island by 17 August. In September Anglo-American forces leaped the Strait of Messina and invaded Italy. Although the Italian government and forces surrendered, the Germans under the command of Field Marshal Albert Kesselring waged a tenacious defense of the peninsula that lasted until nearly the end of the war.

In the Pacific limited Allied offensive action began. Operations on Guadalcanal were completed in early February. New Georgia, New Britain, Makin, and Tarawa all fell; Bougainville was invaded and counterattacks liberated much of New Guinea. Of particular importance was the death of Adm. Yamamoto, whose plane was shot down by American pilots flying from Guadalcanal. American signal intelligence code breakers discovered that the admiral would be visiting forward bases in the Solomons and the timing of his itinerary. American P-38 twin-engine fighters ambushed the admiral's aircraft and the chief strategist of the Japanese navy was lost.

Meanwhile the long-awaited Allied heavy bombing offensive against Germany began in earnest. British bombers had raided deep into Europe under the cover of darkness almost since the beginning of hostilities. The Americans entered the war with excellent heavy bombers (B-17s and B-24s), an accurate bombsight (the Norden), and a different bombardment theory. While the British bombed large areas at night to avoid heavy losses to enemy air defenses, the Americans advocated precision bombing during daylight hours.

During 1942 and early 1943, the American bomber force slowly increased in Europe. Equipment arrived with green crews, and training was intense. Short missions over the continent began on 17 August 1942. The long-range American bombing effort began in earnest late in the summer of 1943, first against the Romanian oil refineries at Ploesti and then against Germany itself.

The strategic bombing campaign reached a thundering climax during 1944. During "Big Week," which began on 20 February, armadas of Allied aircraft bombed German aircraft plants and challenged the Luftwaffe to fight. Although the bombing raids did significant damage, the telling blow was in air-to-air combat, as Allied escort fighters directly battled hard-pressed German fighters. In mid-May some bomber missions were allocated to attacks on German synthetic oil production, the source of much aviation fuel for the Luftwaffe. Although the results were not

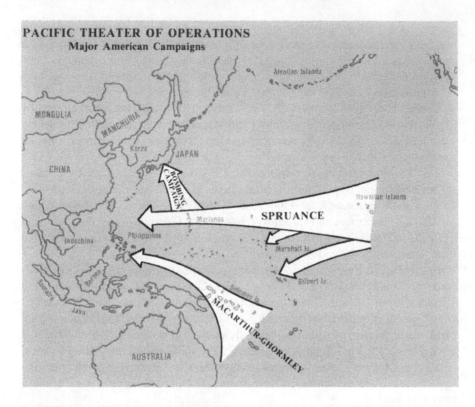

apparent for several months, synthetic oil was the key to total air supremacy for the Allies. Without fuel, the Luftwaffe could not defend against the Allied air offensive, and the destructiveness of the bombing raids mounted rapidly. The vital center had been found, the air battle was won, and the stage was set for the invasion of the continent.

The Germans believed the invasion would be launched directly across the Channel at the Pas de Calais. The Allies nurtured this notion through elaborate deception schemes while preparing to invade across the beaches of Normandy. Although some difficulties were encountered, particularly on Omaha Beach, the invasion went well and a lodgement was quickly achieved. However, it soon became obvious that the planners had worried more about securing a beachhead than they had worried about subsequent operations. It is difficult to imagine fighting territory more unsuited to mobile warfare than the Normandy hedgerow country. The Allies could not take advantage of their superior mobility and quickly became involved in a heavy slugging match that yielded slow progress.

Finally, after a climactic battle in July near Saint-Lô, the Americans breached the German line and poured through the gap into more open country. General Patton's newly created Third Army led the charge across the breadth of France in an armored dash reminiscent of the German blitzkriegs of earlier years. Suddenly, the front was far more fluid and a general Allied advance quickly began. By 19 August American elements were across the River Seine, and Paris was liberated on 25 August. Meanwhile on 15 August, the Allies launched a second invasion, this time on the southern French coast near Saint-Tropez. The troops were quickly ashore and moved north. German resistance in France was in shambles.

In March 1945 the Allies were across the Rhine in force, and the victorious armies fanned out across Germany. Resistance to the Soviet advance continued to the bitter end while resistance in the west rapidly disintegrated. The Germans capitulated on 7 May 1945.

In the Pacific 1944 was a year of relentless Allied advance. American troops seized Kwajalein, Eniwetok, and Saipan by midyear. Allied operations continued successfully on New Guinea, and a Japanese thrust from Burma into India was defeated. By year's end, American bombers were within range of the Japanese home islands, and the stage was set for the destruction of the Japanese will and ability to continue the war.

After midyear Guam and Tinian quickly fell. Finally, on 20 October, General MacArthur fulfilled his promise and returned to the Philippines. The U.S. Sixth Army landed on the island of Leyte and began a long struggle to liberate the islands. The landings provoked the remnants of the Japanese navy to an almost suicidal mission to destroy the American fleet supporting the landing and thus to isolate the troops ashore. The Japanese attack surprised the American forces in Leyte Gulf and caused considerable damage and confusion. However, the Battle of Leyte Gulf was the gasp of a dying fleet and a disaster for the Japanese. In widely separated actions, the Japanese lost four carriers, three battleships, six cruisers, and 14 destroyers. For all practical purposes, the Japanese fleet ceased to exist.

While the campaign for the Philippines continued, another facet of the war began. On 24 November 1944 the first B-29 raid against the Japanese home islands was launched. Using bases in the Marianas, the pace of attack began to build but with mixed results. The new aircraft caused some problems as did the long flight times. Additionally, finding good weather over Japan was difficult. Precision bombing was made difficult because of the jet stream encountered over Japan and because much of the Japanese industrial base resembled a cottage industry—dispersed and difficult to find. As a result, on 9 March 1945 Gen. Curtis LeMay stripped down his B-29s, loaded them with incendiary bombs, and ordered them on a low-level night bombing mission over Tokyo. The raid destroyed nearly 16 square miles of the city, killed 84,000, and injured 40,000 others. Nearly 1 million people were made homeless. American losses were negligible. The die was cast; low-level firebombing would be the principal aerial tactic. By war's end 178 square miles of Japanese cities had been completely destroyed in firebombing raids.

As the B-29s bombed the heart out of Japan, the ground and naval offensives continued, culminating in the campaigns for Iwo Jima and Okinawa. In spite of suicide attacks by Japanese planes, the islands were secured. The jumping-off points for the invasion of Japan were in American hands. Meanwhile the campaign against Japanese shipping, the naval blockade of the islands, and the firebombing had taken their toll. Finally, atomic bombs were dropped on Hiroshima (6 August 1945) and Nagasaki (9 August 1945). On 8 August the Soviets declared war on the Japanese and invaded Japanese-held Manchuria the following day. Even the most die-hard Japanese saw that continued resistance might mean the death of Japanese civilization. On 14 August 1945 Japan surrendered. The only invasion required was peaceful as occupation forces entered the Land of the Rising Sun.

Better State of the Peace

When the war finally came to an end with the formal capitulation of Germany and Japan, the victorious Allies were faced with the task of re-creating the peace. Mindful of the failure of the Versailles conferees to reach a solution guaranteeing an enduring peace, the three major victorious Allies faced the same essential tasks that had confronted the preceding generation. What to do with the losers? How to build an international structure that would allow peace to last? How to restructure a political map torn asunder in the world's greatest bloodletting?

The prerequisites to forming a better state of the peace had been achieved. German and Japanese hostile ability and willingness to continue had been over- come, although in different ways. In the case of Germany, her armies had been effectively destroyed and her industrial web lay in ruins; hostile ability had been decisively overcome. As the Allied armies marched inexorably toward one another in central Germany, hostile willingness (cost-tolerance) evaporated as well. In the end, Germany lay prostrate, more physically defeated and vulnerable than had been the case in 1918.

Japan was a somewhat different proposition. Because of the way the island-hopping campaign against the Japanese Empire was conducted, Japan still retained significant military power even at the end of the war. Certainly Japan's navy lay mostly at the bottom of the ocean and its air force had been reduced to nonexistence; but the Japanese army remained, though scattered at garrisons throughout eastern Asia. Japan's hostile ability was tottering, dispersed, and incapable of being linked together because of the destruction of her naval and air assets, but the Japanese armed forces were not decimated in the way that Germany's were.

What broke decisively for the Japanese was the willingness to continue to resist. Air power, so widely acclaimed as the solution to the European problem, proved the key element in convincing the Japanese leadership that going on was futile. With its navy and air force destroyed, Japan's home islands had been subjected to a merciless pounding, particularly through the use of incendiary bombs. The dropping of atomic devices on Nagasaki and Hiroshima was the coup de grace of the strategic bomber's art that finally brought about capitulation.

Two comments need to be made about this process of overcoming hostile will and ability. The first has to do with the Allies' stated policy (at least the American policy) of unconditional surrender. Although it can never be proved conclusively, that goal and its likely concomitant of physical invasion and occupation undoubtedly stiffened hostile will and prolonged the war against both adversaries. In Germany the major fear was the advancing Red Army and the likely retribution the Soviets would exact. In Japan unconditional surrender resulted in the nonnegotiable U.S. demand that the emperor abdicate. The evidence now suggests that had the United States recanted on that one demand, agreeing in effect to leave the emperor as a figurehead like the British monarch, Japan might have surrendered earlier. Once unconditional surrender became the goal, however, it was politically impossible to reverse gears without appearing to change the goals that had activated sacrifice in the first place.

The second comment has to do with the unique contributions of strategic bombing to overcoming Axis hostile will and ability. World War II was, after all, the test case for the strategic bombardment theories that had been developed in the interwar period, and it was also the only instance when those theories have been tested against developed societies possessing the "vital centers" and industrial web for which the ideas were formed.

In some ways the air power enthusiasts' hopes and claims were clearly excessive. As pointed out, strategic bombardment did not make naval and ground operations obsolete or ancillary. Instead air, land, and sea power were interactive. Naval and ground operations in the Pacific provided staging grounds for the bombers that could not have been secured otherwise, for instance. That the most extreme prior claims for strategic bombardment were overdrawn does not, however, necessarily belittle air power's overall contribution.

The important role that strategic bombardment played in the war emboldened its proponents and probably caused them to over-extrapolate from the experience: because bombardment had worked against Germany and Japan, it would work elsewhere. What was lost in the process was an appreciation of the unique characteristics of the two countries: both were highly industrialized countries with peculiar vulnerabilities and dependencies. When faced by opponents that have not shared these characteristics, strategic bombardment has not been so clearly applicable, and the controversy over strategic air power continues to swirl partly as a result.

If hostile ability and cost-tolerance had been overcome, the crucial remaining task was how to structure the peace. That task in turn meant having to deal with the three questions posed at the beginning of this section. The Allies' answers would determine if the failures of Versailles would be revisited in the wake of World War II.

The first question to be answered was how the vanquished powers would be treated. The answer in 1919 had been to impose a vindictive peace, and it had been a failure. In 1945 the Allies' ability to impose was even greater, since both Germany and Japan were occupied. Fortunately, lessons had been at least partially learned.

In both Japan and Germany, democratic political systems were installed (except in the Soviet occupation zone in Germany), a reminder of post–World War I Germany. The difference was that in 1919 the new political system was burdened with a settlement that virtually ensured economic chaos. That chaos was blamed on the fledgling republic and undermined any likelihood that the people would embrace the system. The installation of democracy after 1945 was accompanied by generous programs of economic rehabilitation and development. These programs nurtured recovery and generated support for the imposed political regimes, thereby removing the vestiges of resistance to victor policies.

The Soviet Union did not share in this process. In the Soviet's occupation zone, an essentially vindictive peace was imposed through a harsh military occupation, the imposition of a Communist regime, transformation of the East German economy into a feeder for the Soviet Union, and punitive reparations (even to the extent of literally transporting German industries to the USSR to replace destroyed Soviet factories). This spirit of vengeance was not unlike that of France after the Great War. Given its parallel suffering, the Soviet Union's urge for vindictiveness should not be entirely surprising (although there were additional Soviet motivations as well).

The second question was how to fashion a postwar order that would prevent a recurrence of war. This task largely fell to a working group in the U.S. State Department that began its work before formal American entrance into the war. The State Department personnel began their job by examining what there was about the mechanism erected in 1919 that had caused its failure. Their answer was that the League of Nations had inadequately institutionalized the process of responding to threats to or actual breaches of the peace, a deficiency they sought to remedy in the new United Nations Charter. Under the League of Nations Covenant, no nation had a specific obligation to respond to any crisis, and the League had no permanent armed forces assigned to its jurisdiction. As a result, the League could only issue a call to its membership to respond to a crisis and hope the members would rise to the occasion. If, as happened during the 1930s, the membership did not heed the call, the League was powerless to stop aggression.

The framers of the UN Charter sought to overcome this problem. In drafting the collective security provisions of the charter, they proposed a radical solution. The permanent members of the new Security Council—the major Allies (the United States, Soviet Union, United Kingdom, France, and China)—would jointly take on the role of policemen of the world. Only they would retain significant military force, while all other nations would be disarmed to the level required for maintaining internal order. These disarmed states would simply act as bases for UN troops from the major powers. Thus overwhelming military force would always be available to confront a potential aggressor.

The victorious Allies had limited success in dealing with the second question, but it was the third that ultimately proved most vexing: the shape of the postwar political map. At this point Anglo-American visions (slightly altered) embodied in the Atlantic Charter clashed directly with Soviet security and hegemonic interests. The inability to resolve these differences satisfactorily has provided much of the basis of conflict in the international system during the cold war.

The operative American, and to a lesser extent British, idea was self-determination for all peoples, although this ideal was occasionally compromised in practice in the third world—for example, as in acquiescence in French reimposition of colonial rule in Indochina. The Soviets, obsessed with physical security and the spread of their ideology, insisted on forming a *cordon sanitaire* (buffer zone) of friendly (Communist) governments in the east European countries they occupied. The Soviet position flew in the face of Anglo-American desires but, given the Soviet occupation of the affected countries, the Western Allies were in a position to do little about Soviet actions.

Where there was joint occupation, matters were even worse. Germany was divided originally among the three combatants (France was later given its own zone carved out of the American and British sectors) and cooperation quickly broke down. Western policy was to push for normalization and economic revitalization as quickly as possible, a preference best served by reintegration of the zones in a united Germany. The Soviets, however, desired (as the French had before them) a permanently weakened Germany, and their best bet in this regard was a divided Germany. Ultimately, division became effectively permanent over the issue of currency reforms in the western zones that precipitated a Soviet response in the form of the Berlin blockade. The Berlin Wall and the miles upon miles of no man's land separating West and East Germany were grim reminders of the inability to resolve the issue.

Korea was the other divided country. In return for the Soviet pledge to assist in the invasion of the Japanese home islands in 1945, the Soviet Union received an occupation zone in the formerly Japanese-controlled nation. According to agreements reached during the war, the purpose of the occupation was to disarm remaining Japanese troops, to establish civil order, and to prepare both sections for free, unifying elections in 1948. Instead a Communist regime was installed in the north and an anti-Communist regime in the south.

Was a better state of the peace achieved? If one uses the Versailles yardstick for comparison, the answer is that this peace has certainly been overwhelmingly more successful than that following 1919. There has been no military resurgence of the defeated powers and a third worldwide conflagration has been avoided. In Europe, the center of international violence for the centuries preceding, bloodshed has been limited to Soviet actions in east Europe for more than 40 years. In that sense the peace has been a notable success.

It was not, however, a perfect peace (if such is possible). The formal mechanism for regulating the peace was ineffective as a collective security instrument when the wartime Allies fell out, and the inability of East and West to redraw the political map in a mutually satisfactory manner sowed the seeds for future conflict. The series of controversies over Berlin was a direct legacy of World War II's better state of the peace, and so are the histories of Korea and Vietnam, to which the discussion now moves.

6

KOREAN WAR

If World War II had been the ultimate example of total war, the Korean conflict began the trend back to limited engagement. Begun a mere five years after the conclusion of the Second World War, Korea started a process of adjustment for American military and political leaders. It is a process of understanding and interaction that still continues.

The Korean War was a new experience for the United States and for the international order that emerged from the ashes of World War II. For the United States, the experience was unique in at least three ways. First, it was our first limited war in modern (twentieth-century) times. As such, it represented a discontinuity in experience that required painful learning and adjustment. Second, it was the first significant cold war confrontation and represented a novel challenge to the emerging American role in the world. Third, it was the first major American military engagement not preceded by a formal declaration of military intent by the Congress of the United States, and it began a constitutional and political process that culminated in the War Powers Act of 1973. For the international system, Korea was the first instance of the application of the collective security provisions of the United Nations Charter, although not an application as envisaged by its framers. Each of these points is of sufficient importance to merit elaboration.

Although the United States had fought wars that were limited in terms of the severity with which they were fought (e.g., the War of 1812) or the ease of their accomplishment (e.g., the Spanish-American War), the Korean conflict was the first time the modern U.S. military, developed and prepared for total war and led by an officer corps steeped in the total wars of the twentieth century, was thrust into a situation in which the political objectives were limited. The result was to create friction between the military and political leaderships. It was a disagreement for which neither group was especially well prepared, and the result was probably unnecessary ill will and inefficiency in the conduct of the hostilities.

A large part of the conflict came straight from the pages of Clausewitz. The Prussian had warned over a century earlier that there is a natural tension between military and political leaders. To the military leader, the military goal is to prevail over the enemy in combat, and the result is a tendency to intensify the level of combat. The military naturally concentrates on a concerted effort to destroy the enemy's armed forces in order to overcome hostile ability. Such was the appropriate military objective in the total war environments of the world wars, and it was a military purpose that a leadership trained in the traditional American military style and blooded in total wars could embrace.

The problem was whether such an approach was appropriate to the politico-military task at hand in Korea. Clausewitz had warned that when the military tendency to expand a war overruns the war's political objectives, political leaders must be especially vigilant to ensure that the political objectives remain supreme. To make matters somewhat more perverse, the tendency to intensify seems to operate regardless of whether one is winning or losing; raising the ante appears the universal reaction. The United States had not, at least in living memory, encountered a situation where military and political objectives clashed, but they did in Korea.

An example may help to clarify this point, because it represents a particularly bedeviling phenomenon. The best example surrounds that tragic figure of the Korean conflict, Gen. Douglas MacArthur, and the politico-military role that ultimately led to his dismissal from command.

MacArthur was very much a product of the American military tradition (it can be argued that he was its epitome). He came from the West Point tradition that taught warfare Ulysses S. Grant style (he was even superintendent of West Point between the world wars), and his most significant service had been in the total-war milieu of World War II.

When he came to command the United Nations forces in Korea, he brought that experience with him. His consistent position and predilection throughout his tenure was to expand the action. When he routed the North Koreans following the Inchon landing and chased them back across the 38th parallel (thus fulfilling the original mandate to rid the south of its foreign invaders), his counsel was to continue onward and to destroy what was left of the Korean People's Army (KPA). When he was allowed to do so and was met by the Chinese intervention that sent his forces reeling south, his advice was to up the ante significantly, taking the war directly to the People's Republic of China (PRC). That such an act, discussed more fully later in the chapter, would have altered substantially the nature of the war did not seem troublesome to someone so thoroughly imbued in the American military tradition of apoliticism. The "American Caesar" exemplified how to violate most of Clausewitz's maxims.

In Korea, the United States encountered another aspect of limited war that it did not recognize at the time and that it would also have difficulty spotting when that aspect reappeared in Vietnam. That phenomenon was an asymmetry in the political purposes of the adversaries that translated into different military approaches as well.

For the United States, the original purposes in Korea were strictly limited: remove North Korean forces from South Korea and ensure they could not return. Accomplishing the goal of liberating South Korea required neither the destruction of the enemy's armed forces (although doing so might aid in ensuring they did not come back) nor the surrender of the enemy and occupation of his territory.

The situation was quite different for the North Koreans. Within their more limited physical resource base, their purpose was total: the occupation of South Korea, the overthrow of the South Korean government, and the forceful uniting of the country. Thus, a form of contrast arose that made the war seem all the more perverse. The United States had the physical wherewithal to wage a total war, but the nature of our political objectives made the use of all those resources undesirable. Had our purposes been total, we might, for instance, have used the atomic bomb. In this circumstance such action was deemed inappropriate for a variety of reasons, including the nature of the objective (Korea was not important enough to warrant the expenditure of part of a scarce resource). On the other hand, the North Koreans had total purposes in mind, but they lacked the manpower and physical resources to achieve those goals after the United States and other United Nations members entered the fray.

Korea was also the first major military confrontation in the developing cold war competition between the Communist and non-Communist worlds. Certainly the two sides had confronted one another earlier, as in the Berlin blockade and airlift, but Korea was the first instance of actual, bloody conflict. It represented the first real test of the American policy of containment that had been articulated by the Truman administration during the second half of the 1940s.

The Korean War was viewed initially as the first opportunity to confront and arrest the spread of communism beyond the Sino-Soviet periphery. In the light of the "fall" of China to Mao Tse-tung and the Communists in 1949 and a rising anticommunism in the United States that was moving toward the excesses of McCarthyism, the idea of stopping communism through the resort to force was initially quite popular, both within the United States and elsewhere in the non-Communist world. There would be no Munichs for the masters in the Kremlin.

Korea was important because of its own geopolitical importance (especially in regard to Japanese security) and as a test of wills. For Americans the real enemy in Korea was not the North Koreans, even if it were they and the Chinese whom one was fighting. The real enemy was the Soviet Union, since Americans assumed that the North Koreans were acting as surrogates for the leaders in the Kremlin, who directed and orchestrated Communist activity everywhere as part of a closely coordinated international conspiracy.

We may never know for sure exactly what the Soviet role in the North Korean attack was, although the gradual opening of Soviet archives by the Russian government has improved the chances. Many assume that the decision process goes back to Secretary of State Dean Acheson's speech in early 1950 in which he outlined the containment line in East Asia and omitted all mention of South Korea (some allege that the omission emboldened the Soviets to authorize the attack on the assumption the United States would not react). Evidence for this position is adduced from the absence of the Soviets from the UN Security Council when the vote was taken to make intervention in Korea a UN enterprise. (The Soviets were boycotting the United Nations because of Western refusal to seat Communist China in place of Chiang Kai-shek's Chinese Nationalists. As a member of the Security Council, the Soviet Union had the right to veto the intervention motion had it been present at the meeting.)

It is likely that the Soviets did at a minimum authorize the North Koreans, possibly at North Korean initiation, to make the invasion. Such a minimal interpretation is justified by the heavy dependence of North Korea on the Soviets for war supplies. The North Koreans simply could not have sustained much of an effort without resupply from their benefactors.

Whatever the nature of the relationship, the North Korean invasion was clearly viewed in the United States as part of the East–West, cold war confrontation. At one level, such a challenge could not go unheeded lest the Soviet Union be emboldened to believe American commitments were meaningless; in that sense Korea was a symbol of the whole evolving shape of postwar international politics. At the same time, many in this country and in Europe firmly believed that the Korean action was a feint to divert American attention away from an impending invasion of western Europe. Such was the prevailing military view during the early months of the conflict, and this belief dissipated only slowly as the Soviet invasion failed to occur.

As noted earlier, Korea was the first major commitment of American forces to combat that was not preceded by a formal declaration of war by the U.S. Congress. Certainly there were precedents for sending Americans into combat without such authorization. These included most of the western Indian wars and the campaign against the Barbary pirates, but all the previous actions had been minor and peripheral. When President Truman went before the American people on 27 June 1950 to announce that American assistance would flow to Korea (as well as to Indochina), there was little concern about the constitutional impact of the action. Only as the war dragged on did such concern arise.

The debate that developed is by now familiar. The Constitution, of course, provides that the Congress has the prerogative to raise and maintain armed forces and to declare war, whereas the president is designated as commander in chief of the armed forces. The problem that arose in the heat of Korea was how much control the Congress exercised in how those forces were used. The framers of the Constitution clearly envisaged that all wars would be declared, thereby giving Congress veto power over the commitment of forces. The Korean War was not formally declared, and, therefore, there was no constitutional (as opposed to practical political) need to get congressional approval. The debate as to whether Congress should have a voice in committing troops to such actions was joined only as public support for the war soured. In the aftermath of Korea, the debate continued inconclusively and was not raised again until another American president used the precedent in Vietnam.

Korea was also the first real test of the fledgling United Nations and the collective security system by which it was supposed to ensure the peace. By virtue of the Soviet boycott mentioned earlier, the U.S. government was able to go before the world body and have the action it would have taken anyway sanctioned as a UN operation. Although the vast majority of the fighting in the war was waged by the United States and the Republic of Korea (ROK), technically the forces combating the North Koreans and the Chinese were those of the United Nations.

The Korean War was as close to a real collective security application as the United Nations ever mounted before the Persian Gulf War over Kuwait in 1990–91, but it was a far cry from the action envisaged in the UN Charter. The charter, attempting to overcome weaknesses in the peacekeeping provisions of the League of Nations Covenant, called for a permanent United Nations force composed of the armed forces of the permanent members of the Security Council (the United States, Soviet Union, Britain, France, and China—the victorious Allies in World War II). In this scheme, all other nations were to be substantially disarmed, providing only basing for the forces of the permanent members, who collectively would police the world and put down threats to or actual breaches of the peace. United States–Soviet enmity had, of course, torpedoed the scheme by 1950.

Clearly, the UN force that fought in Korea did not meet the criteria set forth in the charter. A number of countries did respond to the call of the United Nations and did commit troops, but in most cases the contribution was more symbolic than significant. Gen. Douglas MacArthur and his successor, Gen. Matthew B. Ridgway, were technically UN commanders first and American commanders second, but it was always clearly understood that their orders and authority came from the White House and not the "glass house" on Manhattan's East Side (UN headquarters). Certainly the charter never contemplated a UN action in which a permanent member of the Security Council would be the major supplier and supporter of the force opposing the United Nations.

Issues and Events

From an American perspective, the North Korean invasion of South Korea came as an almost total surprise, making it somewhat difficult to track underlying causes for the action. Certainly the gradual deterioration of U.S.–USSR relations in Europe had produced a strain in the international system and had led many to expect overt conflict between the two giants; that such a confrontation would occur on the Korean peninsula, where the United States had no clearly defined interests, was a nearly complete surprise.

If there was a basic, underlying issue that defined the American position in and interest about Korea, it was embodied in the nature of cold war competition and the question of implementing the new American national strategy of containment. The

Korean invasion was the first overt test of the new strategy, the first time that the communist world stepped across the containment line and threw down the gauntlet.

Seen in that light, the situation in Korea took on a symbolic importance well beyond its objective, discrete value to the United States. When the invasion occurred, the majority of Americans had probably never heard of Korea and most of those who had could not have explained why it was in the American interest to defend that harsh land. After American leaders explained the Korean intervention in the context of the crusade against godless communism and as a symbol of American will to compete, the U.S. action had considerable initial popularity (the effort was actually quite popular until the middle of 1951 when the lines stagnated at the 38th parallel).

Korea was both the first test of containment and the opening event in a debate about the nature of American commitment arising from that doctrine that continued throughout the cold war. As the KPA poured south across the border and routed an ROK force pitifully underequipped to meet the onslaught, the question of what containment meant had to be answered. The failure to respond effectively could mean the policy was a hollow shell and invite further Soviet-inspired actions. A direct response would mean the expenditure of American blood and treasure less than five years after the end of the Second World War.

The United States determined that it had no choice but to respond militarily, and despite the rapidity with which it did so, the response came barely in time to turn the situation around. At the same time, the debate was begun about how the United States should respond to probings of the containment line in the future. One side to that debate (generally associated with Democratic presidents) argued our commitment is encompassing and the United States must be prepared to assist the beleaguered with armed force whenever there is an attempt to breach the line. The other side (historically associated with Republican presidents) maintained such a commitment is too expensive and that we should be prepared militarily to defend only those strategic places clearly vital to our interests (e.g., western Europe and Japan). In this view the United States should grant material aid but leave the defense of other areas to indigenous hands.

The proximate events leading to the crisis can be traced back to the latter stages of World War II. As mentioned in the previous chapter, the United States induced the Soviet Union to declare war on Japan in 1945 so that Soviet forces would be available to participate in the anticipated invasion of the Japanese home islands. A significant measure of the cost that the United States was willing to pay for that commitment was the agreement that the Soviet Union would be given an occupation zone in a liberated Korea.

Under the formal terms of agreement, the Americans occupied the territory south of the 38th parallel and the Soviets occupied the area north of that line. The stated purposes of the occupation were to disarm Japanese troops, to restore domestic order, and to prepare each zone for unifying elections scheduled for 1948. Clearly, things did not go according to plan. Instead the Soviets installed Kim Il Sung and the Korean Communist Party in the north and trained and equipped the formidable KPA to a size and equipment level far in excess of agreed upon standards. The United States, meanwhile, helped bring to power the pro-American Dr. Syngman Rhee in the south and established a military force at a constabulary level as the wartime agreements called for.

When 1948 arrived, the United States pulled out of Korea as agreed, but the Soviet Union did not. The elections to unify the country (to be supervised by the United Nations) similarly did not come about, as both sides refused to allow UN observers in their zones and each accused the other (both probably quite correctly) of trying to rig the returns. Korea was a de facto divided state. The only way it could be returned to its historical unified status (there is no strong historical or cultural basis for a division) was through an act of violence.

The North Korean invasion was the major precipitant, of course, of U.S. action. The invasion caught the Americans and the South Koreans by surprise. Because of deficiencies in intelligence collection, both were unaware of a massing of the KPA until the day before the invasion began. When the North Korean forces streamed across the border on 25 June 1950, they quickly routed the unprepared and ill-equipped South Korean forces. (The KPA attack was, for instance, spearheaded by Russian T-34 tanks; ROK forces did not have a single antitank weapon.) As the KPA threatened to overrun the south completely and thus present a fait accompli that would be extremely difficult to counteract, the United States was forced to act quickly if it was to react at all.

President Harry S Truman's reaction was indeed swift and decisive. With information on the nature of the attack still pouring in, the president requested UN action and on 27 June 1950 announced the American intention to come to South Korea's aid. The first American combat troops arrived in South Korea from Japan on 1 July. Unfortunately, these initial forces were garrison troops accustomed to the relative tranquility of occupation duty rather than the rigors of combat, and they fared little better than the ROK troops they had been sent to assist. By the time frontline American troops arrived on the scene, the forces opposing the KPA were cornered with their backs to the sea at Pusan.

Political Objective

As the first modern limited war in the American experience, Korea was the first instance in living American memory of war fought for less than total purposes. From the Civil War forward, Americans had come to regard war as an enterprise whose purpose was to excise some overwhelming evil. Thus war required the enemy's total defeat and capitulation, and Americans were comfortable with that concept. Korea, however, was not that kind of war and because it was not, Americans ultimately became uncomfortable with it and turned against the experience.

A large part of the problem from the American vantage point was that not only was our purpose limited, but it kept changing. The changes that occurred related directly to the state of the battlefield, recalling Clausewitz's useful admonition mentioned earlier. Had there been constancy in the objective being pursued, much of the American bitterness associated with the Korean experience might not exist. In fact, had the United Nations not strayed from its original statement of purpose, the Korean War would have been a brief and successful enterprise that we might well remember today with a sense of considerable pride.

The assertion runs counter to prevailing wisdom about our three-year campaign on the Korean peninsula, and thus requires some explanation and justification. To provide such a rationale requires looking at the question from two angles. The first vantage point is exactly what the objective was, or rather what alternative objectives were available. The second vantage point is examining how and when the objective changed and how those changes affected popular perceptions of the war.

The major source for determining the political objective is the UN mandate given to the forces that went to Korea. Unfortunately, from the original statement and subsequent debate, two not entirely compatible objectives emerge. The first and most basic was to repel the North Korean invasion, to rid South Korea of those invaders, and to allow reinstitution of South Korean control of its territory. The second objective that can be discerned is tied to the earlier UN role in Korea, supervision of the unifying elections. The first objective translated militarily into the need to simply push the invaders out of the country and to establish some reasonable assurance they would not return. Realizing the second objective required pursuing the KPA into North Korea, destroying it as a fighting force, and occupying the north as prerequisites for holding the elections. Clearly, one objective was much more ambitious than the other. (One might add that General MacArthur, after the Chinese "volunteers" intervened, suggested attacking Chinese territory, which would at least implicitly have expanded the political objective even more.)

When United Nations forces were first committed to combat, the situation in Korea was desperate. The South Koreans and their American allies were pinned down within the Pusan perimeter, and the question was whether sufficient force could be brought to bear quickly enough to keep them from being pushed into the sea. In that circumstance, initial objectives were modest: the goal was to relieve the situation and to rid the Republic of Korea of its invaders (the first objective).

The battlefield situation, however, soon provided the impetus for change. MacArthur's brilliant and unanticipated landing at the port of Inchon, combined with a breakout from the Pusan perimeter, caught the KPA in a classic hammer and anvil maneuver, crushed its ability to resist, and sent it reeling in disarray back across the border. United Nations forces followed the KPA to the border, paused, and weighed their options.

The major question was whether to pursue the North Koreans into their country and totally destroy them. From a military point of view, as MacArthur continuously argued, the task appeared a mere mopping-up exercise, and the original UN mandate on unification provided the justification for the counterinvasion. Moreover, MacArthur dismissed as idle bombast repeated warnings by Chinese Foreign

Minister Chou En-lai that the new People's Republic would not stand idly by should the Americans press toward Chinese territory. MacArthur reasoned that China was still too weak from its recently concluded civil war to field an army of any size or capability.

In the end President Truman agreed to broaden the objective, and he authorized the invasion of the north. With the perfect vision of hindsight, the decision was a monumental mistake that doomed the war effort to historical ignominy in the popular mind. The reason for this, of course, was MacArthur's miscalculation about the Chinese and their impact on the military situation. When the Chinese "snuck" more than 300,000 troops across the border, ambushed the UN troops, and sent them reeling back south of Seoul, the situation changed radically. In January 1951 a regrouped UN force counterattacked, broke the combined KPA and Chinese force, and sent it fleeing back into the north. This time, however, the UN stopped at the 38th parallel, and appeals to go north again (which was militarily possible) fell on the deaf ears of political leaders who had become deeply suspicious of the military advice they were receiving. (Some observers have argued that, given the disarray of the enemy, the decision not to invade in 1951 was as disastrous as the decision to invade in 1950.) With the decision made not to go north, the KPA and Chinese were able to regroup and the war became a static engagement along the 38th parallel that dragged on for two more years until President Eisenhower's threat to use nuclear weapons brought the enemy seriously to the negotiating table.

In the end the result exactly met the original objective. North Korea's political goal of uniting the country by force had been thwarted, and the UN political objective of freeing the Republic of Korea had been achieved. The irony was that the goal had been achieved before, in September 1950, when the KPA had been routed and sent fleeing home. At that point the United Nations had considerable leverage, in the form of the threat to invade, to force the Kim Il Sung regime to negotiate the same settlement. The war could have been over rapidly, Americans would have won quickly and decisively, and the bulk of the forces might have been home by Christmas to be greeted by a hero's welcome. The Korean conflict then might have been remembered as one of America's finer hours, perhaps equatable with the Spanish-American War in terms of length and success.

The decision to expand the purpose destroyed that possibility. When the liberation of the North and unification became the goal, then the act of liberating the South no longer constituted victory in a military or a political sense. When circumstances forced the United Nations to readopt the original goal as the objective, the American public no longer accepted accomplishment of this objective as the definition of winning. Moreover, when negotiations began in 1951, with the KPA and China safe and regrouped behind the 38th parallel, the United Nations no longer had the leverage to force a favorable peace settlement. China and North Korea were able to stall coming to terms for two years while casualties continued and American frustration heightened. The tragedy is that it was all unnecessary; Korea is the story of opportunity lost because of inconstancy in pursuing the objective.

Military Objectives and Strategy

The Korean War is a nearly perfect case study in the relationship of military objectives to the fortunes of battle. The war also illustrates the difficulty of trying to set obtainable and supportable military objectives in a limited conflict, difficulties not found when prosecuting unlimited war. The nature of total war makes the military objective obvious; the enemy's ability to resist must be destroyed. Hence the Allied objective in World War II was total military defeat resulting in unconditional Axis surrender. America's first limited war in the nuclear age presented a far more complex and frustrating situation.

United Nations military objectives in Korea fluctuated directly with the fortunes of the war itself. In the early days of the conflict as the disorganized and surprised South Korean and American forces reeled back under heavy pressure from the invading North Korean army, the immediate military objective was simple survival. The invaders had to be slowed to buy time for the buildup of American forces. Had the defenders not been able to maintain a foothold on the peninsula, an invasion to liberate the overrun territory would have been much more difficult. Without the Allied foothold at Pusan, the North Koreans might have consolidated their conquest, regrouped and resupplied their army, and prepared effective defensive positions. The psychological blow to the South Koreans might also have had an immense impact.

Fortunately, the defensive perimeter at Pusan held, and the UN command could move on to the first real military objective, a counteroffensive to drive the invading forces from South Korea. This UN military objective was limited, circumspect, geographic, and aimed at the political objective of liberating South Korean territory. No mention was made by the United Nations of punishing North Korea, destroying the North Korean army, or forcing North Korean surrender. At the time the United Nations established this objective, no one was sure that enemy troops could be driven out of South Korea. The picture was bleak. Clearly, this limited objective was meant to be nonthreatening to the Chinese and Soviet mentors of the North Koreans. The specter of a wider war and the fear of escalation formed the background for decisions about military objectives.

Success, however, tends to increase expectations and open new possibilities. After MacArthur's master stroke at Inchon and the breakout from the Pusan pocket, the North Koreans fled north in total disarray. The time seemed ripe to clear Communist forces from the entire peninsula. After some deliberation this mission was ordered, encouraged by MacArthur's assurances that the Chinese would not dare enter the war, and that if they dared, he would destroy their intervening forces. Thus the ease and speed with which UN forces achieved their original military objective gave rise to new and expanded military objectives in support of expanded political objectives. In the game of international military poker, the holder of the high hand had just raised the bet.

It seemed to be a good bet. The North Korean army quickly ceased to be an effective fighting force as it fled north in confusion toward China. MacArthur's troops followed, advancing so rapidly that their own organizational and logistic structure began to deteriorate. When the Chinese called the U.S./UN bet by striking in surprising force, these factors contributed to the subsequent UN retreat.

As UN forces streamed back south in reasonably good order, the immediate objective changed again to survival. Some wondered whether any positions could be held on the peninsula because of the overwhelming numerical superiority of the Chinese forces. The Chinese were, of course, stopped, but not before they had invaded South Korea, captured Seoul, and inflicted severe losses on UN forces.

In this new situation it appeared that UN forces would find it very difficult to clear the peninsula of the numerically superior enemy without significantly widening and escalating the war. As a result, the military objective again changed. The objective reverted to the original purpose of driving the invaders out of the south, and then holding on to the status quo ante bellum. To some this was merely a return to the original objective that, given the original conditions, constituted a significant (if limited) military victory. To many, however, such a limited objective seemed an unconscionable compromise. The rallying cry of the discontented was General MacArthur's statement that "There is no substitute for victory."

UN forces regrouped, fought their way north, reliberated Seoul, and inflicted murderous losses on the Communist forces. Communist forces were in disarray, but because the limited objective of restoring South Korean territorial integrity was met by driving the Communists north of the 38th parallel, the UN advance halted along that line. There both sides dug in and the war settled into a bloody stalemate. The combatants were either unwilling or unable to escalate the war in an attempt to achieve complete military victory.

The strategy to achieve the objectives sought was very familiar. Military strategy mimicked World War II. Tactics on the ground were very similar to those of that war, particularly those used in fighting in the "narrow places" of World War II. A military strategist in the Italian campaign would have found himself at home with the military strategy used in Korea. For example, MacArthur staged his version of the Anzio landing at Inchon, only with much greater success.

American airmen attempted to conduct strategic bombing against North Korea as they had in World War II. However, since North Korea had only a limited industrial base, strategic targets were quickly exhausted, and air assets turned to more profitable tactical targets. Rather than being a producer of war materiel and thus vulnerable to strategic bombing, North Korea was a funnel for war materiel produced elsewhere. Such a situation led to enormous frustration because the sources of enemy war supplies and manpower were in China and the Soviet Union, both of which were off-limits to U.S. bombers. American pilots attacking the bridges spanning the Yalu River (the boundary between North Korea and China across which Chinese troops and supplies flowed) were instructed to attack only the southern (North Korean) halves of the spans. If there was any danger of hitting the northern

(Chinese) halves, they were told to abort their missions. And thus was born the idea that "we fought with one hand tied behind our back."

The limitations placed on UN military activities caused significant rancor. To one degree or another, the issue and its debate led to the downfall of General MacArthur and to the downfall of the man who fired him, President Truman. But the crux of the matter was that the United States and the United Nations wanted to risk no wider war. MacArthur's counterinvasion of the north (with U.S./UN approval, of course) had widened the war significantly and led to the Chinese intervention. Further widening the war to attack targets in China could have led to a Soviet-American confrontation and the possibility of nuclear war. Such a circumstance had to be avoided, much to the frustration of those who sought traditional military victory.

Political Considerations

There were two major sets of political concerns that affected the conduct of the Korean conflict. The first was American and centered around continuing public support for the effort. The second had to do with the international scene and specifically focused on the Soviet Union and its role in the Korean adventure.

Domestically there were two distinct phases of public opinion toward the action. The initial phase, which roughly equates with the period when the war was mobile and fluid, was marked by a high level of public support. Motivated by the high level of domestic anticommunism and a crusader's zeal for freeing a beleaguered people from the yoke of "godless communism," Americans endorsed the decision to enter the Korean fray. Their support remained constant throughout the early phases of the war as the armies chased one another up and down the peninsula, and whether the United Nations appeared to be winning or losing at any given point did not seem to influence that support much.

The beginnings of erosion more or less coincided with the 1951 decision not to pursue the enemy back across the 38th parallel, and that erosion was progressive during the remainder of the conflict. In large measure the war lost support because it appeared to lose meaning. Overshadowed by the notion of uniting the country, the original objective no longer seemed like winning; the United States had apparently abandoned winning as its purpose. At the same time, there was the specter of the intransigent North Koreans and Chinese engaging in seemingly endless and pointless negotiations while some of the heaviest fighting of the war raged over barren hills that had numbers rather than names. Casualties continued to mount with no end in sight and for no visible effect. As part of American anti-Communist xenophobia, which was moving toward its crescendo in the Army-McCarthy hearings, the failure to liberate the north became ignominious and equatable with the "sellout" of Nationalist China in 1949.

Since the war's continuation appeared pointless, it took its domestic political toll. MacArthur, rightfully relieved of command for disobeying his commander in chief, was lionized as a hero and became a serious, if unsuccessful, contender for the 1952 Republican presidential nomination. President Truman, who had planned to run for reelection, concluded that the war had made him too unpopular to stand a reasonable chance of winning and withdrew from consideration. The military hero of World War II, Dwight D. Eisenhower, meanwhile swept into the White House in a landslide, with much of his popularity based on his promise that "I will go to Korea" and end American involvement.

The adverse public reaction to the Korean War was, or should have been, a harbinger regarding public attitudes toward American engagement in limited wars. If such was the case, however, it was not evident in public assessments after the fact. Rather, postmortems tended to focus on the failure of American arms and on the inadvisability of becoming involved in future land wars on the Asian continent. With the debate so directed, the problem of whether the American public could or would support a long military engagement for limited political purpose was avoided. Even the admonition to avoid Asian land wars was forgotten a little over a decade later.

The major international political consideration during the Korean War was the Soviet Union. This concern took two forms. On the one hand, there was the question of what the Soviet role had been in authorizing or ordering the invasion, to which allusion has earlier been made. Gathering anti-Communist hysteria tended to place the most sinister interpretation on that question.

If one started from the assumption, as most Westerners did, that the Soviets directed the operation, the second question was what they were up to, and there were two possible interpretations. One was that the Korean invasion was simply a probing action intended to test Western, and more specifically American, resolve. In that case reaction was warranted to avoid setting the wrong precedent, but the situation was not otherwise terribly ominous.

The other interpretation of Soviet motivation was indeed ominous. In this view, shared widely on both sides of the Atlantic Ocean, the invasion was the first move toward general war, a diversionary tactic to draw American attention and forces away from Europe in preparation for a Soviet thrust westward. Throughout the early months of the conflict, this interpretation was widely believed, and it receded only grudgingly in the obvious absence of a Soviet aggression. The sway of the argument, however, is evident in the American reluctance to use nuclear weapons against the North Koreans and Chinese; it was assumed that the arsenal, limited as it was to around 300 bombs, needed to be reserved for use against the Soviets.

Military Technology and Technique

At the time of the North Korean invasion, the American military establishment was in considerable disarray induced partially by postwar demobilization and partially by reorganization of our military establishment under the Department of Defense. But the major cause for confusion was the advent of nuclear weapons. No one knew exactly what to do with them or what their impact would be. To the air power enthusiasts, somewhat perplexed by the mixed results of strategic bombing in World War II, nuclear weapons seemed to bring the bombing theories of Douliet and Mitchell to maturity. At last the decisive destruction of enemy sources of power could be accomplished in a swift and short air campaign.

Air power theorists quickly gained the upper hand in the American military establishment, for few could dispute the decisiveness of the nuclear destruction rained down on the Japanese in World War II. As a result, in the reduced military budgets of the postwar years, the scarce developmental monies available went to the fledgling Air Force. The general American consensus at that time was that air power using nuclear weapons would prevent aggression or, failing in that, would end the aggression through destruction of the offender. As a result, when the North Korean army crossed the 38th parallel, American ground and sea forces were equipped with World War II vintage weapons. And although nuclear weapons were never used in the struggle, the war was fought in the shadow of the mushroom-shaped cloud.

The fear of starting World War III and letting the nuclear genie out of the bottle influenced nearly every decision concerning the conduct of the war. Thus the full potential of American military power was never unleashed, to the anguish of those who saw no substitute for total military victory. The enemy was granted sanctuaries well within range of American air power. Although military commanders pleaded for permission to attack these sanctuaries beyond the Yalu River, permission never came because of the fear of escalation to nuclear confrontation with the Soviets.

The threat to use nuclear weapons may have had a significant impact on bringing the conflict to an end. After his election to the presidency, Eisenhower let it be known through several channels that if the armistice negotiations did not quickly reach fruition, he would seriously consider unleashing America's nuclear might. Such a threat seemed credible because of Eisenhower's military background and because the American Air Force had the means to deliver the weapons. Whether due to Eisenhower's nuclear threats or not (some observers maintain the Chinese never received the threats), the truce negotiators reached agreement six months after the new president took office.

The monies spent on the improvement of American air power changed the nature of the air war in Korea. The Korean conflict saw the first use of large numbers of jet-powered aircraft in battle. American fliers, using the swept-wing F-86 Sabre jet, quickly won air superiority over Korea although the threat of Chinese air power, operating from sanctuaries, was ever present. UN air superiority was so complete that UN ground forces were never seriously hampered from the air, and UN air power attacked enemy ground targets on the peninsula almost at will.

The Korean War also saw the first large-scale use of helicopters. Under development for years, their principal use was in medical evacuation (with a significantly improved survival rate for wounded soldiers), but they also saw duty in a variety of transport roles. Such use only vaguely foreshadowed their eventual use in a variety of important roles in the Vietnam conflict.

The fear of escalating a relatively small war was not the only factor that affected the military techniques used in Korea. Perhaps the dominant factor controlling the conduct of the war was geography. The rugged, mountainous terrain and narrow coastal plains prevented the large-scale armored-maneuver warfare often seen during World War II. Korea was an infantryman's war, a slogging, slugging match up and down the ridges of the Korean peninsula.

On the other hand, the fact that Korea was a peninsula presented several military opportunities. Naval power could be brought to bear, particularly naval air power and amphibious assault techniques. MacArthur saw this clearly and used the amphibious approach to deal the North Koreans their terrible setback at Inchon when they appeared to be on the verge of complete victory.

The peninsular shape of Korea also brought air power into great play, particularly in terms of interdicting the enemy's logistical lifeline. The most concentrated effort to cut the enemy supply lines was entitled Operation Strangle, the same name applied to a similar effort in the Italian campaign in World War II. Strangle was sufficiently effective to seriously affect the enemy's ability to conduct offensive actions. Time after time enemy offensives could not be sustained for lack of materiel. But air power could not completely cut the flow of supplies to the front. When the enemy controlled the tempo of conflict, as he often did when the fighting became a stalemate along the 38th parallel in late 1951, supplies could be slowly stockpiled from those that successfully ran the gauntlet of American air power. Thus the enemy could launch serious offensive actions even though they could not be sustained for long periods.

Military Conduct

Although two-thirds of the population of Korea lived south of the 38th parallel, by 1950 North Korea had a larger, better-equipped, and better trained military establishment than South Korea, thanks to the considerable aid furnished by China and the Soviet Union. The North Korean army numbered nearly 130,000 (augmented by a politically reliable border constabulary of nearly 20,000) and was reasonably well equipped with World War II vintage Soviet equipment, including some 150 T-34 medium tanks and a considerable amount of light artillery. The South Korean army numbered just under 100,000 and possessed no armored forces and a small amount of artillery. Neither North nor South Korea possessed naval forces other than a few patrol boats. The North Koreans boasted a small air force that included just over 100 combat aircraft of various types, while the South Korean Air Force was virtually nonexistent. Thus when the war began, neither side could boast a large, modern military establishment, but the north had a considerable advantage in both numbers and equipment.

When the invasion began in the early morning hours of 25 June 1950, the North Korean plan was to make a quick thrust south through the Uijongbu Gap to seize Seoul (a communications and transportation hub as well as the capital city) and

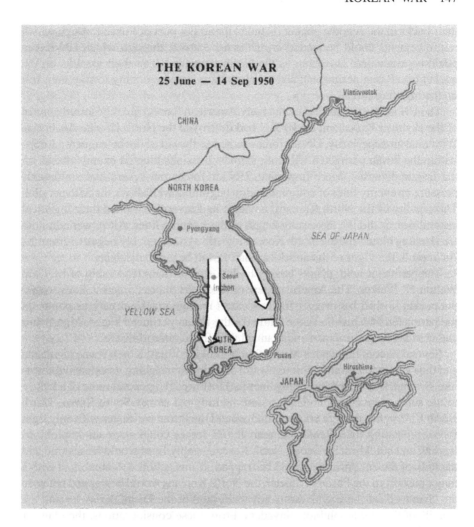

THE KOREAN WAR
25 June — 14 Sep 1950

Vladivostok

CHINA

NORTH KOREA

Pyongyang

SEA OF JAPAN

Seoul
Inchon

YELLOW SEA

SOUTH
KOREA

Pusan

Hiroshima

JAPAN

then quickly overrun the remainder of the south. The plan was nearly successful. The ill-prepared and ill-equipped South Koreans fell back in total disarray. Seoul fell in just three days. To meet the emergency, General MacArthur, the American theater commander, sent the American 24th Division directly into the fighting from its relatively sedate garrison duty in Japan. After arriving on the peninsula, units of the division fought a series of desperate delaying actions as they attempted to slow the enemy advance south from Osan to Taejon. (It was during the confusion and desperation that the apparent slaughter of several hundred South Koreans by American soldiers occurred at No Gun Ri.)

By 5 August UN forces had been forced back into a rectangular pocket, its front roughly following the line of the Naktong River. Although UN forces had

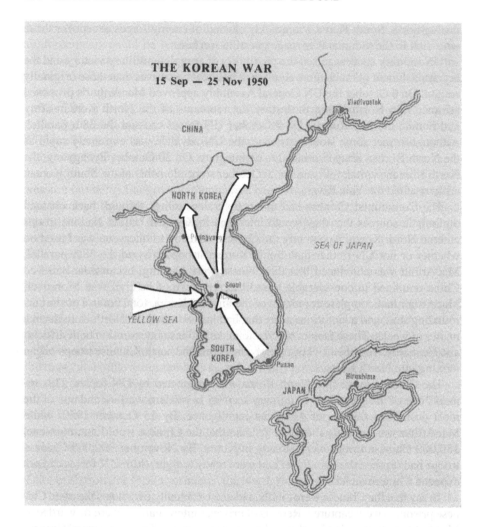

able. Finally, one suspects that no one, particularly at MacArthur's headquarters, wanted to find any Chinese forces. Whatever the reasons, the American and UN intelligence failure was complete and led to nearly catastrophic results.

The Chinese counteroffensive began on 25 November 1950. The Chinese planned to turn the interior flanks of UN forces (which were split into eastern and western forces by the mountainous terrain) and then trap each of the isolated forces in pockets against the seacoast. With this accomplished, the Chinese forces could quickly sweep south and clear the remainder of the peninsula.

In the east some Marine Corps elements of X Corps were quickly surrounded near the Choshin Reservoir, while other elements were simply overrun. MacArthur realized that the scattered units of X Corps were in danger of defeat in detail and

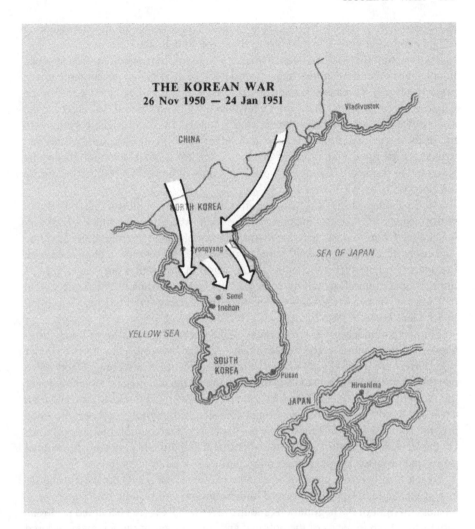

THE KOREAN WAR
26 Nov 1950 — 24 Jan 1951

CHINA

VIadivostok

NORTH KOREA

Pyongyang

SEA OF JAPAN

Seoul
Inchon

YELLOW SEA

SOUTH
KOREA

Pusan

Hiroshima

JAPAN

ordered the evacuation of the corps by sea, a feat accomplished with great skill by 24 December. The remainder of UN forces fought a delaying retreat and by year's end occupied stable defensive positions along the 38th parallel. The Chinese offensive slowed and finally ground to a halt because of logistic difficulties exacerbated by concentrated UN air attacks.

By New Year's Day 1951 Chinese forces were resupplied and reinforced to a strength of about one-half million men, and they resumed their offensive all along the front. General Ridgway was now in command of all forces in Korea under the overall theater command of MacArthur in Japan. Ridgway's tired forces slowly retreated south. Seoul fell again to the Communists on 4 January. The Chinese advance continued, but growing logistic difficulties quickly slowed its momentum as UN resistance increased. Finally, the Chinese attack stalled along a line running roughly from Pyongtaek in the west to Smachok in the east.

To this point the war had been characterized by rapid movement. In just seven months the contending armies had covered the length of Korea nearly three times. First, the North Koreans streamed south, then UN forces advanced north to the Chinese border, and then the Chinese advanced south past Seoul. In each case the rapid advances had stretched the attacker's supply lines and logistic systems to the limit. In each case the defenders had fallen back on shorter supply lines and waited for the opportune moment to counterattack. The first seven months of the Korean War markedly resembled the ebb and flow of battle in North Africa during the early years of World War II.

On 25 January 1951 Ridgway began a methodical counteroffensive that met with considerable success. In spite of occasional savage counterattacks, Ridgway pushed the Chinese north and by 19 April had established a strong defensive line slightly north of the 38th parallel. In the meantime, MacArthur had clashed with President Truman over the conduct of the war and the limitations placed on UN military operations. The result was the sacking of MacArthur and his replacement by Ridgway. In turn, Ridgway's vacant position of commander in Korea was filled by Lt. Gen. James A. Van Fleet.

For the remainder of 1951, the contending forces fought a series of bitter struggles with limited success. The front lines moved back and forth a few miles on either side of the 38th parallel with neither side gaining decisive advantage. Both sides had dug in across the peninsula and the war took on the stalemate characteristics of the Western Front in World War I. In a sense, there were no flanks to turn and both sides could only resort to bloody frontal assaults. Although the seacoast provided inviting open flanks, the Chinese were not capable of major amphibious operations. On the UN side, the return to the limited political objective of restoring the original status quo obviated the possibility of amphibious operations in the north.

By late October truce negotiations (under way since early July but used primarily as a propaganda forum) were moved to Panmunjom and resumed with more seriousness. In light of the negotiations and the fact that UN forces occupied positions satisfactory both in terms of the political and military situation, Ridgway ordered Van Fleet to cease all offensive operations and assume an active defensive.

While the negotiators argued, blood continued to flow in constant but minor fighting. The war was a stalemate, punctuated by patrol actions, outpost skirmishes, and occasional large-scale, but largely unsuccessful, Communist attacks. Finally, after 18 months of fruitless and bloody stalemate, the negotiators at Panmunjom reached bitter agreement. The fighting officially ended on 27 July 1953.

Better State of the Peace

In the popular mind, the Korean War is generally considered either a failure or, more charitably, the absence of a "victory." In Korea the United States did not bring its adversaries to their knees and force their capitulation, an outcome that Americans had come to expect. Rather than an imposed peace, the conflict ended at the negotiating table between unvanquished opponents. The outcome was not like the Japanese surrender aboard the USS *Missouri*, and it was nowhere nearly as satisfying.

Judging whether a better state of the peace was achieved is a matter of deciding whether the political objectives of the United States were realized. The difficulty of making that assessment in the case of Korea is that, as noted earlier, the objectives changed. In turn, the contrasting political objectives translated into different military objectives in terms of enemy hostile ability and will. By one set of objectives, we won. By the other, we did not. However, we can say the North Koreans lost. They failed to unite South Korea under their leadership, and they paid a terrible price for their attempt to do so.

Although UN forces twice came close to overcoming North Korean (and the second time Chinese) hostile ability, in the end that ability was not destroyed. The main reason for this, of course, arises from the fact that the United Nations did not pursue the adversary back across the 38th parallel the second time, when the enemy probably could have been broken. This failure has been widely criticized, but its effect was to allow the KPA and Chinese to regroup and replenish their forces, which were intact at the war's end.

Adversary hostile will was largely overcome in the end. President Eisenhower's threat to use nuclear weapons if the opponent did not agree to an armistice effectively overcame hostile will defined as the willingness to continue (cost-tolerance). The prospect of nuclear devastation was a larger price than either China or North Korea was willing to pay. Similarly, the terms of the peace, which included a divided Korea, forced the North Koreans, however grudgingly, to relinquish the political purpose for which they had initiated violence in the first place, the forceful uniting of the Korean peninsula under Communist rule.

Whether this outcome represents fulfillment of American political purpose or not depends on what that purpose was. If the purpose was the original aim simply of reestablishing the status quo wherein South Korea remained a sovereign and independent state, then the political objective was clearly met. To achieve that goal did not require overcoming hostile ability, although doing so might help ensure the long-term viability of the Republic of Korea. All that was truly required was to force the KPA out of South Korea and to get the North Koreans to agree not to come back, and that was done.

The political objective was not achieved if that purpose was the goal set during the invasion of North Korea: uniting the country and holding elections under UN auspices. To accomplish that end did in fact require the overcoming of enemy hostile ability, since an extant KPA could be expected to oppose such elections on the grounds that they would eventuate in an anti-Communist Korean peninsula. Such a goal was symmetrical with the North Korean aim, and if that was the objective, it can be said that both sides lost.

One can, quite obviously, disagree about whether a better state of the peace was accomplished depending on what the objective was. In the process, however, one must also raise a question about what kind of political objective is satisfying to the American people. Americans do not generally look back upon the outcome of that clash with pride, even though it is possible to argue that it was a success. A reason for this attitude may be that although the goal was achieved, it was not an adequate goal in American eyes. It was not the kind of political objective for which the American public is willing to make a sustained sacrifice. One can debate the proposition put forward earlier that Americans would remember Korea positively, in much the same way we are likely to remember Desert Storm, if the goal of liberating the South had been accomplished in a few months with minimal sacrifice in terms of blood or treasure. It may also be that Americans are willing to support a sustained conflict only if the ends appear grand enough, something like the second objective.

7

VIETNAM WAR

The Vietnam War is one of America's greatest military traumas. It was a bewildering affair from start to finish, and we are still trying to understand what happened to us in that corner of the world. Analysis has evolved; while the war was going on and shortly after our involvement ended, commentary was highly subjective and mostly vituperative, seeking to lay blame for blunders made. That tone has changed somewhat as a second generation of analyses, more removed from the passion of the occasion, attempts to present a more balanced, less emotional treatment of the events. The twenty-fifth anniversary of the war's end has stimulated another surge of remembrance and questions that continue to this day.

The beginning of American involvement in the Southeast Asian conflict is difficult to establish. Our first formal commitment to the fighting there came on 27 June 1950, when President Harry S Truman included assistance to the French in Southeast Asia as part of his message dispatching American fighting forces to Korea. The beginnings of our involvement could be dated earlier. We could choose 1945 when Ho Chi Minh declared the independence of Vietnam, and members of the American military delegation in Hanoi saluted the new republic. We could go back even further to the actual conduct of World War II, when the American Office of Strategic Services (OSS), the predecessor of the Central Intelligence Agency (CIA), collaborated with Ho and his embryonic forces to subvert the Japanese occupation. Thus, there are many possible choices for the date of the start of American involvement.

The choice for an ending date is far more limited. Regardless of when one says our involvement began, it came to a halt with two events. In March 1973 the last American combat troops left the country, leaving behind only a skeletal military advisory presence. Even that presence ended when the last Americans scrambled aboard helicopters at the embassy in Saigon on 30 April 1975, as the capital fell to the advancing North Vietnamese army (NVA) amidst panic and confusion. In total, involvement spanned a quarter of a century (1950–75) or 30 years (1945–75), depending on one's perspective.

American commitment of combat troops to the war's prosecution was also the longest in our experience. Military advisers entered the country officially in 1961 (nonuniformed personnel preceded them in the late 1950s), and by 1963 the number of advisers had climbed to more than 17,000. Officially they were noncombatants, but many performed combat roles. In terms of formal fighting commitment by American soldiers, our part of the war spanned eight years, from the introduction of the first Marine brigade in early 1965 to protect the American air base at Danang, from which combat flights were being conducted, until the final withdrawal of the last combat units in 1973. This length of engagement compares to our second longest war, the American Revolution, which lasted in terms of real warfare for about six years (not counting the military standstill from the Siege of Yorktown until the peace treaty was signed and may be passed in Afghanistan).

The state of American public opinion about Vietnam can be divided into two phases punctuated by the Tet offensive of 1968. Between the introduction of combat forces in 1965 and that event, opinion was divided, with no clear-cut majority in either support or opposition (except for such specific demographic groups as 18- to 24-year-olds, a majority of whom opposed). During this period, however, the United States appeared to be making military progress in the war, as measured by the "body counts" reported on the evening network news and optimistic statements from government officials and even newsmen. As noted, organized opposition tended to be limited to large universities and to center around activities such as the Vietnam teach-ins.

Tet dramatically changed opinion. The reason was simple enough: The Vietcong and the North Vietnamese army launched a major offensive throughout the country (including action within eyesight of the hotels in Saigon where most of the American press corps resided) that belied the optimistic reports and supposed enemy losses. The initial American reaction was shock and was probably best depicted by television news anchorman Walter Cronkite's purported response to the first film of the attack: "Just what the hell is going on here?" Shock quickly gave way to disillusionment and cynicism. The media in particular believed they had been deceived and came to suspect most of the favorable information.

The result of Tet was politically and militarily ruinous to continued public support for the war effort. Gen. William C. Westmoreland was replaced as commander of American forces by Gen. Creighton W. Abrams. By the end of March, President Johnson had announced his intention not to run for a second term, ostensibly so that he could devote his total attention to resolving the war.

Realistically, the only political goal that could be espoused in that presidential year was to get the United States out, and all candidates jumped on the bandwagon of disengagement in one form or another. The tragedies that occurred at the 1968 Democratic national convention and later at Kent State and Jackson State universities only punctuated the pathos.

Another unique aspect of Vietnam was more strictly military in nature: the war was the first major conflict in which the United States confronted a highly dedicated, nationalistic force employing unconventional, guerrilla warfare strategies and tactics. That the United States was basically unprepared to fight the kind of war we encountered was ironic, and it speaks poorly of our collective memory. Certainly Vietnam was not the first guerrilla war in which we engaged. As argued earlier, American adoption of unconventional warfare techniques, especially by the revolutionary militia, helped lead to our independence. In the nineteenth century, the U.S. military was confronted on several occasions with these kinds of operations, notably the Seminole War, several of the western Indian wars (e.g., the campaigns against the Apaches), and the Filipino Insurgency. In the twentieth century, Gen. John J. Pershing encountered the same kind of foes in Pancho Villa and his supporters. The lessons that these experiences might have suggested were largely ignored as our European-style armed forces hit the beaches of Vietnam.

Issues and Events

Vietnam's strong tradition of nationalism is both ancient and finely tempered in the flames of centuries of combat against foreign invaders. The history of Vietnam's struggles can be tracked back to the Trung sisters' insurrection against the Chinese, the ancient enemies of Vietnamese nationalism, in A.D. 40. The modern history of the struggle for Vietnam begins in the 1860s, when the French colonized the region. By 1883 the French controlled all of Vietnam and in 1887 established the Indochinese Union, which included both Vietnam (divided into three parts) and Cambodia (Laos was incorporated into the Indochinese Union in 1893). Vietnamese efforts to cast the French out began almost immediately after the French arrived and continued to greater or lesser degrees throughout the remainder of the nineteenth century and into the twentieth.

French rule spread the seeds of its own downfall. In many respects French rule in Vietnam was colonialism at its worst. However, one change the French brought about stands above all others as the major reason for unrest among the Vietnamese. When the French came on the scene, they found a society dominated by small landowning peasants. When they were forced to depart less than a century later, they left behind a society of landless peasants dominated by an absentee landlord oligarchy.

Ho Chi Minh, who would become the leader of the Vietnamese independence movement, departed Vietnam as a young man in 1911 and did not return until World War II was under way. In the meantime, he gained fame as a leader, in exile, of the nationalist movement. Ho first appeared on the political scene when he attempted to petition the Versailles Peace Conference for Vietnamese independence in 1919. Later Ho joined the French Communist Party, became a party functionary in Moscow, and in 1930 formed the Indochinese Communist Party.

When Ho returned in 1941, Vietnam had been occupied by the victorious Japanese who had, for convenience, left the Vichy French colonial administration in power. Ho's task was to fight both the Japanese and the French. As the war drew to a close, the political situation quickly became even more muddled. In March of 1945 the Japanese attempted to gain public support in Vietnam by ousting the French administration and having Bao Dai (the figurehead emperor of Vietnam who had served under French and Japanese supervision since 1932) declare the independence of Vietnam under the auspices of the Japanese. However, the war was quickly ending and on 18 August 1945, the Japanese turned over governmental power to their principal adversaries in Vietnam, Ho's Vietminh forces. Bao Dai abdicated shortly thereafter and Ho declared his version of Vietnamese independence on 2 September 1945.

In the meantime, Allied leaders at the Potsdam Conference (July 1945) had agreed that the Japanese in Vietnam should be disarmed by the British (in the south) and the Nationalist Chinese (in the north). Ho feared the presence of Chinese troops might lead to permanent Chinese control, yet he had insufficient power to oust them. Thus, he was forced to reach an accommodation with the French to replace Chinese troops with French forces, a most difficult choice between the lesser of two evils in Ho's eyes. In return, the French were to recognize Vietnam as a free state within the French Union and to negotiate about Vietnam's future.

The Chinese left Vietnam and the French returned in accordance with the agreement. However, the Paris negotiations over Vietnam's future quickly broke down because once the French regained control, they had no intention of granting Vietnamese independence. The United States, to whom Ho turned for support on the basis of President Franklin D. Roosevelt's commitment to the Atlantic Charter, stood by ambivalently; we deplored the French return as a violation of self-determination, but recognized that Ho was a Communist as well as a Vietnamese patriot.

After French warships bombarded Ho's supporters in Haiphong, Ho withdrew his forces to the countryside in December 1946 and established a rural guerrilla resistance base. The French controlled the major cities and some of their environs while the Vietminh expanded and consolidated their control of the countryside. On the Vietminh side it was a war of ambush and avoidance of decisive defeat. For the French, punitive expeditions to search for and destroy the enemy in the countryside seemed either to hit only thin air or to be ambushed with tragic results.

During this period (1946–50), the war did not attract much attention in the United States. That changed with the fall of China, the North Korean invasion of South Korea, and the appearance along the Vietnamese border of Mao's Communist Chinese forces. Ho's logistical problems were simplified by a steady supply of modern weapons and munitions from the victorious Communist Chinese. The French, still desperately weak after World War II and suffering from considerable political disarray at home, were in real difficulty.

Suddenly, Indochina appeared to Americans to be part of a coordinated worldwide Communist effort that had to be opposed. President Truman thus included aid to the French in his message committing U.S. troops to Korea. The French then attempted to fight a modern mobile war on an American shoestring. The shoestring grew (by 1954, the United States was footing three-quarters of the tab for the war), and U.S. leadership increasingly viewed Ho as no more than a front for the worldwide Communist effort. American recognition that the war in Vietnam was at least partially an anticolonialist struggle to create a unified Vietnamese nation was lost in the tumultuous aftermath of Mao's victory in China, the Korean War, and the strident anticommunism of the McCarthy era.

French military and economic exhaustion coupled with political instability led eventually to the 1954 Geneva Accords that halted hostilities, divided Vietnam at the 17th parallel, and called for elections within two years to determine the form of government for a reunited Vietnam. Ho was in control of the north while the ever available and compliant Bao Dai, who had been returned as a figurehead by the French as an alternative to Ho in 1949, became chief of state in the south. With American assistance, Bao Dai chose as his prime minister an anti-Communist Catholic nationalist, Ngo Dinh Diem.

Both Ho and Diem began consolidating their power bases as refugees flowed north and south to their government of choice. Two de facto states were emerging in Vietnam, and for Americans it was difficult not to draw an analogy with Korea. Both new states turned to their natural allies for aid and received it. Ho negotiated aid agreements with both China and the Soviet Union in 1955. In the south the United States added Vietnam as a protocol state to the new Southeast Asia Treaty Organization (SEATO). America moved quickly to supplant France and became the chief supporter of South Vietnam. Thus, the United States almost immediately began supplying aid to Diem and agreed to train the South Vietnamese army while Diem moved quickly to suppress various dissident sects that threatened his control. In July 1955 Diem (with full U.S. support) rejected the notion of a national referendum because he feared that Ho, as a national hero, would surely triumph. In October Diem defeated Bao Dai in a popular referendum in the south and thus became chief of state of the newly declared Republic of Vietnam. Diem continued to consolidate his power by cracking down on dissident groups, particularly former Vietminh and their supporters.

By 1957, with the encouragement of Hanoi, Vietminh-led insurgent activity against the Diem regime had begun in the south. As the years passed, the guerrilla activity increased, as Hanoi supplied arms and equipment. Many of the southern Vietminh who had resettled in the north infiltrated back into the south to assist the insurrection.

American involvement expanded gradually. President Dwight D. Eisenhower authorized the first military assistance programs, including the sending of military advisers to the country, and President John F. Kennedy expanded that commitment, including the dispatch of the American Special Forces (Green Berets) to the country. The military situation was not going well. Increasingly, the reason attributed for the deteriorating military situation was President Diem, and he certainly deserved much of the blame.

Although Diem had done a remarkable job in consolidating his power, he still did not enjoy a broad popular base of support. Aloof and distant, he had difficulty forming strong bonds with the people. Effective land reform was not accomplished (it was often thwarted by corrupt officials in Diem's regime), which tended to alienate the landless peasants. Diem's Catholic faith alienated the majority Buddhist population. Diem's government, centered on his family and close Catholic associates, was riddled with corruption. He was petitioned for reform but either ignored the requests or responded with further repression. As a result, the insurrection gained support and momentum, particularly in the countryside.

American aid continued to flow and military advisers continued to arrive. In February 1962 the United States formed the Military Assistance Command, Vietnam, to command and coordinate U.S. military efforts. Meanwhile, in the central highlands, the Vietcong (a derisive name applied to the insurgents by the Saigon government) were forming into battalion-size units as the insurgent momentum accelerated. By the end of 1962, the American commitment had deepened and more than 11,000 advisers were "in country" and were often participating in combat operations.

In spite of U.S. aid, training, and advisers, the military situation continued to deteriorate. Diem's army was led by an officer corps often promoted on the basis of loyalty to Diem rather than military ability. Senior officers often suffered from a warlord mentality—using their troops for personal gain and avoiding too much success in the field for fear that such success would cast them as threats to Diem. As the army became more politicized, it was often more interested in Saigon political intrigues than in combating the enemy. When the army did take to the field, it was often beaten by numerically inferior enemy forces.

By the middle of 1963, the situation was becoming desperate. Buddhist demonstrations against the Catholic-dominated government grew in size, frequency, and intensity. Diem countered with more repression and bloody attacks on the Buddhist temples. In the face of the deteriorating situation both in the field and in Saigon, a group of South Vietnamese army generals staged a coup and killed Diem. However, rather than relieving the situation, the assassination brought on a period of political instability lasting until mid-1965. In 1964 alone, seven different governments were in power in Saigon.

The United States and South Vietnam began small-scale operations against North Vietnam in 1964 to bring the war home to the North Vietnamese and to convince them to cease and desist. South Vietnamese commando teams harassed enemy coastal installations, and American ships cruised in the Gulf of Tonkin (but outside North Vietnamese territorial waters). On 2 August 1964 North Vietnamese patrol boats attacked the USS *Maddox*, a destroyer on an intelligence-gathering mission, and two days later a second attack allegedly occurred against the USS *C. Turner Joy* (whether that attack actually occurred is hotly contested). In response, President Johnson ordered retaliatory air raids against North Vietnamese naval bases on 5 August. On 7 August Congress approved the so-called Gulf of Tonkin Resolution giving the president broad powers to act in Vietnam. The resolution set the stage for direct U.S. combat involvement, the retaliation was the first overt American military act in the war, and the precedent was set for the future.

If there was an underlying, pervasive issue common to the entire sweep of U.S. involvement in Vietnam, it was the containment of communism: the determination not to allow the expansion of another Communist regime. Different presidents expressed this commitment in different ways. Eisenhower, for instance, believed that the failure of the United States to stop Communist expansion in Vietnam would lead to the fall of all Southeast Asia to communism (the domino theory). Truman and Johnson placed more emphasis on the consequences of not stopping an aggressor early, for fear that the failure to do so would encourage further aggression (the analogy with Britain and France at Munich in 1938). Richard Nixon and his vocal national security adviser, Henry Kissinger, later emphasized the importance of honoring commitments so that future as well as current victims of aggression would accept American constancy.

To be sure, the emphasis on containment as a basic issue changed over time, as American foreign policy changed toward the world generally and especially toward our adversaries. In the early period of involvement when anticommunism was at its zenith in this country, containment was a strong, broadly supported policy. When Truman included support for the French in Indochina (a part of the world about which most Americans had heard only vaguely) in his speech on Korea, it was accepted without a raised eyebrow. But times change and so does policy. By the middle 1960s (and certainly by the end of the decade), American policy toward the Soviet Union had shifted in rhetoric (if less in substance) away from the confrontational tenor of containment to the more cooperative language of détente.

Unlike other American military experiences, there simply was no single, dramatic event that drew us into the war, no grand casus belli (the closest candidate is the Gulf of Tonkin incident, which is hardly equivalent, for example, to the North Korean invasion of South Korea). Rather, the theme of events in Southeast Asia is one of gradualism and incrementalism, which trapped decision makers in a maelstrom of ever-widening and deepening involvement by bits and pieces. It is a story not of duplicity or stupidity (as is sometimes portrayed), but of individuals caught in circumstances in which the individual decisions they reached, each of which seemed the best alternative at the time, had the unintended cumulative effect of slowly and gradually dragging the United States deeper and deeper into the fray.

The entire situation is not without irony. Five presidents and their advisers wrestled with the problem of Vietnam, and each was baffled by it. In each case, there were alternative approaches that could be taken, but none seemed attractive. In most instances, there were three things that could be done when a crisis emerged (as it regularly did). One alternative was to cut our losses and get out. The universally recognized consequence of this approach was the quick demise of the Republic of Vietnam, an outcome that each president deemed to be ideologically and politically unacceptable. At the other extreme, each crisis could be met with direct insertion of

American combat forces, an alternative deemed equally unacceptable until Johnson finally succumbed to it (although in a gradual, incremental way). The third alternative, and the one deemed the lesser evil most of the time, was the incremental way, doing just a little bit more. The irony is that most of the time, those who made the incremental decisions had very little hope that their choices would prove effective or decisive; the other alternatives just seemed worse.

The irony of incremental decision making worked in another way as well. After the initial decision to support the French, the decisions that each president made would have been impossible or unnecessary had it not been for those of his predecessors. In turn, each president's choice of the incremental alternative meant that subsequent presidents were likely to be placed in the same position. The effect of each decision was cumulative, with two results. Since each decision enlarged the American investment in the Vietnam outcome, it became increasingly difficult to cut losses. Our South Vietnamese "clients" recognized this American self-entrapment and realized the great difficulty we would have extricating ourselves (which we periodically threatened to do). As a result, U.S. leverage over the South Vietnamese did not expand and in some cases contracted as our efforts grew.

With these general comments in mind, we can turn briefly to the critical but incremental decision path. The first decisions, of course, were made by Truman: the promise to grant and the gradual enlargement of economic and military assistance to the French fighting the Vietminh. Eisenhower, building on this initial investment, came next. His official rejection of the Geneva Accords laid the groundwork for American military and economic support for the Diem government. Kennedy followed the lead by increasing the volume of aid to the beleaguered Diem regime and by introducing the first uniformed combat advisers into the country. Johnson made the big plunge by introducing combat units into the country and, with the agreement of the military, gradually increasing their numbers. Nixon brought the process full circle through Vietnamization, which allowed the United States slowly to extricate itself, but at the cost of geographic expansion of the conflict into Cambodia (Kampuchea) and the intensification of the conflict in Laos.

Political Objective

Making sense of Vietnam is difficult because the war was at a minimum a three-actor event (if one does not ascribe an independent purpose to the Vietcong, which one can do certainly after the early 1960s), wherein each player had different purposes. For the United States, the war was limited in terms of U.S. purposes, if not always in firepower. For the North Vietnamese, the war was one of total political purpose that commanded the complete resources of the people, and the purpose was hardly less desperate for the leadership (if not necessarily the total population) of South Vietnam. Vietnam was, in other words, a war of asymmetrical purpose: the outcome was clearly more important to America's adversaries than it was to Americans.

The political objective of the Democratic Republic of Vietnam (DRV), embodied in Ho Chi Minh, was the unification of Vietnam under its rule, by force if necessary. As such, it was a total and indivisible goal, just as American independence had been in the eighteenth century. Moreover, it was an objective that was maintained constantly from 1945 until its final achievement in 1975.

The goal of the various governments of South Vietnam was to avoid being absorbed by the North. Because overarching Vietnamese nationalism (where it existed) was identified with Ho Chi Minh, the leaders in the South could not embrace the idea of unification under their own control; the support base was not there. Instead, their objective was defensive. Unfortunately, the objective was not overwhelmingly popular nor compelling among the South Vietnamese. Partly this was because the war was more than a simple invasion; it was also an internal insurgency (that part of the war, especially in its early going, conducted by the Vietcong). There were at least four other reasons why the South Vietnamese objective was never accepted.

The first reason was that maintaining the freedom of South Vietnam was a defense of artificiality. The agreement that divided the country at the 17th parallel was an arbitrary matter of convenience, not a reflection of prior political reality. Certainly, there was historical rivalry and even animosity between the primarily rural, agricultural southerners and the more urbanized and industrialized northerners. However, nationalism was not North or South Vietnamese; it was Vietnamese.

Second, the government whose enslavement the RVN sought to avoid was headed by the one Vietnamese politician who had widespread support throughout the country. Ho was the embodiment of Vietnamese independence, the George Washington of his country, because of his role in ridding the land of the French colonialists. No one in South Vietnam had that kind of reputation or popularity.

Third and related, those who ruled the government of the South were less than inspiring. The corruption and repression of the Diem regime had a great deal to do with the original formation of the National Liberation Front, and Diem tenaciously resisted attempts by Americans to institute reforms that might have brought support to his regime. The string of incompetent generals who followed Diem into the presidential palace were no more inspiring.

Fourth and finally, South Vietnam's association with the Americans was a problem. To the average Vietnamese, northerner or southerner, the Americans were not a particularly welcome sight, especially when they began arriving in large numbers. Rather, they were viewed by many as just another group of foreign invaders taking the place of the French and thus to be resisted in the same manner.

The exact nature of the American political objective was considerably more complex than the objectives of the indigenous combatants. As noted, the common thread linking American purpose over time was the extension of the containment idea to Vietnam and, by further extension, to the rest of Southeast Asia. In translation from the central tenet of American foreign policy to specific actions, containment meant the United States had the political objective of ensuring that the South Vietnamese political system not be overthrown by force (at least after 1956). Over time, the underlying purpose (containment) was used in different ways to explain why the objective was worth pursuing. The problem with each of the various containment rationales was that each had a counterargument either in factual content or interpretation.

A further thread running throughout these justifications was the question of who was really being contained in Vietnam. There were always four possibilities, and their plausibility was inversely related to the importance of containing them. The most plausible enemy was North Vietnam and its National Liberation Front collaborators; they were all Vietnamese, they were proximate, and unification was their goal. The second candidate was China (a particular fixation with Secretary of State Dean Rusk). China aspired to superpower status, certainly wanted to be recognized as a primary power in Asia, and hence was worth containing. Third, some argued that the United States was really containing Soviet expansion in Indochina. The Soviets were the most worth containing, but their lack of proximity or obvious interests in the area made them the least plausible. Finally, monolithic communism that must be opposed everywhere was a candidate. This basis is countered by the Sino–Soviet split that began in the 1950s.

The American political purpose in Vietnam was never entirely clear to sizable portions of the American population. Until American involvement in the country became overt in terms of an American military presence, the lack of understanding was tolerable. As first advisers and then combat personnel entered the country, the lack of knowledge of the situation (and even the country; a survey in 1966 demonstrated that more than half the American population did not know where Vietnam was) began to be felt. The American people began to ask exactly what the purpose was.

A second problem was whether Vietnam was important enough to justify an American commitment. In the early days, when involvement was limited to economic and military assistance, the question was relatively unimportant because the sacrifice was minimal and unnoticed by most Americans anyway. When the war began to consume larger portions of American treasure and blood, then it became important to determine whether the objective matched the sacrifice.

The worthiness of the objective could be justified only in geopolitical terms since there was no plausible historical or economic basis for the commitment. Thus, the debate centered on containment but in a different context than the Korean War. In 1950 there was no question about whether halting monolithic communism was worthwhile, but by 1964 perceptions had changed. Communism was no longer viewed as monolithic because the Sino–Soviet split had demonstrated it was not. Moreover, a newer, probably more permissive and less sacrifice-oriented group was entering the American adult population, and its response to calls based purely on patriotism was not so certain as that of a previous age. An objective of arguable vitality was inadequate to appeal to all the population, and especially to that portion which would be forced to fight for it.

As time went by, the question increasingly framed by segments of the American public was whether the United States had any substantial interests in Southeast Asia that could be translated into political objectives that demanded protection. As time went by and casualties increased, the public progressively answered the question negatively, and the political leadership in Washington seemed unable to devise a compelling argument that vital American interests were at stake.

South Vietnamese leadership did not make matters any easier. The succession of leaders who paraded across the television screen lacked the broad-based support of their people, and instances of corruption and inefficiency were rampant. The image created was one wherein the side we were supporting did not appear to be much, if any, better than those we opposed, and many of our memories of Vietnam are of those demeaning characteristics.

In these murky circumstances, devising a political objective that would galvanize the American people to the task at hand proved impossible in the long run. Those characteristics of a "good" political objective discussed earlier (e.g., simplicity and moral loftiness) were never successfully attained. More to the point, the objective never translated clearly into a military objective to guide strategy.

Military Objectives and Strategy

American understanding of the nature of the war posed the first major obstacle in formulating an effective military strategy. The United States viewed the Vietnam conflict as a limited war, a conflict fought with limited means for limited political objectives. Although viewed as part of the larger struggle against an aggressive Communist threat to the free world, Vietnam was at the outer periphery of U.S. national interests. The United States could not allow itself to become overly involved because the important struggle would come in western Europe. Like Korea, Vietnam was a sideshow that could easily divert attention and weaken America's ability to resist at the critical points. No matter what other issues were involved, U.S. policy makers viewed the conflict in Vietnam as one more confrontation with the Communists.

Such a view was at best overly simplistic and at worst so far from the truth that our efforts could not help but fail in the long run. The important enemy motivation was nationalism rather than communism. This is perhaps best illustrated by the events that took place after the North Vietnamese had seized control of all Vietnam. Rather than a great victory for monolithic communism, fragile Communist alliances quickly disintegrated as Vietnam invaded Cambodia (ruled by the Communist Khmer Rouge) in December 1978, which in turn led to the Chinese invasion of northern Vietnam in February 1979. Even the Asian Communist "monolith" proved to be little more than a figment of American imagination.

Spurred on by nationalist fervor, the Viet Cong and North Vietnamese waged an unlimited rather than a limited struggle. Their objective, to overthrow the government of South Vietnam and to impose their own control, was unlimited. Although compromise might be accepted in the short term, it would be only a pause in the ultimate struggle. There could be no compromise with the long-term nationalist goal. Populations were mobilized for the effort and every means of battle at the enemy's disposal was used. Only the limitations on the resources available to the enemy preserved the American illusion that this was a limited war.

American understanding of the war was further confused by the complexity of the struggle. In essence, the war was fought at three levels. The first level, on which the American military focused its efforts, was the war against enemy main force units—both Vietcong units and units of the regular North Vietnamese army. The second level was the shadow war against the enemy guerrilla fighters. The third level was the war for the loyalty of the population, perhaps the most important part of the entire struggle.

All three levels were intertwined. Although enemy main force units were somewhat dependent on supplies from North Vietnam, they also depended on the cooperation, or at least the neutrality, of the South Vietnamese population for succor, recruits, and intelligence. This was particularly true of Vietcong main force units. Guerrilla units were almost totally dependent upon the cooperation or neutrality of the population. Without the aid of the people, the guerrillas would have been exposed, resisted, and starved of supplies, and thus could not have operated. All of this points out the importance of the struggle for the loyalty of the population. But to win over the population, it had to be protected from enemy main force units and guerrillas. Thus, U.S. forces faced somewhat of a "chicken-or-egg" problem. If efforts were concentrated on the main forces, the enemy's infrastructure within the population would be left alone to spread, gain support, and supply troops and materiels to the main force and guerrilla units. If efforts were concentrated against the infrastructure, enemy forces in the field might be able to consolidate their positions and further intimidate the people and embarrass the government.

The American position and perception of the war were further complicated and confused by three other factors. First, U.S. forces could not be a decisive factor in the struggle for the loyalty of the population. Actions to win the hearts and minds of the people had to be performed by South Vietnamese to be fully effective. The United States could help with organizational skills, expertise (if there was any real expertise in such a task), and money. A second complication was the difficulty American forces had combating guerrilla fighters on their own terms. The average U.S. soldier was trained and equipped to move and fight in large units and to make use of overwhelming firepower. Few were trained to operate alone or in small groups, to operate with great stealth, and to fight effectively at very close quarters. A third problem was that the American style of war made it difficult to fight enemy main force units and achieve decisive victories. The enemy stood and fought only on its own terms. Seemingly trapped enemy forces were often able to melt away into the mountains and jungles, avoiding decisive defeat because their mobility was based on the foot power of the individual soldier rather than the mechanization typical of large American units.

U.S. strategy was also dominated by what the country, as represented by our political leadership, was willing and able to do in Vietnam, a factor that varied over time. From 1954 until 1964, the United States was willing to send large amounts of aid to bolster the South Vietnamese government and growing numbers of military advisers to train its army.

From 1965 through 1968, American political leadership was willing not only to engage in large-scale combat operations but also to carry the major portion of the war-making burden. But the United States did not leap precipitously into large-scale combat. The American response to the challenge in South Vietnam was gradual and graduated. Rather than being bent on a full-scale war, American political leadership continually sought compromise solutions and attempted to use the gradual escalation of its efforts as a bargaining tool, with very limited success.

From the latter part of 1968 through the end of 1972, U.S. leadership was unwilling to continue its large-scale prosecution of the war. Disillusioned with a long war without apparent progress, Americans wanted out. The most the United States was willing to do was to scale down its participation gradually, to withdraw its troops, and through provision of training and equipment to attempt to leave South Vietnam in such a position that it could defend itself. However, the United States was willing to fight at arm's length through the use of air power. The final chapter, 1972 through 1975, witnessed the total collapse of American will to aid a faltering ally under heavy attack.

It did not become clear to Americans until well after the war that North Vietnam was constantly willing to increase its efforts in the South because the North was fighting an unlimited war. While American willingness to fight rose and fell over time, North Vietnamese willingness never wavered. During the period from 1965 through 1968, the United States believed it was escalating the war, hopefully to a point at which the North Vietnamese would realize that their objective was not worth the cost. From the North Vietnamese viewpoint, the war was already escalated because North Vietnam was engaged in a total war for an invaluable objective.

Finally, American strategy was influenced by the willingness and ability of the South Vietnamese to prosecute the war. In the struggle for the loyalty of the peasants, the pacification efforts of the South Vietnamese were often poorly conceived, badly organized, and haphazardly executed. The government resisted the sweeping political reforms required for success (particularly land reform) until late in the war. In the military struggle, the South Vietnamese displayed varying capabilities. In the early years, a warlord mentality sapped the leadership of the army. Later, as political upheavals shook the South, the coups and countercoups of the highly politicized army diverted attention from the military struggle in the field. By 1965 the South Vietnamese army was on the ropes and ripe for defeat. The American entrance into the war saved the day and gave the South Vietnamese army time to regroup and reorganize. By 1968 the South Vietnamese army had developed a number of first-class fighting units that conducted themselves well during the Tet offensive. Improvement continued through the departure of the American troops.

In the long run, however, the American training and equipment that turned the South Vietnamese army into a credible force contributed to its final undoing. U.S. advisers trained the South Vietnamese to fight in the American style, relying on heavy firepower, unlimited air support, and the logistical system to make it all work. When the Americans left and the logistical pipeline dried up, the South Vietnamese found themselves at a fatal disadvantage.

With all of the foregoing as background, what were the strategies used by U.S. forces? As one would suspect, the strategies changed over time and must be dealt with by time periods. The period from 1954 through 1964, the advisory years, can be dealt with quickly. The objective was to help the South Vietnamese help themselves by equipping and training their forces. It was also a time for testing the various theories of counterinsurgency being touted in the United States. Army Special Forces units were sent to Vietnam and operated extensively in areas far from the political intrigues of Saigon. Other members of the ever-expanding advisory force trained South Vietnamese forces and accompanied them on operations in the field. American airmen trained their South Vietnamese counterparts and often flew with them on combat missions. In 1962 American planes and crews also began spraying herbicides to defoliate the jungle hiding places of the insurgents and, within certain areas, to destroy the crops the insurgents used.

The arrival of large numbers of American ground combat troops in 1965 led to a considerable debate over the appropriate strategy for their use. The original rationale for the insertion of American troops was to protect American air bases in the wake of several inordinately destructive raids by enemy troops. Some officials argued that the American role should be limited to protecting these enclaves rather than becoming deeply involved in a war on the Asian mainland. Others argued that limiting the American role to guard duty was a waste of superior military capability and went against the "aggressive nature" of the American soldier. By mid-1965 the enclave strategy had been discarded and America became fully involved in a ground war in Asia, thus ignoring a long-held Western military phobia. And, as had been predicted by some, the Americans began to take over the ground war, a role quickly and easily relinquished by the South Vietnamese.

A major factor in the decision to widen the American involvement in the war was the military situation in South Vietnam in 1965. The Vietcong were massing in large units, and an increasing number of regular North Vietnamese units were operating in the south. Both enemy forces were on the offensive and the South Vietnamese army was rapidly disintegrating as it lost nearly a battalion per week through battle and desertion. As the enemy offensive gained momentum, district capitals fell at the rate of one per week. In the language of protracted war, the guerrilla war had advanced to the third or large-unit maneuver phase.

Without direct American intervention, an American enclave strategy might have been pointless as South Vietnam collapsed under the enemy onslaught. Thus, as more American troops poured into South Vietnam with a fighting mission, General Westmoreland, commander, U.S. Military Assistance Command,

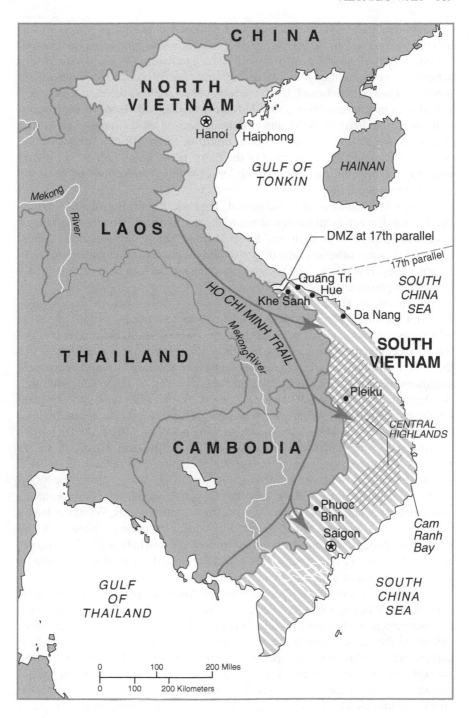

Vietnam (COMUSMACV), mapped out his master plan. The first step was to halt the losing trend and to stop the enemy initiative. Once the crisis had passed, Westmoreland would move on to step two, which envisioned aggressive offensive action to seize the military initiative and destroy enemy forces. Once enemy main force units had been defeated, U.S. forces would enter phase three, in which they would mop up the remaining guerrilla force structure and clean out any enemy units in remote base areas. Meanwhile the American air campaign against the North, named Rolling Thunder, would continue in an effort to persuade the North that it should end its support of the battle in the South. Air power would also play a significant role in the South by providing interdiction and close air support for ground troops.

By the end of 1965 the crisis had passed. The major enemy thrust, aimed at cutting the South in half from the mountains to the sea, had been defeated. Two major American operations (Starlight and Silver Bayonet) had tested American troops, blunted the enemy offensive, and demonstrated the concept of air cavalry in the Ia Drang Valley. It was now time to move on to the second step of Westmoreland's plan and aggressively seize the offensive.

The enemy, its ambitious large-unit offensive blunted, became less aggressive and operated in smaller tactical units. American troops initiated a series of spoiling attacks and "search-and-destroy" sweeps through the countryside, but they occupied no land for any length of time. In the purely military sense there were no strategic points to occupy, yet every point was strategic because a significant portion of the war was waged for the loyalty of the people. Although the Americans could "sweep" an area clear of enemy forces, when they moved on, the enemy returned, forcing the peasants into some kind of accommodation. Although South Vietnamese pacification personnel often followed in the wake of American sweeps, their mission was made doubly difficult when U.S. forces moved on to other operations, leaving the Vietnamese to defend themselves against enemy forces that might return.

Westmoreland's strategy was to keep the enemy off balance with spoiling attacks and to inflict the maximum number of casualties. In other words, Westmoreland's was a strategy of attrition as he attempted to exhaust the enemy's manpower and will to fight through superior American firepower. If overwhelming U.S. firepower could kill the enemy fast enough, North Vietnam would not be able to sustain its support of the war in the south, or so the idea went.

Ironically, the enemy strategy for defeating the American forces was also based on attrition. The Vietcong and North Vietnamese had been on the verge of administering the coup de grace to the hapless South Vietnamese forces before the American forces arrived on the scene. However, they could not hope to inflict decisive battlefield defeats on the well-trained and superbly equipped U.S. forces. They fell back on the concepts of protracted warfare, a kind of warfare unsuited to democracies in general and anathema to impatient Americans. Their strategy was to avoid defeat, harass the Americans (and their allies, of course), prolong the war, and cause as many American casualties as possible. Combined with a well-orchestrated propaganda campaign, the bodies of dead American soldiers returning home would have a devastating effect on the American will to continue the struggle (cost-tolerance).

Thus both sides pursued an attrition strategy, but there were differences. U.S. attrition was aimed at killing a maximum number of enemy soldiers on the battle-field, which became an end in itself. The enemy strategy of attrition was aimed at the morale of the American people. The enemy considered the war a struggle between entire societies while American strategic interests concentrated on the narrower confines of the battlefield.

In early 1968 the enemy attempted to speed up the process of American and allied defeat by fomenting a national uprising among the people of South Vietnam. The tool to accomplish this end was a major offensive against South Vietnam's cities beginning during the Tet holidays. After some initial setbacks under the massive onslaught, U.S. and South Vietnamese forces soundly defeated the enemy. The Tet offensive failed to start an uprising and resulted in a crushing military defeat for the enemy, but it provided the straw that broke the back of American will to continue the struggle.

In the wake of Tet, the American objective in the war clearly changed. Rather than a military solution, the United States sought a way out of the war with mini-mum damage to its prestige. Strategy quickly changed to accommodate this new goal. The United States started the process of Vietnamization—turning the war back over to the South Vietnamese—and reducing U.S. troop levels (although for a while troop numbers increased, reaching their zenith in early 1969). As American troops departed, efforts were made to bolster the South Vietnamese army's ability to stand on its own by providing both training and equipment. American forces continued combat operations on the ground but at a reduced level. Many of these operations, including an incursion into enemy sanctuaries in Cambodia, were aimed at protecting the withdrawal of U.S. forces.

The defeat of the enemy in its Tet offensive also offered a great opportunity for the United States and South Vietnam. The offensive was led by Vietcong forces, and their destruction left a power void in the countryside. In effect, the Vietcong, who had avoided defeat against superior American forces for years, had been destroyed by the enemy's decision to make them the shock troops for Tet. The Saigon government quickly moved to fill the void with massive new pacification programs. In essence, the guerrilla war was won (or more accurately, lost by the Vietcong when they abandoned guerrilla tactics during Tet) and the war for peasant loyalty was being won. But the Americans continued to withdraw.

After a conventional invasion by North Vietnamese forces in the spring of 1972 and its defeat by the South Vietnamese with the assistance of American air power, the North Vietnamese and remnants of the Vietcong signed a cease-fire in early 1973, but only after a final massive U.S. bombing offensive in North Vietnam centered on Hanoi and Haiphong. The cease-fire allowed the United States to leave with some arguable degree of honor. In 1975 the North Vietnamese again invaded and the South Vietnamese army, without the massive ground firepower, air power, and logistical support of the United States, quickly crumbled.

Having discussed the factors that influenced U.S. strategy in Vietnam and traced the changes in strategy over time, it is appropriate to evaluate the strategic choices made by the United States. Had the United States not taken to the field in 1965 to battle directly with the enemy, any subsequent strategic decisions would have been academic as the enemy would have overrun the South. The situation in 1965 was desperate. The strategic choices made after the crisis passed are more legitimately questionable.

Perhaps the biggest error made by American strategists was not realizing that the enemy was fighting an unlimited war. In a sense, the enemy turned President John F. Kennedy's famous inauguration speech on its head. It was the enemy who would "pay any price, bear any burden, meet any hardship" to gain its objective of a united Vietnamese nation. Given such total commitment, a policy of gradualism—slowly increasing the pressure—had little effect except to strengthen enemy resolve and slow their progress to the ultimate goal.

Gradual escalation of the American effort severely reduced General West-moreland's strategic choices. As the American buildup slowly progressed, he did not believe he had enough combat forces or the political backing to launch an overwhelming attack against the enemy (either in the south or the north), or to seize and hold an ever-expanding area that could be effectively and permanently pacified. Short of withdrawing, an attrition strategy seemed to be one of his few choices. And from the standpoint of superior American firepower and mobility, it seemed to be a logical choice.

Unfortunately for the Americans, the choice of an attrition strategy ignored several critical factors. First, it ignored the commitment of the enemy to the cause. Attrition would have had to be of incredible proportions to dent the enemy's resolve. Second, adopting an attrition strategy assumed that we could inflict the appropriate casualties upon the enemy. Time and again, the enemy slipped from the grasp of elaborate operations and eluded the overwhelming American firepower. Although enemy forces suffered enormous casualties, they never approached the attrition level required to bring American victory. Third, by its very nature, a war of attrition is a long and drawn-out affair. This factor played against the American penchant for quick and decisive results. America's characteristic impatience had a great deal to do with its ultimate undoing in Vietnam.

Political Considerations

If the United States had a difficult time translating the political objective in Vietnam into a working military strategy that could achieve American purposes, our government had an even more difficult time devising a translation of those ends into terms the American people could support and sustain. The central reality of the Vietnam experience politically was its growing unpopularity, and both the

Johnson and Nixon administrations labored diligently if unsuccessfully to develop a positive consensus around the war effort.

As we have already noted, it was not the clarity of the political objective that was the problem, as some earlier observers have maintained. The principle of containment rather precisely defined American involvement, and this objective remained the rationale at least until the Tet offensive and counteroffensive. The real problem was that for segments of the population, this purpose was not an adequate reason for sustained sacrifice. Partly this may have been because the containment policy was not obviously applicable to the kind of struggle going on in Vietnam. Containment was, after all, devised to blunt aggressive, presumably Soviet-inspired expansion, and although the North Vietnamese and Vietcong were Communists, they were also nationalists. At the same time, overt national support for containment, which had been high in the anti-Communist atmosphere of the 1940s and 1950s, had flagged by the latter 1960s. This was, after all, the period of dawning détente with the Soviet Union, and the fact that we conducted business as usual with the Soviets throughout the war certainly did not contribute to containment-based fervor (all of SALT I, for instance, was negotiated while American soldiers were fighting and dying in Vietnam).

In this atmosphere, the government labored hard and long to produce adequate justifications for our sacrifice, and a series of explanations was "run up the flagpole" to see if they would prove convincing. At one point, it was the Munich analogy that underpinned our commitment, while at another it was the threat of falling dominos in Southeast Asia. If the worth of containing a minor power like North Vietnam was questioned, American leadership argued that it was really the Soviet Union or the People's Republic of China that we were combating. The longer and harder that government officials tried, the more their efforts fanned an increasingly large and cynical antiwar movement.

The Tet offensive in 1968 was the straw that broke the camel's back in terms of popular support. Before the television extravaganza that the Tet offensive provided, opposition was significant but limited. When Tet appeared to reveal that the progress reported in the war was illusion, overall public opinion turned decisively against the war and forced the objective to change to extrication from the war. Vietnamization was directly attributable to public pressures surrounding the 1968 presidential election.

There were international political considerations as well. One major category of those concerns, similar to the same phenomenon in the Korean War, was Soviet and Chinese commitments to the DRV. Both countries were openly supporting and supplying our adversary, and both had mutual defense arrangements with North Vietnam. The problem from an American vantage point was to keep the hostilities at such a level that those commitments would not force a direct confrontation between the United States and China, or even worse, the Soviet Union. These worries caused the United States to impose limits on the levels of violence, especially in the air war over North Vietnam and particularly on target restrictions for American bombers.

The result was detailed instruction and control of military operations by civilian authorities, motivated by political rather than military considerations.

At the same time, the United States had to be concerned with international opinion about the war. Because we were fighting against another third world country, much of the Afro-Asian world opposed our participation and voiced its objections loudly in such forums as the United Nations. While these were relatively minor irritations, there was also fairly widespread opposition to our participation among our major allies. Whether motivated by basic opposition to the enterprise, a belief in the futility of the entire effort, or the debilitating effects Vietnam was having on the level and quality of our participation in NATO, America's principal allies showed considerable disgruntlement with our efforts. American diplomats were forced to expend a fair level of energy attempting to justify our case.

Military Technology and Technique

Because of the extraordinary length of the Vietnam conflict, the military technology available changed significantly over its course. As weapons and equipment changed on the battlefield, military techniques also changed as the antagonists struggled either to take full advantage of new technology or to avoid its lethal impact.

During the 1950s and early 1960s, the American military focused most of its attention on the Soviet threat to western Europe. Even the vaunted Special Forces were originally designed and trained to foment insurgent activities behind Soviet lines in eastern Europe. The Special Forces' eventual use came in the very different terrain of Southeast Asia with the very different mission of *countering* an insurgency. As to the rest of the world, the American military assumed that being prepared for the worst case in Europe was sufficient to counter a least case in another part of the world. But in Vietnam, the American military found that many of its techniques and weapons were of limited value against a very different kind of enemy in a very different kind of war in a very different part of the world.

During the Vietnam War, the vast technological capabilities of the United States were harnessed to develop and produce an array of highly sophisticated equipment and weapon systems to find and target the elusive enemy, and then to deliver large amounts of firepower on the target with great accuracy. But it was a more mundane piece of equipment—the helicopter—that shaped the character of the American ground war effort. The helicopter became nearly as ubiquitous and certainly as versatile in Vietnam as the jeep had been during World War II.

When American advisory personnel first introduced large numbers of combat helicopters into the conflict during the early 1960s, they proved to be highly effective. In many instances during these early encounters, enemy guerrilla troops broke and ran at the approach of helicopter borne forces. Before long the panic faded, cooler heads prevailed, and the enemy developed tactics to counter the helicopter. Although helicopters offered tremendous mobility and flexibility, they were noisy, relatively fragile, and quite slow. The enemy could hear their approach at a considerable distance and either take evasive action or prepare active defenses.

Thin-skinned helicopters moving slowly or hovering at low altitudes were vulnerable to ground fire and even small-caliber weapons had considerable effect. During the course of the war, the United States lost nearly 5,000 helicopters.

In spite of their vulnerabilities, helicopters offered advantages to ground troops that simply could not be ignored. And so it was that the United States brought to the war a totally new kind of fighting organization, the air cavalry. Such units were designed and structured from the ground up for rapid movement by their own helicopters. Troop transport, artillery transport, supply, medical evacuation, close air support, and reconnaissance were all accomplished by organic assets. The effectiveness of the air cavalry concept was first demonstrated in October 1965 against North Vietnamese regular army units in the Ia Drang Valley. Moving in coordinated "packages" of infantry supported by mobile fire bases (each emplaced and then displaced by helicopter as the battle moved on), the air cavalry successfully countered the foot mobility of the enemy and soundly defeated the NVA forces.

The American infantry forces that the helicopters delivered to the battlefield were the most potent in the history of warfare thanks to their new standard weapon, the M-16 automatic rifle. In the early months of its use, the M-16 was the subject of considerable controversy concerning its reliability in difficult combat situations. Overall, however, the M-16 gave the individual infantryman much more firepower than had previously been available. It was lightweight (8.4 pounds loaded), which meant that the individual soldier could carry more ammunition than ever before without increasing the overall weight of his pack. Because of the very high muzzle velocity of its 5.56-mm (approximately .22-caliber) bullet, the weapon had superior "killing power," particularly at ranges of 100 yards or less. Finally, if needed, the M-16 could spew out fire at the rate of 700 rounds per minute in automatic operation.

As significant as it was in terms of infantry fighting power, the change in standard infantry armament was minor compared with other sophisticated equipment used during the war. Light amplification devices and other detection devices tracked enemy movement through the seismic shocks of their steps as they walked down jungle paths. Once the enemy was found, he could be struck with numerous new weapons including cluster bombs (small bomblets dispensed from a larger bomb to give larger-area coverage) or smart bombs that could be guided to their target with pinpoint precision. Defoliants dispensed from aircraft destroyed jungle foliage and thus robbed guerrillas of their hiding place. Guerrilla areas were saturated with bullets from transport aircraft modified to carry electrically driven Gatling guns that fired over 6,000 rounds per minute as the gunship relentlessly circled the target.

In the air war, the United States used its most sophisticated aircraft, and the enemy countered with sophisticated antiaircraft artillery and thousands of antiaircraft missiles imported from Communist-bloc countries. In turn, U.S. forces countered the antiaircraft threats with electronic jamming devices to foil enemy aiming and guidance systems and with air-to-ground anti-radiation missiles that homed in on the transmissions of enemy radar sights. Success in the electronic war often spelled the difference between victory and defeat in the air.

Vietnam was also the first war to incorporate the large-scale use of computers. Computers were particularly important to the American logistical effort. The massive planning and control problems of moving mountains of supplies and munitions for both American and allied troops 10,000 miles across the Pacific were ready-made for solution by computers. The fact that U.S. troops were lavishly supplied with both combat essentials and creature comforts speaks well for the logisticians and their computers.

The traditional weapons of the guerrillas stood in sharp contrast to the wizardry of smart bombs, light amplification devices, and computers. Although many guerrillas used modern weapons (infiltrated into the country or captured from their enemy), they continued to rely on more primitive but no less effective weapons. Punji stakes (sharpened bamboo shafts dipped in excrement or other infectious substances) hidden along jungle trails penetrated many a soldier's boot with particularly nasty consequences. Guerrillas commonly remanufactured captured or unexploded munitions into ingenious booby traps that killed or wounded the unwary and unlucky. The guerrilla aspect of the war remained a technologically primitive affair.

As in every war, the military technology available affected the way the war was fought. The American position was simple. American lives should be spared by the effective exploitation of modern technology. This very reasonable outlook translated into substitution of American firepower for American bloodshed. U.S. forces relied on overwhelming firepower to aid and save the infantryman on the battlefield. Whether it came from artillery or air power, whether it was used directly on the battlefield or behind the lines to interdict the flow of enemy forces and materiel, Americans became incredible spendthrifts with firepower. U.S. artillery fired an average of 10,000 rounds every day. American aircraft dropped 8 million tons of bombs (four times the total tonnage dropped in World War II). Never had firepower been so one-sided and so lavishly used to save American lives.

Such great firepower requires targets or it is wasted. The objective of American tactics on the ground was to find targets for firepower and then to destroy those targets, if possible at "arm's length." Large-scale sweeps through the countryside, called search-and-destroy missions, had as their object finding the enemy, fixing the enemy in place by cutting off escape routes, and then bringing overwhelming firepower to bear from the air and from supporting artillery fire bases.

The problem was finding the enemy. It was difficult and often impossible to tell friend from foe unless the enemy was ready to fight. Thus, in spite of the offensive nature of the American search-and-destroy tactics, Pentagon analysts estimated that in 90 percent of the ground combat with the enemy, the enemy opened fire first. In other words, the enemy forces initiated the combat and fought only when they wanted to fight. Thus, a major difficulty for the American military was in finding the enemy so that overwhelming firepower could be brought to bear.

Air power was used to destroy and harass enemy logistics in South Vietnam, in North Vietnam, and along the enemy lines of communication through Laos and Cambodia (the Ho Chi Minh Trail). The effectiveness of these interdiction efforts remains a matter of conjecture. On one hand, there is no question that air power made the North Vietnamese logistical efforts extremely difficult and vastly increased the cost of supporting enemy troops in the field. On the other hand, some supplies always got through, and this was the crux of the American problem. An enemy that was difficult and often impossible to find and who would stand and fight only on his own terms controlled the tempo of the fighting. As a result, interdiction efforts had minimal effects because, when short of supplies, the enemy lowered the tempo of combat and built stockpiles from the trickle of supplies that survived the air power gauntlet.

The interdiction effort was also hindered by supplies transshipped from Sihanoukville (Kompong Som), a port in "neutral" Cambodia, to Vietcong and North Vietnamese forces in the southern portion of South Vietnam. Weather also played a role as the monsoons limited air operations during long portions of each year. The relatively primitive North Vietnamese logistic system in itself helped to lessen the impact of American air power. Tons of supplies were transported on the backs of porters or pushed on bicycles along narrow footpaths through the jungle. Such transportation methods were largely immune to the application of air power.

Perhaps the greatest hindrance to the interdiction effort was the inability to strike at the sources of the supplies. Strategic targets, those parts of the industrial web of a nation that produce the wherewithal of modern warfare, simply did not exist to any significant extent in North Vietnam. The "reservoir" of strategic targets lay outside North Vietnam and were off-limits to American bombers. Since the reservoir was out-of-bounds, the next most effective way to interdict the flow would have been to turn off the "spigot"; that is, to mine the North Vietnamese harbors through which the supplies were imported. Such action was politically unacceptable (for fear of sinking Soviet or Chinese ships and escalating the war) until 1972. Thus, massive American air power could only poke holes in the enemy's logistic "hose," and the enemy was largely able to compensate for such losses by increasing the flow of supplies through the spigot.

The American substitution of technology for bloodshed was natural. When one possesses overwhelming technological and materiel superiority, one should use it, particularly if it saves the lives of one's soldiers. This is the rich man's technique. Conversely, the Vietcong and the North Vietnamese fought a poor man's war. Guerrilla techniques, used throughout the war but most in evidence before the slaughter of the Vietcong during the Tet offensive in 1968, were the classic techniques of the weak. Guerrillas strike by surprise at isolated elements of the enemy and then melt away into the jungle or mix with the noncombatant population. As a result, guerrillas are difficult to find and target, and must remain so if they are to survive. The purpose of guerrilla techniques is to negate the superiority of the enemy by not giving the enemy a lucrative target. The political infrastructure that recruits and supports the guerrillas is virtually immune to firepower because it lives among the "friendly" population. A Vietcong cell in Saigon, for example, was not subject to destruction by overwhelming American or South Vietnamese firepower.

North Vietnamese regular army units at times stood and fought against American forces, and each time they did they were soundly defeated. For the most part, however, the North Vietnamese satisfied themselves with protracted-war techniques. Operating from sanctuaries in Cambodia and Laos, they struck at U.S. forces with the purpose of inflicting casualties rather than gaining decisive victories and then melted away into their sanctuaries to prevent their own defeat. The North Vietnamese technique changed to a considerably bolder approach as U.S. combat troops withdrew from the war. In 1972 the North Vietnamese launched a conventional invasion across the demilitarized zone that the South Vietnamese army defeated only with the massive use of the still available American air power. In 1975 American air power was no longer available, and another massive invasion from the north quickly resulted in the fall of South Vietnam. With the Americans and their superior firepower no longer on the scene, the North Vietnamese no longer had to fight a poor man's war.

Military Conduct

As discussed earlier in this chapter, enemy military activity had grown larger and bolder over a number of years, as had the size of the American commitment to South Vietnam. American actions had not reached the point of overt combat intervention by 1964, but it was clear a military crisis was fast approaching. The 2 August 1964 attack on the destroyer *Maddox* (and the disputed attack on the destroyer *C.Turner Joy*) resulted in the near unanimous passage of the so-called Gulf of Tonkin Resolution, which gave the president nearly carte blanche authority to employ American military forces. The United States clearly had unsheathed its sword, and it was obvious that American patience was in short supply.

However, the enemy forces were not deterred by the threat of American responses to their attacks. On 30 October 1964 the Vietcong attacked the Bien Hoa Air Base, killing five Americans and destroying six American aircraft. On 24 December the Vietcong planted a bomb in the Brinks Hotel, an American military billet in Saigon. The explosion killed two and injured nearly 60 others. On 6 February 1965 the Vietcong attacked the American base at Pleiku, killing eight and wounding more than 100. Fed up with enemy actions and armed with the Gulf of Tonkin Resolution, President Johnson ordered a reprisal air raid into North Vietnam, an operation called Flaming Dart. As the enemy attacks continued, so did the Flaming Dart reprisals. Finally, on 2 March 1965, reprisal raids ceased and a continuous bombing campaign (Rolling Thunder) began. Closely controlled by the president, Rolling Thunder was designed to display American determination, to persuade the North Vietnamese to stop their support of the war in the South, and to disrupt North Vietnamese military capabilities. Rolling Thunder continued for three years, the longest and largest (in terms of bomb tonnage) aerial bombing campaign in history.

The fact that Lyndon Johnson thought Rolling Thunder could cause the North to stop its efforts probably displays more vividly than any other event how little we understood the nationalist resolve underlying the North Vietnamese war effort. At the same time, this step up in American involvement meant U.S. air facilities became important targets for the enemy and would henceforth require greatly heightened security that we did not trust the ARVN to provide. To that end, on 8 March 1965 Marine Corps Battalion Landing Team 3/9 set foot on the sandy beaches north of Da Nang and became the first U.S. ground combat unit committed to the war. Its mission was to protect the American installation at Da Nang, the most important base in northern South Vietnam. More Marine combat troops followed as well as logistical support troops.

Although a considerable debate ensued about the use of U.S. troops, the president sided with General Westmoreland, the MACV commander, who stressed that the best defense was found in offensive actions. Once the precedent of putting Americans in the field was established, it became easier to add more in the same incremental manner. As the South Vietnamese position continued to deteriorate, the deployment of American troops increased. By the end of 1965, there were nearly 200,000 American troops in Vietnam; at the war's zenith in early 1969 that number was well over a half million.

The military problem for the United States was how to fight this kind of war. The immediate problem as Americans entered the country in large numbers was how to turn the military tide, which had been running consistently in favor of the enemy. By mid-1965 the crisis point had been reached. The Vietcong large-unit offensive was in full swing and regular North Vietnamese army units were in evidence on the offensive in the south. To alleviate the crisis, American troops began large-scale offensive operations. In mid-August the Marines launched Operation Starlight to destroy a Vietcong stronghold on the Van Tuong Peninsula south of Da Nang; after seven days of bitter fighting among fortified villages with extensive protective tunnel complexes, the Marines reported nearly 700 enemy dead.

In late October General Westmoreland sent units of the First Air Cavalry Division to search out and destroy regular North Vietnamese units in the highlands of Pleiku province. The North Vietnamese objective was to cut South Vietnam in half from the Cambodian border to the sea. Using their helicopters for total mobility of troops, artillery, and logistics, the Air Cavalry searched for the elusive enemy. Heavy contact was made in the Ia Drang Valley in mid-November, and fierce close-quarter fighting lasted from 14 to 18 November. For the first time, Strategic Air Command B-52 heavy bombers loaded with conventional iron bombs were called upon to provide tactical support to the ground troops. The North Vietnamese units were badly mauled and limped back across the Cambodian border, leaving more than 1,300 dead comrades behind.

American troops had been blooded successfully. By the end of 1965, the immediate crisis had passed. American offensive actions had blunted enemy momentum. The Rolling Thunder bombing campaign continued until 25 December, when President Johnson temporarily halted the campaign as a conciliatory gesture to induce the North Vietnamese to sit down at the negotiating table. The effort was unsuccessful and Rolling Thunder resumed on 31 January 1966.

The problem was what to do next. If the enemy had lost the initiative, he had not quit the field. The question was how to bring about the defeat of the North Vietnamese, which meant bringing enough pressure to bear to convince them to abandon their objective of forcefully uniting the country. This effort was hampered, of course, by our lack of understanding of the North Vietnamese objective (from their perspective the war was a civil war) and the tenacity with which it was held. The U.S. effort would be two pronged. On the ground, the method (one is reluctant to call it a strategy) was search and destroy. In the air, the solution was the continuation and intensification of the Rolling Thunder bombing campaign in the North.

With enemy momentum stopped, General Westmoreland's search-and-destroy method was an attempt to seize the military initiative by going out aggressively after the enemy, locating him, and destroying him. Westmoreland planned to inflict such enormous casualties on the enemy that he would not be able to sustain the war effort in the south. During 1966 and 1967 the Americans mounted a multitude of large-scale search-and-destroy operations with such names as Masher, White Wing, Thayer, Irving, Double Eagle, Lanikai, Fairfax, Cedar Falls, Attleboro, Junction City, Malheur, and Pershing. Many of these were massive and complex operations. Operation Junction City, for example, involved 26 battalions (22 American and 4 South Vietnamese) in a complex mission using armor, parachute drops, and helicopters.

The missions were largely unsuccessful for a number of reasons. First, when the enemy was confronted with such an operation (about which his intelligence organs had usually given him advance knowledge), he reverted to guerrilla tactics, breaking up his units and disappearing. Westmoreland's strategy forced him to send his troops where the enemy was (or where he thought the enemy was). All too often, many if not most of the enemy escaped, only to return to the area after the Americans moved on. As a result, other operations at later dates took place over the same ground in pursuit of the same enemy.

Second, this method did not translate into a military pattern that was easy for politicians or the public to follow. American battle maps (and newspaper maps for the home front) portrayed a series of separate and seemingly unconnected operations rather than cohesive campaigns. In a sense, the large-scale operations were not part of a strategy. The objective was attrition, and the operations themselves became the strategy.

Third, it was difficult to measure real success. Throughout this phase, progress in terms of the body count was reported at the "Five O'Clock Follies" and on American television, but where was the end? Body counts became the standard measure of success (since careers were made and broken on the body count, inflation was almost inevitable), and taking and holding territory had no meaning. It is unclear how many of the enemy were killed by American large-unit operations, but the enemy casualty rate was significantly higher than the rate of American casualties. By our standards, it was hard to imagine that the North Vietnamese would or could continue to accept the casualties we were inflicting. They, however, placed a different importance on the objective, making their cost-tolerance much higher than we realized. The ground strategy of attrition, in other words, was not going to work as long as there were able-bodied North Vietnamese who could be put in the field.

The other part of the strategy was Rolling Thunder. As originally proposed by the military, Rolling Thunder was to be a short, intensive bombardment campaign to cripple the North Vietnamese lines of communication to the south and virtually seal off North Vietnam from outside aid. Instead, Rolling Thunder was implemented gradually in an attempt to limit its violence. Thus, although Rolling Thunder was originally designed to achieve specific military objectives, it was implemented in a fashion designed to persuade the North Vietnamese to negotiate rather than to destroy their capabilities. In the final analysis, Rolling Thunder had only limited success in achieving either objective.

Rolling Thunder was at first concentrated in the area immediately north of the demilitarized zone at the 17th parallel and limited to a small number of the targets proposed in the original campaign plan. Between its inception in 1965 and its termination in 1968, the attacks gradually expanded northward, and the target list grew longer. Airmen took great pains to avoid unnecessary civilian casualties and to avoid provoking a strong reaction from the Chinese. The campaign was also halted several times in attempts to get the North Vietnamese to the negotiating table.

Rolling Thunder unquestionably inflicted terrible damage upon the North Vietnamese war effort. The bombing destroyed numerous ammunition depots, oil storage facilities, power plants, and railroad shops. Road and rail rolling stock were decimated. The small North Vietnamese industrial base was virtually destroyed, and over 500,000 North Vietnamese had to be mobilized for repair, dispersal, and transportation duties. Many more were mobilized for air defense efforts.

Although the Rolling Thunder campaign undoubtedly hampered the North Vietnamese war effort, ultimately it was of limited effect. Certainly it never had the decisive results that were hoped for. At least four reasons underlie this failure. First, the manner of implementation almost guaranteed minimal results. Piecemeal attacks on important targets allowed the enemy to make repairs or find alternate means to accomplish ends. Second, North Vietnam was a developing nation with no particular industrial base and thus no significant "industrial web" that could be attacked and destroyed. Third, the only way to attack the basis of enemy supply was to go to its sources, which were the Soviet Union and Red China, or at least to interdict the supply of materiel coming from those sources. For fear of widening the war, targets near the Chinese border were off limits, as were the ports of Haiphong in North Vietnam and Sihanoukville in Cambodia through which supplies flowed. Finally, the North Vietnamese effort was nowhere near as dependent on outside supply as the American effort anyway, so that it was questionable how much effect any aerial bombardment campaign could have. North Vietnam was not World War II Germany.

As the Americans pursued North Vietnamese and Vietcong main force units on the ground and bombed the North Vietnamese at home and on the infiltration routes into South Vietnam, the war in the villages to win the support of the people continued but with little success. South Vietnamese troops and officials were supposed to follow the victorious American troops on their massive sweeps of the countryside, and "pacify" the population. These efforts were hampered by insufficient forces, inefficient management, poor leadership, corruption, and a host of other factors.

The temporary nature of American successes in the big-unit war, the limited success of bombing in the north and on the infiltration routes, and the failure of the pacification campaign were all revealed in the enemy's 1968 Tet offensive. The tone of Americans reporting on the war through the end of 1967 had generally been one of progress and success. For that reason the Tet offensive of January 1968 was a tremendous shock to the American public. During these Buddhist holidays, an estimated 80,000 Vietcong troops launched an all-out offensive against South Vietnamese population centers. Hue was virtually overrun and even the American embassy grounds in Saigon were penetrated. Three dozen provincial capitals, several autonomous cities, and more than 60 district capitals came under heavy attack. The official object of the attack was to foment a massive uprising (a peculiar Vietnamese twist to traditional protracted-war theory) by the South Vietnamese people to throw out both the Saigon government and the Americans.

Militarily, the attack was a disaster for the Vietcong. After some striking initial successes, the attacks were driven back with heavy losses. By the end of February, General Westmoreland claimed that his forces had killed 45,000 of the enemy. The Vietcong, who had spearheaded the attack, were destroyed as a fighting force and were never again a major military factor in the war. Enemy hostile ability was crippled, but American cost-tolerance was broken.

The Tet offensive had a devastating effect domestically in the United States. The scenes of fighting in the streets of Saigon and the virtual devastation of Hue as Americans and ARVN fought door-to-door to dislodge the enemy contradicted the reports about American success and about victory being in sight. An enemy who had been attrited systematically for three years was not supposed to be capable of such an action, and commentators in the print and electronic media were left to speculate about the futility of three years of combat and (to that point) 20,000 American battle deaths. When MACV requested an additional 206,000 troops for the final, climactic push against the enemy, internal debate revealed that such an action was no longer possible. The antiwar movement, long isolated on college campuses, spread throughout the land. On 31 March Lyndon Johnson appeared on national television to announce that he would not seek reelection.

Ironically, the Tet offensive also presented the Americans and the South Vietnamese with an unparalleled opportunity. In essence, the Vietcong had destroyed themselves, and a power vacuum existed in the countryside. In the months and years following Tet, the allies took quick advantage with a well-coordinated pacification program developed and organized in 1967. Civil Operations and Revolutionary Development Support (CORDS) made all parts of the pacification program integral parts of the military command. Combined with the controversial Phoenix program (which sought to destroy the enemy's rural political infrastructure), CORDS largely pacified the countryside by 1972.

For the Americans, the die had already been cast. From the 1968 Tet offensive onward, the American objective was withdrawal with honor while still attempting to preserve an independent. South Vietnam. In mid-1968 General Westmoreland was promoted to chief of staff of the Army and replaced in Vietnam by Gen. Creighton Abrams. Abrams was less sanguine about big-unit actions, although they often still occurred. Vietnamization, the gradual process of turning the war over to the South Vietnamese, became the American strategy. Given the tide of American public opinion, Vietnamization seemed the only possible option for the United States.

From 1969 through 1972 the war continued but with a different tone. The enemy, badly weakened by Tet, reverted to the tactics of protracted war, launching operations out of sanctuaries in Cambodia to inflict casualties but avoiding decisive battles. The allies launched operations against the enemy, but now Vietnamese code names reflected the steady withdrawal of American troops. Training the South Vietnamese army became a top priority, as did providing the new equipment and supplies needed to let the South Vietnamese stand alone. At the end of April 1970, allied forces launched a short campaign against enemy sanctuaries in Cambodia, which touched off additional antiwar demonstrations in the United States. By the end of 1970, American troop strength had been virtually cut in half from its peak in early 1969. In February 1971 South Vietnamese troops struck into Laos to cut the Ho Chi Minh Trail. After some initial success, however, the North Vietnamese counterattacked and routed the South Vietnamese. Training and equipping the South Vietnamese continued, and by the end of 1971, American strength was down to only 140,000, half the level of a year earlier.

The first real test of the newly trained and equipped South Vietnamese occurred in the spring of 1972, when the North launched a major conventional offensive across the demilitarized zone. Although the northerners achieved some initial success, the South Vietnamese forces performed relatively well and, with the help of massive American air power, drove the invaders back. Included in the bombing effort was a massive campaign against North Vietnam code-named Linebacker during which many of the bombing restrictions were lifted. Targets close to Hanoi and Haiphong were attacked and North Vietnamese harbors were closed by mining. New smart bombs made short work of important railroad bridges that had survived many Rolling Thunder attacks.

Meanwhile negotiations that had been continuing since 1968 in Paris took a favorable turn. Secret negotiations between the North Vietnamese and the Americans finally yielded results in the late fall of 1972. However, the Saigon government opposed the agreement since it provided for an in-place cease-fire that did not require the withdrawal of enemy troops. Negotiations broke down over changes proposed by the South Vietnamese.

On 18 December 1972 President Nixon ordered a new bombing campaign against the North, Linebacker II, to convince the North Vietnamese to sign the cease-fire agreement. In a campaign featuring massive B-52 bombing raids, remaining military targets around Hanoi and Haiphong were systematically destroyed. Although enemy air defenses were initially very robust, the North Vietnamese quickly ran out of sophisticated ground-to-air missiles and were, by the end of the campaign, essentially helpless. Finally, after 11 days of pounding, the North agreed to resume talks. The final cease-fire agreement was initialed on 23 January 1973. The American war in Vietnam was over. On 29 March the last American combat troops left the country, and on 1 April the final American prisoner of war was returned.

In early 1975 the North launched another invasion of the South. After some early setbacks, southern military leaders attempted to consolidate their positions by withdrawing troops from the northern provinces of South Vietnam. The withdrawal turned into a panic and the South Vietnamese army fled south in total disarray. Meanwhile Americans debated whether to aid the South Vietnamese. Although promises had been made at the highest executive level, Congress prohibited all support, and the rout in Vietnam continued. Finally on 30 April 1975, Saigon fell to North Vietnamese troops.

Better State of the Peace

As has been mentioned, the main distinguishing feature about Vietnam was that it was the first and only occasion when the use of U.S. military force did not accomplish any of America's purposes. Enemy hostile ability was not overcome; nor was enemy willingness to continue. Rather, for the first time an adversary accomplished its political goals against us; and if North Vietnam did not defeat the United States on the battlefield, it most certainly overcame our cost-tolerance and thus our willingness to continue. By so doing, North Vietnam was able to impose its will on the Republic of Vietnam and force us to accept its policies.

The agony of Vietnam is that we "lost," although it has been the burden of much of our analysis to try to frame what is meant by losing. Vietnam was not a military defeat for the armed forces of the United States, but it was certainly not a military victory either. At the tactical level of individual engagements, superior American firepower consistently carried the day against an enemy willing to endure staggering losses. At the strategic level of overcoming hostile ability, the United States was unable to prevail. In retrospect, a major reason for this was that the North Vietnamese clearly found their cause to be more important than we found ours. For the Democratic Republic of Vietnam, the war was a total contest with unlimited objectives (at least within its means), and it was willing to pursue those objectives with all its energies. For the United States the purposes of the war, to the extent they can clearly be stated, were limited and bounded, and that limitation was eventually reflected in the vigor and resolve with which Americans pursued those ends. This asymmetry of purpose certainly had something to do with our ultimate failure.

There were at least two major political failings in our handling of the Southeast Asian situation. The first and possibly most basic was our failure to adequately comprehend the situation. As several recent accounts of Vietnam have shown, American presidents from Truman forward had Indochina as a constant concern, but most of the series of incremental decisions they reached were made in ignorance of the situation in the country or because other influences, generally domestic in nature, were more important.

Early policy toward the area well illustrates the point. In 1945, when Ho Chi Minh declared the independence of the country, the American attitude was generally supportive. Ho had, after all, collaborated with the United States during the war, and President Roosevelt had been explicit in his determination that the area would be allowed to determine its own future.

Ho Chi Minh did not change between 1945 and the beginnings of American aid to the French in 1950. He had been both a patriot and a Communist in 1945 and he still was in 1950. What was different, of course, was the international situation and the way the United States viewed itself in its changed environment. Wartime cooperation between the United States and the Soviet Union had been transformed into an all-encompassing confrontation, and this altered environment was dramatically demonstrated by the North Korean invasion of South Korea in June 1950. In that light, Americans could not countenance dealing positively with a Communist of any ilk. Thus, if Ho was viewed as more patriot than Communist in 1945, in 1950 the perception was just the opposite and that meant he was the enemy. That he also represented the nationalist sentiment of a large portion of the Vietnamese population was almost beside the point. Moreover, the situation in Europe had also changed. In 1945 the United States was not overly concerned about alienating a prostrate France (especially given President Roosevelt's well-known animosity toward Charles de Gaulle). By 1950, however, France was an important NATO ally, faced a difficult internal political dilemma in the form of a popular Communist party, and appeared both in need of and worthy of American assistance in the worldwide fight against communism.

The important point to note about all this is how little any of these circumstances had to do with the situation in Vietnam itself. This lack of correspondence between the objective situation and the bases on which American decisions were made haunted the United States throughout its Vietnam involvement. Certainly there were those both within and outside the government who had expertise and understanding, and many within the government recognized that the prospects for decisive success were remote. Nonetheless, those who saw the situation as either hopeless or not amenable to a military solution and who early on counseled disengagement were not heeded. One wonders and can only speculate what the nature of American involvement in Vietnam might have been had the sole criterion for decision making been an accurate assessment of the politico-military situation.

The second political element in our failure, which contains a military component as well, was in deciding what parts of the problem were amenable to political as opposed to military solutions. A war fought on the model of Mao Tse-tung's war of national liberation strategy is as much, if not more, a contest for the hearts and minds of men as it is a military struggle. A war for the hearts and minds of men is a contest over loyalty, and it is primarily a political contest in which the adversaries, through deed or promise, seek to gain those loyalties.

While this seems obvious enough, it is not at all clear that American policy recognized or appreciated this distinction and translated it into action. The task of winning hearts and minds is not something for which military force is especially well suited. Military force may provide the necessary shield behind which political conversion occurs, but the military itself has no unique capabilities to perform this role. And yet, that is exactly the way the military was used repeatedly during the war, while the South Vietnamese governmental officials who were the only ones who possibly could engage in civic action too often remained on the sidelines. The failure of the South Vietnamese to win the hearts and minds of the population may have made any American success impossible.

In understanding this failure, the military cannot be totally exempted from blame. If the war did not result in a conventional military defeat, it certainly did not result in a military victory (even if we came close to such an outcome during the Tet counteroffensive). Moreover, the professional military did not exactly distinguish itself in its ability to recognize its lack of progress and to innovate appropriately in the face of this quagmire. Why was this so?

The most obvious answer is that our military as an institution simply did not understand the nature of the war or how to fight it. At one level, this may have been the result of our enemy's successful mixture of guerrilla insurgency and conventional war tactics as prescribed by Maoist doctrine. At another level, however, it may have been that the military did not comprehend how to fight against the kind of mixed guerrilla and mobile warfare with which we were faced.

The latter point should come as no great surprise to anyone familiar with American military history. Despite the use of guerrilla tactics by the revolutionary militias during the American Revolution, the U.S. Army has never shown particular talent or enthusiasm for unconventional forces. Our history of combating irregular, unconventional forces has never been particularly distinguished. Moreover, the military has always resisted preparing for this kind of conflict. Possibly the greatest failure of the United States military in Vietnam was in not recognizing and admitting this frailty to political authorities. Had the services said "we're not sure" rather than "can do," different decisions might well have been made.

The United States failed miserably in Vietnam. There was a better state of the peace after the North Vietnamese captured Saigon in May 1975, but it was their better state and not ours. That is the unique and bitter legacy of Vietnam. What lessons may be learned from that tragic outcome and how those lessons can be applied to the future are discussed and assessed in the next chapter.

8

PERSIAN GULF WAR

Operation Desert Storm, or the Persian Gulf (or simply Gulf) War, was the first major employment of American military force after the Vietnam War. In important ways, it was a transitional military experience for the United States. The war served as the culminating point of the period of national adjustment to what was generally considered to have been the American failure in southeast Asia. Desert Storm was the debut of the new professional American military based in the all-volunteer concept on the world stage, and it marked the return of the United States to geopolitical assertiveness after a decade and a half of introspection and readjustment of America's place in the international order. American leadership in the international effort to reverse Saddam Hussein's invasion, conquest, and annexation of tiny, oil-rich Kuwait served as a launching pad for the emergence of the country to the role of the world's sole remaining superpower in the 1990s.

At the same time, the Persian Gulf War was also the opening event of a movement to American concentration of military effort on the countries on or near the oil-soaked littoral of the Persian Gulf that remains a major focus of U.S. defense concerns, a presence that has been constant ever since. When the war ended, the United States lingered by providing protection for Iraq's Kurdish minority and Shiite majority from revenge of the Saddam Hussein regime. Operation Provide Comfort (later Northern Watch) in the Kurdish regions and Southern Watch in the Shiite south provided "no-fly zones" patrolled by American, British, and (for a time) French air forces; their missions did not end until the overthrow of the Iraqi regime in 2003. Al Qaeda, the primary antagonist in the ongoing contest against international religious terrorism, was an offshoot of the Afghan resistance to the Soviet occupation of that country in the 1980s, in which the United States played a supporting role. Both were part of the continuing web of American military presence in the region and helped create the context for understanding the Gulf War.

The American people and the American military had staggered away from the calamity of Vietnam. Combined with the impending Watergate scandal, Vietnam helped undercut the faith and trust of Americans in their government. The professional military retreated, shunned by a population reluctant to be subjected to military service and deeply suspicious of the government and the media. All three links in the Clausewitzian trinity (the bond between the people, the military, and the government) lay fractured.

The period after Vietnam can be seen as a process of healing the fractures caused by the Vietnam War, as both the citizenry and professional military that serve it reassessed and adjusted the role of military force in American society. That adjustment, in turn, was influenced by three different factors: the Vietnam experience and adjusting to its meaning, the changing attitudes of the American public toward the military, and the use of military force, and the end of the cold war.

The primary, immediate legacy of Vietnam was a turning inward of both the American people and the military. The American people wanted nothing of foreign military involvements: "no more Vietnams" meant, at least temporarily, no more military engagements. Many Americans registered their disdain for the military by avoiding service in the new All-Volunteer Force (AVF). They also let the politicians know in 1976, the first presidential election year since the fall of Saigon, by electing a president who promised an absolute reduction in military spending.

The military became introverted as well. Many of the best and brightest left military service; those who remained asked, Why had the catastrophe occurred? Most felt betrayed. On reflection, however, two things happened. First, the armed forces dedicated themselves to professionalism to overcome the general perception of their incompetence. The tool for this was the same AVF concept originally adopted to insulate Americans from involuntary service; instead, being a member of the AVF became a badge of professionalism. Second, their analysis brought them back to Clausewitz and the trinity; in the future, support from the American people would be sought, even demanded, in advance of committing American forces into harm's way.

The revival of the American military in the 1980s paralleled changing attitudes in the American electorate. As Vietnam faded in the public mind, so did the overt animosity toward the military fade in most minds. The change, a renaissance in support for an active military role, was symbolized—even energized—by the presidential career of Ronald Reagan. In 1976, Reagan sought the presidency on the conservative, promilitary platform that propelled him into the White House in 1980. In 1976, he could not wrest the Republican Party nomination from Gerald Ford. In 1980, the same combination of patriotic values and a pledge to reverse the "unilateral disarmament" of the Carter years produced a landslide victory. Reagan had not changed; the people's attitudes had.

The result was a decade of prosperity and prestige for the American military unprecedented in peacetime that began and grew in the 1980s. Money was available, high quality recruits streamed into the recruiters' offices, and morale soared. The military was back, and when it was called upon in 1990, it could respond.

The period between Vietnam and the Persian Gulf War thus provided the military a comparative respite in which to regroup and regenerate itself. During the final 15 years of the cold war, the U.S. was called upon to use force only in modest ways: attempting to rescue the hostages held in the American Embassy in Tehran and saving American medical students at risk in Grenada, for instance. Desert Storm was its first major post-Vietnam challenge.

The Gulf War also highlighted a geographical transformation in the focus of American military effort that continues to the present. Its roots lie in events and dynamics that precede Iraq's invasion of Kuwait in 1990 but were not overwhelmingly obvious at the time. Two 1979 events, in retrospect, were particularly important.

The first was the Iranian Revolution. On January 16, 1979, the American-backed Shah of Iran fled his country, abdicating the Peacock throne and setting in motion events that resulted in a radical Islamist government in that country. Americans remember this period principally because of the seizure and kidnapping of the American embassy and its personnel in Tehran, but the deeper significance was geopolitical. Prior to 1979, the Iranian government had secured American access to Persian Gulf oil, acting effectively as a surrogate for American interests and minimizing the need for direct American presence. After the revolution, Iran no longer served as an ally, but as an adversary. The result was the need for a direct, permanent American military presence; the Persian Gulf became an American pond.

Over Christmas 1979, the Soviet Union invaded and occupied Afghanistan, a country in which the United States had no discernible previous interests. Fearing that Soviet intentions went beyond the land-locked, inhospitable Afghan terrain—possibly to the Persian Gulf—the United States determined an interest in frustrating Soviet ambitions and began clandestinely to offer monetary aid and military support for the Afghan *mujahidin* resisters. When the Soviet Union abandoned Afghanistan in 1989, remnants of that resistance became, among other things, the core for Al Qaeda and the Taliban.

American participation in Desert Storm was a watershed in American military activity in the Persian Gulf region. Sending over 425,000 Americans to the region ended the post-Vietnam self-imposed quiescence in the uses of American military power. It also created the precedent for future American military power in the area. Before Desert Storm, the United States was only potentially a military factor in the region, and an unknown factor at that—one reason Saddam Hussein invaded Kuwait, after all, was his misperceived belief that the memories of Vietnam would inhibit an American response. After Desert Storm, the United States emerged as the dominant military force in the region, a status that not all in the area find entirely comfortable.

Along with the aftermath of U.S. assistance to Afghan rebels in the 1980s, the Desert Storm experience thus serves as the basis from which much subsequent U.S. involvement has evolved. The Gulf War was the result of an action by Iraq (invading Kuwait), and it ended with Iraq's forceful removal from Kuwait. The war did not, however, remove Saddam Hussein from power as punishment for his transgression, and the nonrepentant Iraqi leader remained an international irritant whom many in the United States believed needed toppling, a wish they realized in the Iraq War. Part of the rationale for the 2003 invasion was based in alleged ties to international terrorism that had its roots in the Afghan adventure of the 1980s.

Issues and Events

The underlying issues and events leading to the American involvement in the effort to evict Saddam Hussein's Iraqi army from Kuwait are both simple and complex, depending upon the level at which one investigates them. In one sense, the pre-cipitating event—the lightning conquest of the tiny, oil-rich state of Kuwait by the Iraqi armed forces—is straightforward. Moreover, with their conquest complete, there was no physical barrier to the Iraqis continuing to plunge farther south and grabbing the Saudi oilfields on the Persian Gulf as well.

A deeper understanding requires looking at two different items. The first is the period leading up to Iraq's invasion, including both Iraqi attempts to gain financial assistance from Saudi Arabia and Kuwait, and American attempts to woo the Iraqi government and hence bring it into the international mainstream. The second is the rea-son Saddam acted militarily as he did, rather than pursuing alternative objectives.

Iraq had a set of grievances with Saudi Arabia and Kuwait that arose from Iraq's "protection" of Arab interests in the Iran–Iraq War. That war pitted Arab Iraq against Persian Iran, and Saddam Hussein saw it as a gallant crusade against the Persian infidels and as a sacrifice that would bring him thanks and glory among fellow Arabs throughout the region. Some even believed he hoped the war would catapult him into the position of leader of the Arab world, a designation unfilled since the death of Egypt's Gamal Abdul Nasser in 1970.The eight-year struggle ended in 1988 and left Iraq exhausted militarily, politically, and economically. Iraq had borrowed over $30 billion during the war, mostly from the Saudis and Kuwaitis, and could not repay it. Saddam Hussein requested cancellation of the loans; he was refused. Next, he requested additional credits to help rebuild an economy shattered by the fighting. This request also had political bases: Hussein's continued popularity and stability required a return to normality. His requests were not honored, and his ambitions were unfulfilled.

Finally, Saddam requested that the price of oil be raised and production reduced as a way to maximize Iraqi oil revenue. Not only was the request denied, but also Kuwait allegedly engaged in "slant drilling" from Kuwaiti soil into the Rumalia oil fields claimed by Iraq, in effect stealing Iraqi oil. In the months preceding the invasion, Iraq threatened consequences if Saddam's requests were not honored; no one, including the United States, paid any attention.

Iraq had a new suitor during this period—the United States. The U.S. had reopened relations with Iraq in 1985 (Iraq broke relations with the U.S. in 1958), in effect tilting the Americans against Iran in their war. When the war ended, rela-tions continued to warm and expand; ultimately Iraq diverted U.S. grain credits to military purchases—a scandal that has become known briefly as "Iraqgate." The

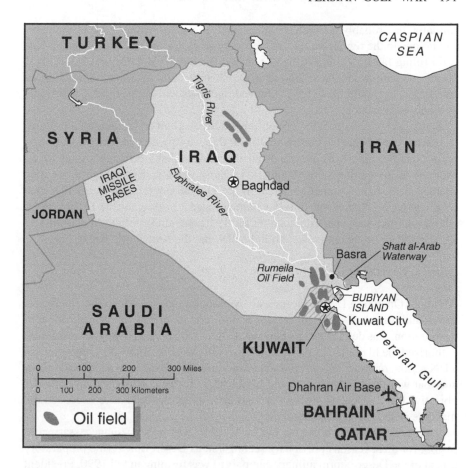

purpose was to try to curry influence that could steer Iraq into the mainstream of international life.

In the process, the United States averted its gaze from Saddam's more atrocious behavior, such as gassing Iraq's own Kurdish population in 1987. Prior to the fall of the Shah of Iran, Saddam Hussein had been the pariah of the region; after 1979, he looked better compared to the Ayatollah Ruhollah Khomeini, but it was not because he had changed. Had he bullied Kuwait in 1978, the United States would have believed him because of the perception he was an evil ruler; it chose not to in 1990. Two weeks before the invasion, April Glaspie, U.S. ambassador to Iraq, assured Saddam that the U.S. had no interest in oil negotiations among the Arabs or in boundary disputes. When asked after the war why the United States did not foresee the impending invasion, she replied, "We didn't believe he was that stupid."

When he contemplated military action, Saddam had three options: simply occupy the Rumalia oil fields and annex those fields; invade, conquer, and annex Kuwait; or invade Kuwait and move onto the Saudi fields. Had he simply seized

Rumalia, he probably would have succeeded: the slant drilling gave him a case, and no one in the region would have felt very sorry for the arrogant al-Sabah Kuwaiti ruling family. Similarly, had he seized the Saudi oil fields, the Saudi royal family probably would have negotiated a cash settlement to ensure withdrawal. In both cases, he would likely have achieved his goals without triggering a decisive international reaction.

Saddam chose, instead, the one option certain to elicit an international military response. The UN had no choice but to act: this was the first and only time to that point a member state had conquered and annexed another UN member state. However, the disintegrating Soviet Union no longer posed a veto threat in the Security Council to protect its sometime client. If the UN did not reply in a case like this, why have a UN? At the same time, Saddam violated the unwritten law of the region, which said that Arab states do not alter the 1919 boundaries of the Arab world by force. Saddam broke both rules; both the UN and the Arab world had to react.

Political Objective

Although Operation Desert Storm was overwhelmingly an American enterprise, it was authorized as a United Nations action, and its objective, clear throughout the duration, was defined by UN Security Council Resolution 678. That resolution authorized the liberation of Kuwait and the restoration of Kuwaiti sovereignty. As a UN action, *that was the only political objective that could have been authorized*, and that was the objective accepted by the U.S. To have authorized any broader objective (such as overthrowing the Iraqi government) would have exceeded the mandate stated in the UN Charter.

This question of the objective has been the subject of some unnecessary confusion, for three reasons previously seen. First, as Desert Shield (the preparatory aspect of the Desert Storm military campaign) was forming in fall 1990, President Bush began referring to Saddam as a "Hitler" figure. This was meaningless, loose rhetoric, the implications of which were probably not well thought out, but which was intended to rally support for the enterprise. Since "Hitlers" are to be removed, however, some felt the objective included getting rid of Saddam. This was never the case at the time.

This limitation was important as the crisis unfolded and was the result of the nature of both rationales for the response. Under its Charter, the UN can authorize actions intended to restore the status quo ante, which in this case meant evicting the Iraqis and restoring Kuwaiti sovereignty. To do more—such as invading Iraq and overthrowing its government—would have stretched the UN mandate beyond limits most of its members would have supported. One aggression, in other words, did not justify another. In the case of the Arab states, the Iraqi transgression could not be punished beyond restoring Kuwaiti borders without attacking Iraqi soil, and doing so would make the invading states guilty of the same infraction for which Saddam Hussein was being punished (changing the 1919 borders by force). These limits help explain the frequent assertion that had the United States decided to pursue the Iraqis in 1991, the coalition would have dissolved around them as other coalition members refused to exceed the mandates on which their participation in the war was based.

Second, the ease of the 100-hour ground war made it seem easy to "finish" the job by rolling on to Baghdad, grabbing Saddam, and destroying his supply of nuclear, biological, and chemical (NBC) weapons in the process. This might have been militarily possible, but, as already argued, it would have transformed the purpose of the war in much the same way as occurred in Korea the first time the allies got to the 38th parallel. Third, subsequent UN efforts (authorized by Security Council Resolution 687, which also created the cease-fire that ended the war) to disarm Iraqi NBC capabilities met with some frustration. Some have subsequently maintained that if the United States had "done the job right" in the first place, there would have been no problem. To repeat, however, the UN could not order an invasion of Iraq; Resolution 687 was directed at Iraqi violation of international treaty obligations (the Non-Proliferation Treaty), not the invasion of Kuwait.

The confusion had two Korea-like aspects. In Korea, the original and eventual objective was to restore the status quo; the UN mandated the same end in Kuwait. In Korea, the objective expanded when the North Korean Army was broken in the initial counteroffensive; the alternate objective of freeing the peninsula to hold unifying elections was made possible by military success. In the Persian Gulf War, military success made the removal of Saddam seem possible; in 1991, restraint prevailed. However, in both cases flirtation with broader objectives left the attainment of the original objective seem a bit hollow and unfulfilling. That broader objective, of course, never died and was rekindled in 2003.

Military Objectives and Strategy

Changing military objectives and strategies divided the Persian Gulf conflict into two distinct but closely related phases, Desert Shield and Desert Storm. Desert Shield, as the name implies, was almost entirely defensive in nature. The only exceptions were the United Nations economic sanctions, enforced by Coalition naval forces, which were intended to force Iraqi withdrawal from Kuwait. Desert Storm was purely offensive, designed to drive the Iraqis from Kuwait by force of arms, to destroy Iraq's offensive military capability, and thus to create a better state of peace in the war's wake.

Desert Shield

Following the invasion of Kuwait on 2 August 1990, hurried conferences involving the president, Secretary of Defense Richard Cheney, commander of Central Command Gen. H. Norman Schwarzkopf, and other key officials reviewed military options. They quickly decided to deploy U.S. forces to Saudi Arabia, if invited by the Saudi government. The objectives of the deployment, quickly dubbed Desert Shield, were fourfold. The first was to deter the Iraqis from further aggression beyond the Kuwaiti-Saudi border where their advance had stopped, perhaps only temporarily. Second, if the Iraqis were not deterred, it would defend Saudi Arabia, and by inference, the United Arab Emirates. Third, the forces would enforce the economic sanctions dictated by the United Nations Security Council Resolutions 661 and 665. Fourth, the Central Command commander would put together an effective military force among the Coalition members. It would not be easy to attain any of these objectives.

The crux of the problem was to get sufficient forces to Saudi Arabia quickly enough to defend the kingdom. If the Iraqis attacked quickly, before adequate forces were in place, Schwarzkopf had to have a defensive strategy that would both effectively defend and still retain the air and seaport facilities to continue the buildup. This "window of vulnerability," the amount of time it would take to get sufficient forces in place, could be as long as 17 weeks, according to Central Command Operating Plan 1002-90.

Schwarzkopf based his initial concept of operations on an enclave strategy that would trade space for time. The enclaves would be the ports and airfields on the Gulf coast, principally Al-Jubayl and Dhahran. Holding these areas would be essential to continuing the buildup. Saudi land forces would provide a screening force along the Kuwaiti border to slow an Iraqi advance, but the major combat power available to the Coalition in the early stages of the buildup would be Coalition air power.

Without adequate ground forces on the scene, heavy reliance on air power was an obvious choice. The terrain in the region was ideal for rapid armored movements— an Iraqi strength—but it was also the most ideal possible terrain for air operations—a U.S., and hence Coalition, strength. The relatively flat desert terrain meant that there was no place for Iraqi ground forces to hide from Coalition air power. Further, the nature of armored warfare placed great reliance on mobile resupply of munitions, fuel, and other expendables—ideal targets for air power ranging deep behind advancing Iraqi armor. Schwarzkopf understood that air power would be an advantage of the Coalition even after sufficient ground forces were in place.

While the Saudi ground forces and Coalition air power slowed an Iraqi advance, trading space for time, U.S. ground forces would defend the enclaves and the buildup would continue. The Saudis expressed some reservations about this strategy. Their natural instinct was to defend all of Saudi Arabia, fighting for every inch of Saudi soil. Fortunately, the Iraqis did not attack and the differing approaches to the problem were not put to the test in battle.

Desert Storm

Although Desert Shield was defensive, plans began to take shape almost immediately for offensive operations should the Iraqis fail to withdraw from Kuwait under the economic pressure of the United Nations sanctions. The earliest effort was a broad conceptual plan called Instant Thunder, an obvious play on the name of the Rolling Thunder air operations in the Vietnam War. The contrast between the two operations was stark. Twenty-five years earlier, the Rolling Thunder campaign had emphasized slow, graduated pressure on the enemy. Instant Thunder visualized an aerial blitzkrieg designed to destroy 84 strategic targets in Iraq during the first week alone. The entire concept plan included four phases, the first three using air power (land and sea based) almost exclusively. Phase I consisted of the aforementioned strategic air campaign designed to isolate the Iraqi leadership from their fielded forces, to destroy their ability to command and control their forces effectively, and to destroy Iraqi nuclear, biological, and chemical warfare research, production, and storage facilities. Phase II sought air supremacy over Kuwait and its immediate environs, the Kuwaiti theater of operations (KTO). Phase III shifted attention to the Iraqi ground forces in Kuwait, particularly the elite Republican Guard units, and to an interdiction campaign to isolate the Kuwaiti battlefield from effective reinforcement and resupply from Iraq. Phase IV was a combined air and ground offensive to drive the Iraqis out of Kuwait.

Over the months of Desert Shield operations, this concept plan was expanded and refined but still retained its basic thrust and flavor. The final plan had five objectives: (1) to neutralize the Iraqi command capabilities, (2) to eject the Iraqis from Kuwait, (3) to destroy the Republican Guard, (4) to destroy the Iraqi missile, nuclear, chemical, and biological warfare capabilities, and in the aftermath, (5) to assist in restoring the Kuwaiti government.

The four-phase strategy of Instant Thunder continued, but with some adjustments in priorities and nomenclature. The Phase I strategic attacks now also included attacks on the Republican Guard and high-value interdiction targets. Phase II remained relatively unchanged while Phase III became "Battlefield Preparation," but otherwise remained relatively unchanged. Phase IV, the ground war, changed significantly over time, as will be discussed.

In reality, the military objectives looked almost as much toward the aftermath of the struggle as they did toward the current conflict. Destroying the Iraqi nuclear, biological, and chemical warfare capability had at least as much to do with possible future conflicts in the region as it had to do with Desert Storm. The destruction of other military production facilities also looked more toward the future than the present. Once Desert Storm actually began, it would almost certainly be over before any factory production could have an impact on the war. Clearly, a better state of peace was on the minds of the planners.

The first three phases of the aerial campaign were those initially to be put in final form, although target sets and tactics were continually updated and expanded as additional forces became available during the buildup. Ground operations posed a more complex planning problem. Although Coalition air power quickly moved to Saudi Arabia and its nearby waters, heavy armored forces took considerably longer to arrive. In the meantime, there was great pressure for General Schwarzkopf to devise the fourth phase—the ground offensive portion of the strategy. Unfortunately, when the planning began, sufficient forces were not available to wage the kind of quick, decisive ground campaign that would limit Coalition casualties. Nor were projected deployments scheduled to result in the numbers of ground forces General Schwarzkopf believed he needed.

Bowing to pressure, Schwarzkopf and his staff devised a high-risk offensive plan that called, essentially, for a frontal assault on the extensive fortifications the Iraqis were building along the Kuwaiti-Saudi border. In early October, General Schwarzkopf reluctantly allowed a member of his staff to present the high-risk ground plan (along with the now expanded air plan) to the president. But he insisted that the briefing include the caveat that an additional heavy corps was required to guarantee success in the ground offensive.

The weaknesses in the high-risk plan were clear to the president and his advisers. By 31 October, chairman of the Joint Chiefs of Staff, Gen. Colin Powell, notified Schwarzkopf that his forces would be increased significantly—well beyond the heavy corps he had asked for—and that General Schwarzkopf should begin planning seriously for offensive operations including an expanded ground campaign to liberate Kuwait.

Iraqi dispositions dictated the final ground offensive plan. To nearly everyone's amazement, the Iraqis were building formidable defensive barriers only along the Kuwaiti-Saudi border. The Iraqi-Saudi border remained unfortified and almost unmanned. The classic strategy would be to strike at the Iraqi weakness (i.e., across the undefended Iraqi-Saudi border), outflanking the Iraqi forces and rapidly advancing to the Euphrates River to trap the Iraqi forces in Kuwait.

The problem, of course, was that the Coalition had concentrated its forces in a defensive posture in front of the Iraqi forces in Kuwait. To strike across the undefended Iraqi-Saudi border would require a giant "left hook" to the west. However, positioning the forces to the west would undoubtedly warn the Iraqis, who would also move their forces and build defensive barriers. General Schwarzkopf's solution to this dilemma was unique and daring.

The solution required total air supremacy and a complete pin-down from the air of all Iraqi ground and air forces. This strategy would essentially blind the Iraqis to any Coalition movements, and even if discovered, would make it next to impossible for the Iraqis to respond effectively. Only after the air campaign had begun would General Schwarzkopf move his powerful armored striking force to the west for the left hook into Iraq. The logistical task, however, would be enormous and tightly constrained by time.

Once the ground attack began on the left flank, other Coalition forces would attack into the teeth of Iraqi defenses now seriously weakened by air power and artillery-battlefield preparation. The effect would be to keep the Iraqi forces in place while the flanking action moved forward in the west. This pinning action would be complemented and reinforced by a potential amphibious assault from the Gulf, an operation very publicly rehearsed (for Iraqi benefit) during Desert Shield.

One of the military problems facing the diverse Coalition was the reluctance of one Arab state to attack another—even in the face of Iraq's naked aggression against Kuwait. Schwarzkopf soothed Arab sensitivities by assigning them to attack into Kuwait (thus liberating, rather than attacking, an Arab neighbor) and designating the Arab forces as those who would actually liberate Kuwait City.

As General Schwarzkopf planned the ground war, he emphasized to his commanders the importance of trapping and destroying the Republican Guard units. It was not enough for the great left hook to trap the elite Iraqi force. His commanders were to turn to the east and crush it. Conventional wisdom held that the Republican Guards were both the central prop of the Iraqi regime and the potential nucleus of possible postwar Iraqi military power. In this sense, General Schwarzkopf s plans for the ground war addressed both the present conflict and the state of the peace that would follow.

Political Considerations

Within the complex of actions taken to reverse Iraq's military action were purely political concerns. From a domestic viewpoint, the main concern was how to gain congressional acquiescence to a military action. Internationally, the problem was how to form a coalition of states that would make the liberation of Kuwait seem more like an international than an American action. The two were related because the success of the Coalition required American participation (425,000 of the Coalition total of 675,000 were American), which in turn politically required congressional approval.

The campaign to gain congressional acquiescence to military action hinged on how long the economic sanctions that were the first UN action against Iraq (Security Council Resolution 660) should be allowed to work before military force was employed. Those who wanted a maximum application of economic sanctions also believed that when war came, it would involve many—some estimated thousands—American casualties. The debate was heavily partisan; most of the opponents of early use of force were Democrats, led by Senator Sam Nunn, chair of the Senate Armed Services Committee; most of the supporters were Republicans loyal to President Bush.

Ultimately, proponents of the war prevailed, although not by much. Congress voted to authorize the use of force on 12 January 1991, four days before the air campaign began. The vote was close in the Senate: 42 Republicans and 10 Democrats voted in favor; two Republicans and 45 Democrats voted against, a slim margin of 52–47. In the House of Representatives, 164 Republicans and 86 Democrats voted for the measure; 179 Democrats, three Republicans, and one independent voted against, a margin of 250–183. The authorization took the form of a resolution in support rather than a formal declaration of war.

Internationally, the formation of the Coalition that eventually numbered 25 countries ranging from those in the Arab world to the first and third worlds was the major imperative. Led by Secretary of State James Baker, who darted between national capitals, an impressive array was assembled.

Not taking away from Baker's accomplishments, the composition of the Coalition is less amazing in retrospect than it seemed at the time. The coalition consisted of countries drawn to involvement by each of the rationales for the response to the invasion: UN principles and the sanctity of the 1919 borders in the area. Thus, those who supported the UN came because not to have done so would have gutted what little credibility the organization had at that time; instead the action energized the world body. At the same time, Saddam's violation of basic tenets of the Arab world reasonably guaranteed Arab participation as well. In some ways, it would have been more surprising had the coalition not formed.

What should be reiterated, however, was that the Coalition existed for the purpose of restoring Kuwaiti sovereignty and nothing more. To have invaded more of Iraq would have forced the UN to withdraw support. Similarly, the Arab members could justify punishing Saddam for his transgression; the occupation and conquest of Iraq by Westerners (principally the Americans) would have been viewed as an unacceptable Arab humiliation. To have upped the ante would have made the action almost exclusively American. General Schwarzkopf stated publicly after the war that had the United States gone further, it would have been us and "maybe" Great Britain alone.

Military Technology and Technique

In the nearly two decades between the inglorious departure from Vietnam and the stunning victory in Desert Storm, the U.S. military establishment journeyed from agonizing catharsis through robust rejuvenation. Both parts of the journey affected military technology and technique.

The humiliating withdrawal from Vietnam led to an almost immediate search for scapegoats, with fingers pointing in all directions. A spate of books written by former senior military commanders blamed the political leadership, while others accused the military leadership of incompetence. It was not until near the end of the 1970s that more objective and balanced analyses concluded that there was plenty of blame to go around.

A crisis in confidence shook the U.S. military for a decade following the withdrawal from Vietnam. At least three factors fueled this crisis. The first was the aforementioned search for scapegoats and the accusations of military incompetence. Second, the end of conscription changed the U.S. military to a smaller professional force. Third, U.S. superiority in military technology had seriously eroded. While the U.S. was heavily engaged in the jungles of Southeast Asia, many weapons system modernization programs were delayed or canceled. Meanwhile, the Soviet Union, still perceived as the paramount threat, continued to modernize its forces rapidly.

The perception of incompetence was a mixture of both fact and myth. While it was true that the U.S. military was poorly prepared for the kind of war it faced in Southeast Asia, it did not necessarily follow that the U.S. was poorly prepared for or incapable of prosecuting a modern, high-tech conventional or nuclear conflict.

On the contrary, the tactics, technology, and techniques used in Vietnam were ultimately unsuccessful because they were often better suited for conventional warfare. This fact was lost in further accusations of incompetence that followed the military disaster in Beirut and the less than sterling military performance in Grenada, to name two instances.

Considerable skepticism greeted the end of conscription and the birth of the first professional army since before World War II. Analysts of every persuasion wondered if the military services could attract enough men and women. Beyond worries about sufficient numbers were worries about the quality of those attracted to service. Would they be capable of handling the sophisticated technology of modern warfare? Would they be capable of effectively leading and managing the military in peace and war?

Ultimately, the answers to these questions were in the affirmative. In retrospect, the key to recruiting sufficient numbers of high-quality personnel seems to have been a sense of exclusivity. The services set higher standards for personnel quality than had been the case during the years of conscription, and flaunted these standards. Typical was the Marine Corps recruiting slogan noting that they were looking for a "few good men," with emphasis on the "few." Combined with clever promotion of traditional service values and the lure of high-tech "adventure," the services were consistently meeting their personnel quantity and quality requirements as the 1980s began.

The military technology problem was both disturbing and real, particularly in the area of conventional weaponry. Analysts refer to the 1970s as the era of the "hollow force" both because of the general reduction in force following the withdrawal from Southeast Asia and because of the outdated equipment used by that smaller force. In the early 1970s there was serious need in the ground forces for a new main battle tank, a new armored personnel carrier, and many other types of equipment. In the air, where technology is king, both the Air Force and the sea services were using fighter aircraft originally designed in the mid to late 1950s, and the principal Air Force long-range bomber was even older.

The technological rebuilding of the American military began in the early 1970s (for example the Navy's F-14 and the Air Force's F-15 fighters became operational in 1974), but shifted into high gear as the 1980s began. The results of the massive rebuilding program were spectacular, and produced the remarkable technology used in Desert Storm.

The technological revolution of the American military was so vast and reached into so many areas of military operations that it cannot be discussed comprehensively in this survey history. However, at least four areas of development were so important that they merit discussion because of their impact on Desert Storm: precision-guided munitions, night vision devices, space systems, and low-observable (stealth) technology.

Precision-guided munitions had the most visible impact on Desert Storm. During the month-long air campaign that preceded the ground war, television news organizations broadcast hundreds of pictures showing guided munitions striking their targets with almost unbelievable accuracy. Taken from aircraft targeting systems and distributed at military news conferences, these grainy, black-and-white video pictures were the only images available of à war that ranged hundreds of miles into Iraq.

Precision-guided munitions had their modern introduction during the latter stages of the war in Vietnam. They were used successfully to attack high-value targets (bridges being the most notable example) that were both difficult to hit and heavily defended. In the intervening years, these systems were steadily improved and became much more commonplace in the U.S. arsenal. Although their best-known use is from aircraft dropping "smart" bombs, other systems also used precision targeting and guidance techniques. Cruise missiles used variations of the precision technologies that allowed attacks on heavily defended targets without putting air crews at risk. In Desert Storm, the Navy used the Tomahawk cruise missile successfully in this role. These systems accounted for about 7 percent of the bombs used in Desert Storm.

Precision munitions guidance technologies were not limited to aircraft and missiles. Among ground forces, other variations on the precision theme became commonplace for U.S. forces. For example, the U.S. MIA1 "Abrams" main battle tank was equipped with a sophisticated fire control system that included a laser range finder, a ballistic computer, and a gun stabilizer. Its accuracy approached the "one shot, one kill" ideal, even while firing on the move.

The effect of precision munitions was extremely significant in Desert Storm. In the air, they redefined the meaning of "mass." During World War II, hundreds of heavy bombers were sent to strike targets that a few well-placed bombs would have destroyed. Unfortunately, well-placed bombs were not the norm even with the best bombsights then available. Thus thousands of bombs were dropped in the hope that some would strike the target at the critical points. "Mass" required hundreds of bombers and thousands of bombs. In Desert Storm, "mass" often meant one aircraft or one cruise missile with one bomb or warhead striking the target at exactly the right point to destroy it.

The impact was to allow the Coalition air forces to destroy more targets faster— and the results staggered the Iraqis. Saddam quickly lost his ability to control and supply his forces, and Coalition air power then turned its attention to destroying deployed Iraqi ground forces, while continuing to pound strategic and interdiction targets. Historically, air forces conducted such operations sequentially. In Desert Storm they were essentially conducted as parallel operations, thanks to the effectiveness of precision-guided munitions. The shock effect on Iraqi forces was overwhelming, as evidenced by their lack of resistance and desire to surrender when the ground war began.

On the ground, the impact of precision munitions and sophisticated firing systems was also significant. Fast-moving armored forces gutted what was left of the Iraqi armored forces, often at great distances, often with one-shot, one-kill efficiency. The advantage of those who have such systems over those who do not is almost incalculable.

Night vision devices also had their debut in Vietnam. By the time of Operation Desert Storm, they had—in one form or another—become commonplace for U.S. forces. No longer would the dark of night become a time of relative calm in combat, nor would the dark offer a place to hide. Night vision devices made combat a round-the-clock proposition. Further, they offered enormous advantages if the enemy did not have the same capabilities. If you can see in the dark and the enemy cannot, you prefer to engage in the dark; the dark offers you safety but none to the enemy.

Combat aircraft contained some of the most sophisticated night vision systems. For example, FLIR (forward-looking infrared radar) and LANTIRN (low-altitude navigation and targeting infrared for night) systems allowed Coalition aircraft to use the cover of darkness for protection, to find precise targets on the ground, and then to attack them with precision munitions. For many pilots, night was the preferred time to fly combat missions during Desert Storm.

Although secrecy veils many of the military capabilities of space systems, it is clear from published accounts that space systems played a major role in Desert Storm. Among other things, spaced-based systems provided both warning of Iraqi Scud missile launches and vital weather data; they were essential for communications both within the theater of operations and between the Gulf region and other parts of the world.

Of particular interest concerning space systems was the success of the Global Positioning System constellation of satellites. The terrain in Saudi Arabia and much of Kuwait and Iraq is featureless desert. Finding one's way is difficult at best, yet crucially important in a war of rapid movement. At night, the situation was just that much worse. The Global Positioning satellites, combined with a small hand-held receiver, allowed personnel to find their position within a matter of a few feet.

Low-observable technology, generally referred to as stealth, was exemplified by the Air Force F-117 aircraft. The F-117 was first used in Panama with limited success. In the Gulf War, it had outstanding success. It flew primarily at night to maximize its stealthy characteristics—it could not be seen in the dark with the naked eye and was very difficult to detect on radar. The aircraft's success in delivering precision munitions in heavily defended areas was notable both for its precision and because none of the stealth aircraft received any battle damage.

Stealth aircraft did not significantly affect the general outcome of Desert Storm. They did, however, make the attacks on heavily defended areas more effective and far less costly to the Coalition. The importance of stealth technology lies in its foretaste of things to come. The general trend in aircraft design will be to make them less and less vulnerable to radar detection. Perfection of such techniques could cause major changes in the nature of aerial warfare in the future.

The general impact of technology upon the technique of modern warfare, as demonstrated in Desert Storm, was at least threefold. First, it increased the intensity of combat. Modern combat has become a round-the-clock proposition with no respite. Second, technology has made combat much more efficient. Precision munitions make it possible to do much more much faster, with fewer assets. The side benefit is less collateral damage—fewer unnecessary casualties and less unnecessary destruction. Third, the gap in capabilities between those who can exploit modern military technology and those who cannot is growing ever more significant. Precision munitions, 24-hour capability, space systems, stealth technology, and many other technologies provide the possessor with much more than marginal improvements in military capabilities. Properly used, they provide an overwhelming advantage.

One other factor deserves mention. Dismayed at perceived interservice bickering, rivalry, and lack of cooperation, the U.S. Congress passed landmark legislation in 1986. The Goldwater–Nichols Defense Reorganization Act forced greater cooperation between the armed services on a number of fronts. Perhaps most important for Desert Storm, the legislation gave much more power to the chairman of the Joint Chiefs of Staff and to the commanders in chief of the various unified commands. One of those unified commands, U.S. Central Command, had overall responsibility for Desert Storm.

Central Command's commander, Gen. H. Norman Schwarzkopf, was able to command and control his disparate forces with far greater authority than had been the case previously. The result was vastly improved coordination between land, sea, and air forces that further compounded the problems faced by the Iraqis. For example, with Schwarzkopf's new authority, his air commander, Air Force Lt. Gen. Charles Homer, could centrally control all U.S. air power (and in practice, all Coalition air power) using a single daily Air Tasking Order. Unlike Vietnam, where there were as many as six different air wars in progress at any one time, in Desert Storm there was one highly coordinated air campaign. The results were devastating to the Iraqis.

Military Conduct

The initial order to deploy U.S. combat forces to the Gulf was issued on 6 August 1990. By 9 August, Air Force F-15 fighters from the continental United States were flying combat air patrols along the Saudi northern border, and the first 82nd Airborne Division ready brigade from Fort Bragg, North Carolina, was establishing a defensive perimeter around the airport at Dhahran.

The overriding fear was that the Iraqis would quickly move south from Kuwait after pausing to regroup. Early arrivals among the ground troops feared they might be little more than "speed bumps" in the path of the Iraqi advance. But the Iraqis continued their pause and the Coalition buildup quickly gathered momentum, closing the "window of vulnerability" without incident.

The rapid deployment of massive air, land, and sea forces to the Gulf region was a logistical feat without parallel. Airlift forces transported the most urgently needed personnel, equipment, and supplies on what came to be called an "aluminum bridge" from the United States and some European locations to Saudi Arabia. Other heavier equipment stored aboard prepositioned ships at the Indian Ocean island of Diego Garcia quickly arrived at Saudi ports. The bulk of the heavy armored equipment arrived later via a massive sealift from the United States and Europe.

At sea, Coalition naval forces clamped a tight blockade on Iraqi shipping to enforce the UN-imposed economic sanctions. While politicians and diplomats argued over how long the sanctions should be given to force an Iraqi withdrawal, General Schwarzkopf made good use of the time by training his forces in the harsh desert environment, designing the complex command and control arrangements, planning for an offensive campaign, and addressing problems caused by the clash of Western and Middle Eastern cultures.

Command and control of such a diverse coalition presented exceptional challenges. Political and cultural differences became of extreme importance in holding the fragile Coalition together. Eventually a dual chain of command evolved with General Schwarzkopf controlling the non-Arab forces, and the Saudi commander, Gen. Khaled bin-Sultan, controlling the forces from the Arab states. Although General Schwarzkopf and his Saudi counterpart sat as technically equal partners in command, the U.S. commander was clearly the senior partner. Fortunately, the war went well and the strains of conflict never put this delicately devised structure to a severe test.

In many respects, General Schwarzkopf played a role similar to that of General Eisenhower during World War II. As supreme allied commander in the European theater, Eisenhower became a de facto diplomat, holding the Allies together, at times with considerable difficulty. In some ways, General Schwarzkopf's job was more difficult because of the clash of Western and Arab cultures.

The culture clash manifested itself in many ways. Western female military personnel essentially offended Muslim traditions by their very presence. Christian religious services had to be downplayed or concealed. Alcohol was forbidden. In spite of these problems, compromise, coordination, and understanding minimized difficulties and the Coalition endured.

Israel posed the most serious potential threat to the Coalition. If for whatever reason the Israelis joined the attack on Iraq, the Coalition might quickly disintegrate. To many Arab states in the Coalition, Israel remained the supernumerary enemy even in the face of Iraqi aggression. The Coalition expected the Iraqis to provoke the Israelis, particularly if the war went badly. The tool of provocation would likely be Iraqi Scud missiles, which had sufficient range to strike Israel. Too inaccurate to be an effective military weapon, the Scud could still cause significant casualties when launched against population centers. It was a terror weapon, but useful for Saddam's purposes.

Iraq ignored the UN ultimatum to withdraw from Kuwait by 15 January 1991. Live television announced the commencement of Desert Storm to the world from Baghdad at 3 A.M. (Baghdad local) on 17 January. A CNN news team, broadcasting via satellite, displayed fantastic pictures of the night sky over the Iraqi capital filled with antiaircraft fire. Impressive as the antiaircraft appeared to be, it was almost random—no central direction, no good targets. The Iraqi air defense radars and control centers had already been destroyed and the aircraft overhead were F-117 stealth fighter-bombers—virtually invisible to any remaining Iraqi detection systems. Iraq had lost the air war in the first minutes of the struggle. CNN and the other electronic media were broadcasting the beginning of the most successful air campaign in history.

Spectacular as the television pictures were, they did not reveal the crushing results achieved by Coalition land and naval air forces during the first three phases of Desert Storm. Strategic targets were destroyed quickly. It rapidly became very difficult for Saddam to communicate with and effectively control his deployed forces. Known military production facilities, particularly those associated with nuclear, biological, and chemical weapons, were hit hard. Air superiority was achieved within hours and total air supremacy within days.

The Coalition achieved the objectives of the air campaign so rapidly that the second and third phases of the campaign began quickly. Rather than a sequential execution of the campaign, Coalition airmen were able to execute the phases almost in parallel, magnifying the shock effect on the Iraqi war machine. Overwhelming amounts of Coalition air power enhanced by the efficiency of precision-guided munitions made parallel execution possible. Careful planning, superior technology, and clever tactics kept Coalition losses far below what many military pundits had projected. During all four phases of the war, Coalition air forces lost only 39 fixed-wing aircraft in combat plus eight in non-combat operations while flying well over 100,000 sorties.

As expected, the Iraqis quickly began launching Scud ballistic missiles toward Israel in the hope of provoking an Israeli reaction that would strain and perhaps divide the Coalition. The Israelis refrained from entering the conflict, and the Coalition diverted considerable air resources to "Scud hunting" in an effort to end the threat. Although mobile Scud launchers proved to be elusive targets, the Iraqi missile offensive proved to be of no significant consequence.

As the air campaign continued during January and February, the full weight of the air offensive came crashing down on the heads of the deployed Iraqi forces. The aerial onslaught pinned the Iraqi forces down, isolated them, and cut them off from food, water, munitions, and fuel. As the date set for Phase IV drew closer, these troops came under more severe and nearly constant attack. For the Iraqis, if there was an aircraft overhead, it belonged to the enemy. And there were enemy aircraft overhead around the clock. General Schwarzkopf estimated that the air campaign reduced most of the Iraqi forces near the Saudi border to 50 percent effectiveness or less. Further, it had significantly damaged the Iraqi fortifications and mine fields behind the Kuwaiti-Saudi border. The battlefield was well prepared.

Meanwhile, as the air campaign blinded and pinned down the Iraqis, General Schwarzkopf moved his heavy striking forces far to the west, away from the forti-fied Kuwaiti border area and to the nearly undefended Iraqi border. By 24 February all was ready. The air war had stunned, isolated, and devastated the Iraqi forces. Coalition ground forces moved into position for the great "left hook" without being detected.

At 4 A.M. on 24 February, the ground war began. Massive Coalition armored forces swept over the Iraqi border toward the Euphrates River and Basra. The advance was spectacularly rapid. Coalition forces brushed aside limited resistance as they raced to close the trap. To the east, Coalition forces slashed into the teeth of the Iraqi defenses on the Kuwaiti border and breached them with remarkable ease. They raced toward Kuwait City.

The air campaign and the shocking tactics of the fast-moving ground forces had done their job. The Iraqi forces, particularly those deployed far forward, were far more interested in surrendering than in fighting. Dazed, hungry, thirsty, and often deserted by their leaders, they surrendered in such numbers that they were difficult to handle and get to the rear efficiently.

In 100 hours, it was all over. At least one Republican Guard formation, still willing to fight, attempted to slug it out with the Coalition forces, but was manhandled by the better-equipped, -led, and -trained Coalition armored forces. As other Iraqi forces attempted to flee Kuwait, they were under continual assault. Reports of carnage on what was dubbed the "highway of death" leading away from Kuwait City brought pressure in Washington to end what appeared to be a wholesale slaughter of the Iraqis. The result was a cease-fire at the 100-hour mark of the ground offensive. Significant Iraqi forces escaped complete destruction, but they were insufficient to maintain Iraq as any sort of regional threat for the foreseeable future.

In the war's aftermath, some criticized the termination of the conflict before the Iraqi armed forces were totally crushed. The Iraqi forces that remained effective were no threat to Iraq's neighbors, but were sufficient to quell both Shiite and Kurdish rebellions in Iraq and keep Saddam in power. Although the president seems to have chosen the 100-hour mark for a cease-fire because it had a "ring" to it, the decision was far from frivolous. The UN objective had been accomplished. Kuwait was liberated. The remaining Iraqi forces were no longer a regional threat, and continuing the war would have risked the lives of more Americans for what appeared to be only marginal advantages.

Better State of the Peace

Whether the better state of the peace was achieved in the Persian Gulf War depends on what that better state was defined to have been before the war began. Here again, the answer seems to be that the initial objective—freeing Kuwait and reinstalling the al-Sabah regime—was clearly realized, meaning the better state of the peace was clearly and overwhelmingly achieved.

The situation has a striking resemblance to the Korean War in that respect. In Korea, the objective of a better state of the peace defined as restoring South

Korean independence was initially acceptable and easily achieved. Only when it was expanded to unifying the Korean peninsula and that objective was frustrated did the original objective lose its persuasiveness as a worthy and achieved end. In Desert Storm, some of the same occurred. Evicting Iraq was an acceptable objective until it proved easy and the road to Baghdad seemed relatively unimpeded; then, the advocacy of "finishing the deal" by overthrowing Saddam Hussein entered the mix of objectives. That expansion was, of course, rejected, so the Korean analogy does not apply exactly. Knowing there was a prospect of attaining more, however, left a sour taste in the mouths of some.

As a result, the Gulf War in retrospect seemed less than triumphant to many in the American public for two reasons. The first and most obvious was the continuation in power of a notably unrepentant Saddam Hussein in Baghdad. Although removing Saddam was not an objective as the war effort was being mounted, that limitation was never publicly stated at the time. Moreover, George H.W. Bush's demonization of the Iraqi leader left the implication that his removal was indeed desirable, a point reinforced by Bush's call to the Iraqi people to rise and overthrow Saddam on 15 February 1991.

This latter declaration created the second, and ultimately most sour failure of the war effort. When Bush called on the Iraqis to rise, he presumably was exhorting internal Sunni Baathist elements within Saddam's cohort base to rise against him. They did not and realistically could not have been expected to do so; any dissidents had either been exiled or executed by Saddam. Instead, the Kurds in the north and the Shiites in the south rose to arms.

The Bush administration wanted neither to succeed, because it feared an Iraq fractured into three countries as the result would not provide an adequate counterweight to Iran. Neither revolt succeeded, and because some of the Iraqi military was left intact at the end of hostilities, it could be employed to crush the rebels.

The U.S. responded with Operations Provide Comfort (renamed Northern Watch in 1997) for the Kurds and Southern Watch for the Shiites. By the terms of these operations, the Iraqi government was prohibited from exercising authority over enclaves of these groups in sovereign Iraqi territory. The agent of enforcement was the American military, augmented by the British and French.

This part of the better state of the peace was apparently not anticipated when the war began, but it is important for two reasons, each of which tarnishes the success of Desert Storm. First, it created the precedent that infringement on territorial sovereignty in a good cause is justified, a direct repudiation of the organization of the international system since the Peace of Westphalia in 1648. There is no public evidence that the Bush administration even considered this when it organized the operations. Second, it created a perpetual obligation for the United States. Put simply, the Kurds and Shiites were safe from Saddam's revenge for as long, and only as long, as the Americans are willing to enforce the violation of Iraqi sovereignty. This obligation remained until the United States invaded Iraq and overthrew its government in 2003.

9

AFGHANISTAN WAR

Both why and how war is fought have changed in the new century. If the 1990s was a decade of small and limited military involvement for the United States and others, the use of force has become broader in the 2000s. Two events—the end of the cold war (including the demise of the Soviet Union) and the rise of international religious terrorism exemplified by the 11 September 2001 attacks against the United States—have been the most visible symbols of change. Each event was distinct and different, but both contributed to an increased militarization of the American view of dealing with the world. The wars in Afghanistan and Iraq are the result.

The end of the cold war changed the power balance in the world fundamentally. During the cold war, the military power of the United States and its allies was counterbalanced by the military might of the Soviet Union and its allies. That balance, accentuated by the civilization-threatening prospect of nuclear weapons, was the world's most important military reality and created restraint and caution in the contemplation and actual use of military power.

When half the balance disappeared in the early 1990s, thinking and planning about the dynamics of military power changed as well. American military power was no longer checkmated by countervailing Soviet military power. Instead, the United States was the "sole remaining superpower," the center of a "unilateral moment" in history, even the mightiest world power since the Roman Empire. The implications were intoxicating to some, suggesting the ability and utility of American armed force to mold the world in America's image. Fresh from the exhilarating success of American arms in liberating Kuwait (see Chapter 8), perceptions of that utility underpinned military activism in the form of humanitarian involvements in the Balkans and elsewhere during the remainder of the 1990s. American power seemed inexorable in its reach for righting situations around the world in America's favor.

The terrorist attacks of 9/11 created a compelling rationale for applying that force to a direct threat to vital American national interests. The attacks produced a righteous rage among Americans (and others worldwide) against the perpetrators—Al Qaeda and its leader, Osama bin Laden. The result was the defiant declaration of a "global war on terrorism" (GWOT). Afghanistan and Iraq have been that war's principal theaters.

U.S. military actions in the early twenty-first century can be only fully understood and appreciated by referring to both influences. The successes of the 1990s predisposed leaders in the United States toward a belief in the efficacy and efficiency of the military instrument to solve its problems: American military forces could sweep aside the Al Qaeda-shielding Taliban regime in Afghanistan, and the "shock and awe" power of the Americans could produce "regime change" in Iraq. American military predominance not only made these goals achievable, but applying force would be reasonably easy and painless. These judgments have proven overly simplistic and ultimately false.

It all began in Afghanistan, where the Taliban (the term means "students," in reference to the followers of Mullah Mohammed Omar) refused to relinquish the Al Qaeda perpetrators of 9/11. An understandable sense of vindictiveness guided the American action to topple the Taliban and thus wipe away that barrier to getting to bin Laden; dealing with the nuances and difficulties of a post-Taliban Afghan state were largely ignored in the process. In Iraq, misleading and ultimately false representations of the ongoing threat posed to the United States by the Saddam Hussein regime overwhelmed sober predictions about the maelstrom into which the United States was plunging with the ebullient expectation of a brief, painless, and decisive outcome.

This chapter examines the Afghanistan example of these dynamics; Iraq is the subject of Chapter 10. The two experiences are and have been very different, because they have occurred in very different countries that have posed unique and intractable barriers and problems. In both cases, the United States either did not fully understand the unique difficulties that imposing its military might on these countries would create, and, particularly in the case of Afghanistan, there is not overwhelming evidence to indicate it understands those problems today. Iraq is working its way toward an ending that will be described as successful only by the more generous evaluators; the prospects in Afghanistan are, if anything, less optimistic.

Afghanistan is an ancient country that has been a point of contention for centuries, largely due to its strategic location along what the U.S. government refers to as the "land bridge" for trade between eastern Asia and the West. As well, north-south routes from the Middle East to the Asian subcontinent traverse what is now Afghanistan; among those who passed through the area and left a mark was, for instance, Alexander the Great.

What is now Afghanistan first achieved independence as a country in 1747 when Ahmad Shah Durrani consolidated the Afghan Empire, which encompassed modern Afghanistan and neighboring areas in what is now Pakistan, stretching as far as the Persian Gulf. Both before and after the emergence of Afghanistan as an identifiable entity, the country had been the victim of numerous invasions, and during the nineteenth century, it was the subject of the so-called "Great Game" between the British and Russian empires, for which it acted as a buffer. During the nineteenth century alone, there were two inconclusive Anglo-Afghan Wars. A third Anglo-Afghan War ended in 1919 with the full assertion of Afghan independence, a condition it has not relinquished since. The Russians (as the Soviet Union) invaded Afghanistan in 1979 and for a time occupied the country, but withdrew in 1989. A common theme from this history has been one of invasion and temporary conquest of Afghanistan followed by the repulsion of the invaders by the fierce tribes that are the major political units of Afghanistan. Generally, the tribal group that has led these repulsions of foreign invaders has been the Pashtuns, some of whom form the backbone of the current opposition to the Afghan government and the NATO occupation. No foreign country has successfully gained control of and ruled Afghanistan for a lengthy period, especially in the modern age.

Afghanistan has been and continues to be a country of antagonistic ethnic groups organized into tribal associations. The largest and historically most dominant of these groups is the Pashtuns. When the term Afghanistan first began to be used, it was not uncommon for the descriptors Pashtun (also sometimes called Pashtoons or Pathans, among other spellings) and Afghan to be used interchangeably. Today, the Pashtuns are the largest ethnic group in the country with around 40 percent of the population, followed by Tajiks, Uzbeks, and others. Before the Soviet invasion of the country in 1979, they represented a majority in the population, but since nearly 85 percent of the estimated 6.2 million Afghans who fled the country during the Soviet occupation were Pashtuns, their majority disappeared. The Pashtuns live throughout Afghanistan but are concentrated in the southern and eastern parts of the country, are mostly rural dwellers, and are very fiercely independent and resistant to outside presence and influence.

Straddling the boundary between Afghanistan and Pakistan, the Pashtuns are the largest of Pakistan's minority ethnic groups and the main opposition to the Punjabi majority, which, with few exceptions, has ruled Pakistan since it gained independence in 1947.

There are about 41 million Pashtuns living in Afghanistan and Pakistan. They are the majority population of the territories on both sides of the border, including the so-called Federally Administered Tribal Areas (FATA) of Pakistan, where Al Qaeda has found a haven. Indeed, the Pashtuns do not accept the boundary between the two countries as legitimate. It is based on something called the Durand Line, which was an artificial border concocted in 1897 by a British colonial official (Sir Mortimer Durand) to separate Afghanistan from the British Raj. Because it cuts across historic Pashtun lands, it frustrates the desire of many Pashtuns to create their own homeland (Pashtunistan), a desire not unlike that of the Kurds to create a Kurdistan out of parts of Iraq, Turkey, Iran, and Syria. At the time of the partition of the Asian subcontinent in 1947, Pashtuns in what became Pakistan desired but were denied the option of becoming independent or of joining in the formation of Pashtunistan. In this respect, the fate of the Pashtuns is similar to that of the Kashmiris on the border between Pakistan and India. At any rate, it is resistance in the Pashtun tribal lands on both sides of the border that forms the greatest barrier to the American effort in Afghanistan.

One other characteristic of the Pashtuns should be mentioned: their traditionalism and resistance to change, a characteristic shared by many other groups in the country. In fact, Afghanistan experienced an attempt at modernization not unlike the abortive White Revolution that occurred in Iran between 1954 and the fall of the Shah of Iran in 1979. The attempt at reform and modernization in Afghanistan began in 1919 with the ascension of Amanullah to the Afghan throne and lasted for about a decade. The general theme of his reforms was to adopt the "best" of western ideas and to incorporate them within an Islamic, Afghan context. Some of the reforms he attempted were western in content (e.g., education and equal treatment for women) and were offensive to many traditional—and especially rural—Afghans, including the rural Pashtuns in their tribal areas, and were reversed after Amanullah lost power. The reforms attempted by Amanullah, among other things, created a rift between the Afghan countryside and the urban educated elite that remains a potent factor in Afghan politics. This is worth noting because the Taliban, a principal American opponent in Afghanistan, is composed largely of Pashtuns, although by no means do all, or even a majority, of Pashtuns support the Taliban. The rural-urban split also helps explain why there is not great Pashtun support for the president of Afghanistan, Hamid Karzai, who is a respected former Pashtun warlord but who is widely considered a part of the urbanized elite who has betrayed his heritage by coalescing with non-Pashtun minorities to the perceived disadvantage of his fellow Pashtuns. Indeed, understanding the Amanullah period (1919–1929) may be instructive in understanding the current situation; there are parallels between Amanullah and Karzai and between the forces that overthrew Amanullah and the Taliban.

On the American side, interest and involvement in Afghanistan was minimal until relatively recent times. During the 1950s, the United States made a modest investment in economic assistance to the country, but its imprint was neither large nor lasting. Two events in 1979—the Iranian Revolution and the Soviet invasion of Afghanistan—changed that situation, however, and helped to create the setting for the current American involvement in that country.

The Iranian Revolution was the first hammer blow that entangled the United States in the region. Prior to 1979, the direct American presence, especially with military forces, was minimal. The only real American interest there was (and arguably still is) in secure access to Persian Gulf oil, and the Iranian government of Shah Reza Pahlevi served as a surrogate for the United States in protecting that interest. This was accomplished through an arrangement by which the United States provided economic and military assistance and advice to the shah's regime in return for the dedication of some of Iran's military might to guaranteeing that oil continued to flow to the West—and notably the United States. With access to oil so assured, American dependence on Persian Gulf oil grew during the period leading to 1979 and after.

This bargain provided short-term benefits but long-term negative consequences. American assistance was vital to the Shah's White Revolution, an attempt to modernize and westernize the country as a way to reassert the power of the Persian (Peacock) Empire in the region. The negative side effect was to create a strong reaction among the conservative, largely rural, and fundamentalist Islamic masses in Iran that, when mobilized by the Shiite clergy, led to the Iranian Revolution. One of the chief effects of the revolution was to replace the pro-western regime of the shah with a fervently anti-American theocracy that remains in power today. Another was to lead to the breakup of the American-assisted military and to replace it with a pro-revolutionary force quite opposed to U.S. interests.

In a geopolitical sense, the major impact of the Iranian Revolution was to end that country's role as the American military surrogate in the region. After the revolution, there was no regional military force capable of or willing to guarantee American access to Persian Gulf oil. Instead, the United States would have to shoulder that burden itself. This change was recognized when President Jimmy Carter asserted what became known as the Carter Doctrine in his 1980 State of the Union Address; in it, he declared that free access to Persian Gulf oil was an American vital interest that we would defend with military force. Operationally, the change resulted in the formation of an American Rapid Deployment Force (RDF), which has become the Central Command with military responsibility for the Persian Gulf (including Iraq and Afghanistan).

The Soviets invaded Afghanistan in December 1979, adding a sense of urgency to the new geopolitics of the region. Their apparent intent was to prop up communist rule in Afghanistan, which had been imposed by Afghan communists in 1978 but was faltering amid in-fighting between factions within the Afghan Communist Party. The invasion was also almost certainly motivated by a desire to bring Afghanistan into the Soviet orbit, a goal that went back at least to the nineteenth-century Russian-British competition for influence as part of the "Great Game." From a western viewpoint, the most ominous prospect was that if the Soviets gained secure control of Afghanistan, they would be within 300 miles of the Persian Gulf and potentially in a position to interrupt oil flowing through crucial naval "choke points" like the Gulf of Hormuz.

The Soviet occupation of Afghanistan, of course, turned out to be a disaster that helped contribute to the demise of the Soviet Union. The Afghans resisted their rule, as they always have opposed invaders. That resistance came in the form of loosely coordinated tribally-based forces augmented by foreign Islamic volunteers from throughout the Muslim world. The resulting movement of so-called *mujahidin* gradually sapped Soviet resources and will, especially when the United States came to the aid of the Afghan fighters by providing monetary assistance and weaponry (notably *Stinger* missiles that were used with great effect against Soviet helicopters). Like the British and others before them, the Soviets were demoralized by the experience and were forced to withdraw in 1989; their failure and the disillusionment of their returning veterans helped seal the fate of Soviet communism.

The experience also enmeshed the United States in Afghan affairs in ways for which the country is still paying. American assistance to the *mujahidin* had always been instrumental: the goal was to create problems for the Soviets rather than to secure any particular American interest in Afghanistan. When the Soviets left in 1989, the United States rapidly terminated its assistance program and left Afghanistan to fend for itself (a formal agreement to this end was signed by the Soviets and the Americans in 1991). What ensued was a period of instability and civil war among competing *mujahidin* factions that continued until 1996.

Two legacies of American assistance to the *mujahidin* that continue to haunt us are the Taliban and Al Qaeda. Both were products of the war against the Soviets: the Taliban emerged from Pashtun elements in the *mujahidin*, and Al Qaeda arose from foreign fighters who had been part of the resistance. In 1996, the Taliban seized power in Afghanistan by winning the internal civil war and overthrowing an extremely corrupt government. Not long after they came to power, bin Laden and his cohorts came to Afghanistan, where they were embraced by Mullah Omar and their former *mujahidin* Taliban allies, an embrace strengthened by a shared fundamentalist Islamic ideology. This embrace included the extension of the Pashtun code of hospitality to Al Qaeda, a reason the Taliban refused to turn over their "guests" to the United States after 9/11.

Issues and Events

The clash between the United States and Afghanistan was not, like many of America's other conflicts, the result of long simmering or gradually accumulating underlying causes or issues. As already noted, there was little history of U.S.-Afghan relations of any kind prior to the twenty-first century, and what existed generally was of a lower order of magnitude or derivative in its nature. The United States had briefly participated in aid programs to Afghanistan in the 1950s, and the Soviet invasion had activated an American counterresponse in the form of assistance to the *mujahidin,* but that provision of assistance was motivated out of anti-Soviet rather than pro-Afghan sentiments and calculations. Once the United States helped negotiate the terms of the Russian withdrawal in 1989, Afghanistan returned to its accustomed place far down the list of U.S. interests and concerns.

Within the U.S. government itself, there was a growing awareness of Afghanistan during the 1990s as that country experienced growing instability and civil war, from which the Pashtun-dominated Taliban emerged victorious. That concern, however, was not deemed particularly threatening to the United States, but more as an anachronistic sideshow as the fundamentalist Taliban imposed an increasingly tight, repressive hold on Afghan society. As Afghan society plunged back toward traditional rule and values, it seemed and was an anomaly in a globalizing world economy and political environment.

Taliban rule became more than a curiosity when Al Qaeda took up residence in the country and, among other things, began appropriating former *mujahidin* camps and training facilities (some of which had been financed by the CIA during the 1980s) for the training of its terrorist cadres. Al Qaeda, which had its shadowy beginnings somewhere around 1988, had a symbiotic relationship with the Taliban building upon their alliance against the Soviet occupation and shared ideology. When it began its campaign against the United States in 1993 with its first, abortive attack on the World Trade Center in lower Manhattan, the anti-terrorist community within the U.S. government began to take notice and perceive a threat. As terrorist incidents against Americans overseas increased during the 1990s (the August 1998 attacks against the U.S. embassies in Dar es Salaam, Tanzania, and Nairobi, Kenya and the attack against the USS *Cole* in a Yemeni port in October 2000, for instance), concern with Al Qaeda and their apparent safe haven in Afghanistan occupied progressively more attention among anti-terrorism experts.

Most of this concern, however, was not passed along forcefully to the American people until the awful terrorist attacks of 11 September 2001 became the event that would impel the United States and Afghanistan into conflict. While it may have been hyperbolic to suggest, as it was at the time, that the attacks "changed everything," they certainly did rivet attention on Afghanistan because of the association of that country with the perpetrators of the 9/11 disaster.

The initial American response to the attacks was to seek revenge against Al Qaeda, which was quickly identified as the perpetrators of the atrocity. Their location in Afghanistan since 1996, when they were thrown out of the Sudan at the request of the U.S. government, was known and was being monitored. Although Afghanistan was embroiled in a civil war between the Pashtun-dominated Taliban and a Northern Coalition comprised mostly of minority ethnic groups opposed to the Taliban, Al Qaeda was harbored in the eastern and southern parts of Pashtun-dominant Afghanistan that were controlled by the Taliban. Thus, a request to apprehend the Al Qaeda leadership and turn it over to international or U.S. authorities was a request that could be made without a direct impact on the civil war.

Mullah Omar, of course, refused the American request on the grounds that there was not sufficient evidence that Al Qaeda was behind the attacks. Underlying this rationale (it is unclear whether the Taliban believed in Al Qaeda's innocence) was the Pashtun code of hospitality, which militated against turning on their guests and which acts as a barrier to Pashtun acquiescence in or cooperation with efforts to dislodge Al Qaeda from tribal lands to this day. The Taliban refusal to extradite bin Laden and his cohorts, however, placed the United States and the Taliban government of Afghanistan on a collision course. The result was the U.S. intervention in Afghanistan's civil war, which led to the overthrow of the Taliban in November 2001 by the Northern Coalition, and the installation of Hamid Karzai in December 2001. In January 2002, the first contingent of foreign "peacekeepers" entered

Afghanistan, and in May of that year, the United Nations extended the mission of the International Security Assistance Force (ISAF) to December 2002. American participation in and leadership of ISAF meant the United States was a partner in the effort to ensure the Taliban did not return to power. When it became clear that the Taliban was neither destroyed nor had accepted their exclusion from rule, the United States became part of the ongoing civil war—which it characterizes as an insurgency—a conflict that has effectively become an ethnic civil conflict between the Pashtun Taliban and the anti-Taliban (and to some extent anti-Pashtun) government.

Political Objective

When the United States first contemplated involvement in Afghanistan in 2001, its political objective was clear and virtually universally accepted both within the United States and the international community. That objective was the eradication of international religious terrorism and meant the capture, punishment, and elimination of the Al Qaeda terrorist organization (and especially its leadership) that had organized and committed the attacks against New York and Washington, D.C. Initially, it was a limited, discrete, and achievable objective that met the requisites for a "good" political objective with a level of clarity unusual in contemporary politico-military situations. Indeed, assaulting Al Qaeda may have been the clearest and most universally supported political objective for which the United States has fought since the decision to destroy the fascist states in World War II.

The goal was limited because it did not represent an assault against or need to overthrow any government, since Al Qaeda itself was (and is) a subnational political actor that does not represent any state. That characteristic of the objective changed, of course, when the Taliban refused to cooperate in the apprehension of the terrorists, because attaining that objective therefore entailed interaction with the government of Afghanistan. Originally, there was a discrete objective in that it was bounded and focused on the single purpose of isolating and disposing of a single entity, which, at the time, was largely congregated in one geographical location (unlike its current worldwide presence). With the cooperation of the Taliban government, or the provision of adequate dedicated force when that cooperation was not forthcoming, it was (or at least was believed at the time to be) an achievable goal: capturing and destroying Al Qaeda in the Afghan mountains would have dealt a lethal blow to international religious terrorism. Moreover, it served both of the functions of a political objective: it provided the basis for attaining public support for the effort, and it provided initial guidance for how to achieve it.

This political objective enjoyed and, in large measure, continues to serve the purpose of galvanizing political support, because it met all the criteria of a good political objective identified in the introductory chapter. Destroying Al Qaeda was a simple and straightforward goal that could serve as an unambiguous rallying cry that everyone understood. The atrocious, vicious nature of the attacks meant the attackers were easy to portray as immoral and reprehensible, making their elimination a morally lofty endeavor. Because these actions might represent the tip of the iceberg of a continuing campaign that threatened American lives elsewhere, the elimination of the opponent was vital to American interests (even a survival interest) and, since these attacks could occur anywhere, it was a goal that was important to all Americans. Destroying bin Laden and Al Qaeda represented a clear and compelling moral imperative not witnessed since the destruction of Hitler and the Nazis over a half-century earlier.

The problem, however, came in translating the political objective's clarity into guidance for achieving it, a difficulty discussed in greater detail in the next section. Had the Taliban either captured and turned over Al Qaeda themselves or acquiesced in allowing the Americans to do so, this would not have been a problem, but they did not. As a result, implementing the political objective meant devising a way to compete with and defeat an enemy in a hostile and difficult environment. It also meant enlarging the political objective beyond that of simply destroying Al Qaeda.

Two dynamics, of course, strongly influenced this change in objectives. The first and most obvious was the Taliban refusal to cooperate in dealing with Al Qaeda, which made them an obstacle that had to be dealt with before the main objective could be achieved. The second was the conceptual widening of the campaign against international religious terrorists to include Iraq and the subsequent diversion of both intellectual energies and resources away from Al Qaeda in Afghanistan and Pakistan (essentially Pashtunistan). One result of these dynamics was to contribute to the failure to achieve the original objective swiftly and decisively, when Al Qaeda eluded the American effort to trap and squelch it in the Tora Bora Mountains of eastern Afghanistan in the winter of 2001–02. In trying to carry out that operation, the United States became enmeshed in deciding the outcome of the civil war in Afghanistan that was being waged in 2001 between the Taliban and the Northern Coalition.

The decision to overthrow the Taliban became the second political objective in Afghanistan. Its accomplishment was logically prior to attacking Al Qaeda, since the Taliban government provided the shield behind which Al Qaeda hid. In addition to overcoming that obstacle to the primary goal, the forceful removal of the Taliban from power also was intended to serve a second function, which was to issue an unambiguous warning to any other government that might harbor terrorists: such a course of action would be met with a fierce American military response that the government attempting the protection would not survive.

By coming to the aid of those who sought to overthrow the Taliban and physically assisting in that overthrow, the United States developed another political objective in Afghanistan that continues to this day: support for an Afghan government that rejects Taliban rule. Operationally, this second objective has come to be defined in terms of creating a stable, preferably democratic Afghanistan that both provides a bastion against Taliban resurgence and, more recently, provides assistance in combating the production and distribution of narcotics from Afghanistan. This second objective has operationally superseded the first (destroying Al Qaeda) in importance, since it is generally believed that a stable Afghanistan is a necessary prerequisite to isolating and destroying Al Qaeda by depriving it of a compatible staging area. Additionally, a stable and prosperous Afghanistan would also demonstrate the virtue of cooperating with the Americans in their war on terrorism.

Because of the nature of the Afghan government and the resurgent Taliban movement, this third political objective has effectively made the United States the major opponent of the cause of many Pashtuns in Afghanistan and, by extension, Pakistan, a goal which it never intended and which places it at cross-purposes with Afghan history. It also introduces a source of confusion into perceptions about what the American objective is, whether it should be supported, and whether it is feasible.

The political objective of a stable, democratic Afghanistan, while noble in the abstract, does not so obviously meet the laudatory characteristics of the anti-terrorism objective. As a goal, it is an arguably total objective, since it requires the utter defeat of the largest ethnic group in a country where ethnic divisions are the most basic datum of power. It is an open-ended rather than discrete objective in that it is not at all clear that the Pashtun nationalism and religious conservatism on which the Taliban feed could easily or permanently be overcome (it has not been throughout Afghan history). It is thus an objective the feasibility of which is open to considerable question, and it is not clear that it provides either a clarion call for American public support or a blueprint to guide military achievement. The critical point here is the extent to which Pashtun opposition to the Taliban can be exploited. As is the case in these situations, virtually all the Taliban may be Pashtuns, but not all Pashtuns are Taliban. A successful counterinsurgency requires that there be enough opposition to the Taliban among Pashtuns to make it possible to isolate and overcome the Taliban.

Moreover, the goal of a stable Afghanistan does not obviously meet the criteria of a good political objective. While it can be stated simply and unambiguously (prevent the pro-Al Qaeda Taliban from regaining power), it is a much more complex goal. What, for instance, constitutes the rule of law and democracy in an ethnically divided Afghanistan? Must the representatives of the Pashtuns, since they constitute over 40 percent of the population, be included at proportional levels? Does that entail somehow incorporating the Taliban as at least partial representatives of the Pashtuns? The announcement in October 2008 that the Afghan government and NATO would pursue talks with "moderate" Taliban leaders seems to acknowledge that possibility. If the Taliban/Pashtuns are to be included in the government, should the United States continue to try to subdue them militarily? Also, the Pashtuns have grievances against the government because of the government's alleged mistreatment of Pashtuns from the northern and western parts of the country who have fled to the south and east, bringing tales of horror with them. How does the U.S.-led coalition deal with this impediment to a politically inclusive stable Afghan democracy?

While the goal of a stable, democratic Afghanistan sounds morally lofty, the steps that might have to be taken to achieve it could tarnish some of the appearance of moral clarity. Like Iraq, Afghan society is, as noted, a very diverse, complicated patchwork of ethnic and tribal loyalties. The kind of strong central government the United States hopes will emerge would be a notable exception in Afghan history and almost certainly require substantial tradeoffs and compromises that would dull the luster of the entire enterprise. The evolution of the process of democratization in Iraq offers some harbinger of the fate of the similar experiment in Afghanistan. Moreover, much of the opposition on which the Taliban feeds comes from the extraordinarily corrupt nature of the current Afghan government and the resulting opposition to it.

It is also not entirely clear that the emergence of a viable and stable democracy in Afghanistan is an important interest of the United States. The only real objective interest the United States has in the country is the eradication of international religious terrorism. Beyond that, are Americans affected (or do they perceive themselves to be affected) sufficiently to justify what is being done there? If the emergence of a stable Afghanistan is the *sine qua non* for cradicating Al Qaeda, then it may be worth it. But what if it is not? It is at least arguable that a democratic Afghanistan would not necessarily lead to the eradication of Al Qaeda (they can still operate out of Pakistan, for instance) and that the pursuit of Al Qaeda would not necessarily be any easier in a democratic versus an authoritarian Afghanistan.

The case of the Afghanistan National Army (ANA) reflects these problems. The development of a national army is seen as a necessary means to restore order to the country, and one justification for continued NATO operations is to provide a shield behind which that army can be assembled, trained, and readied to stabilize the country. Developing such an internal stabilizing force, is central to the counterinsurgency approach General David Petraeus took in Iraq and has transferred to Afghanistan.

While hardly anyone opposes an enhanced ANA in principle, creating one is a different matter. As always, the devil is in the details. The first concern is the ethnic composition of the ANA. While official depictions are notably vague in this regard, a truly national Afghan army would presumably be ethnically representative of the country from which it comes. That would mean that a substantial portion would be Pashtun. Does the ANA reflect the ethnic diversity of the country? Official explanations are evasive on the subject (with apparent American acquiescence, the ethnic composition of the ANA was not reported after 2005), but there have been suggestions that the ANA is disproportionately non-Pashtun, with a heavy Tajik flavor. It is, in other words, more reflective of the Northern Alliance coalition that overthrew the Taliban/Pashtuns than the country as a whole (indeed, one Pashtun objection to the army is that they are underrepresented in it). Thus, a *truly* national Afghan army would have to have many more Pashtuns throughout its ranks if it were to be the kind of force that could bring peace to the country. As long as it looks like an anti-Pashtun coalition, it will probably translate into Pashtun opposition and will mean an ongoing Taliban problem.

Solving the ANA problem raises a conundrum. Afghanistan needs an ethnically-representative national army to combat an ethnically-based insurgency or civil war, but it cannot have such an army, because the ethnic opponents of the regime are not represented (or are underrepresented). At the same time, if it were possible to form a truly representative ANA, then all ethnic groups would be included and there would be less of a need for a national army, because the civil war would have less reason to continue. Since the first condition is what actually exists in Afghanistan, this leaves two options for those—especially outsiders—trying to influence the outcome. Either they can continue to support the development of an ethnically unbalanced ANA that is effectively the government side of a civil war, or it can insist on a cessation of the civil war to allow for inclusion of the outsiders into the ANA. The latter action requires diplomatic activity to create a cessation of hostilities that the government, which likely prefers fighting the insurgents to reaching power-sharing compromises with them, probably opposes.

The final problem with the political objectives is that they may be contradictory in application. Support for the Afghanistan adventure has been high because it is equated with the first objective of eliminating Al Qaeda, and support for the goal arises from the contribution that a stable Afghanistan would make to demonstrating the sensiblility of cooperating with the United States and destroying Al Qaeda. But the two objectives may not be complimentary due to the Pashtun factor. Given that Al Qaeda resides in largely Pashtun territories on both sides of the Durand Line, Pashtun assistance or at least acquiescence in finding and defeating Al Qaeda is almost certainly necessary—the implication is a pro-Pashtun policy. Unless the current Afghan government changes its views on wider Pashtun participation, however, efforts to bolster that government to stabilize Afghanistan will be at least implicitly anti-Pashtun, making more likely their continuing opposition to the first objective.

Conflating the defeat of Al Qaeda and the Taliban is thus a much more problematical set of goals than is generally suggested in discussions about what the United States is trying to accomplish in Afghanistan. This conceptual morass becomes even more difficult when one tries to sort out the political objectives and try to translate those into military objectives and strategies.

Military Objectives and Strategy

After the shocking 9/11 attacks in New York and Washington, DC, Osama bin Laden and his Al Qaeda organization were clearly the target of American retribution. The same was true of the Taliban leadership of Afghanistan, which had played host to Al Qaeda and its terrorist training camps since 1996. From those training camps Al Qaeda had grown to an international organization with operatives and organizations in many parts of Asia and Africa. In the fall of 1998, the United States had launched cruise missile attacks against the Al Qaeda training camps in Afghanistan in response to nearly simultaneous car bombings of the U.S. embassies in Dar es Salaam, Tanzania and Nairobi, Kenya. Both attacks were traced to local members of Al Qaeda. The limited response with standoff weapons proved to be little more than a timid gesture that failed either to destroy or deter Al Qaeda, or to weaken Taliban support for the terrorists. Moreover, in the wake of the 9/11 disaster, only much more severe measures would satisfy the popular demand for retribution. Further, the Bush administration declared that action against Al Qaeda and the Taliban in Afghanistan would be only the first step in a long, global struggle against terrorists worldwide.

With the foregoing in mind, the fundamental military objective quickly became obvious: the destruction of Al Qaeda in Afghanistan including its training facilities and arms caches, with particular emphasis on capturing or killing its leadership. Concurrently, the military action would be abetted by political/economic action to seize Al Qaeda assets or those of various front organizations around the world. In a very real sense the objective was to both starve and cut off the head of the terrorist "snake" in the hope that the disparate parts of the organization scattered across the globe would weaken, wither, and die.

A second and closely related objective for military forces, the destruction of the Taliban government, also came into focus when, on 21 September 2001, the Taliban government, which had so brazenly given succor to Al Qaeda and had openly supported Al Qaeda's terrorist training operation, refused U.S. demands that they hand over Osama bin Laden. Although destroying the Taliban government was not an absolute prerequisite for the destruction of Al Qaeda, there were at least three powerful reasons for doing so. First, the destruction of the Taliban government would gain the support of the Northern Alliance, a loose collection of non-Pashtun tribal warlords that was waging a civil war against the Taliban regime. Support from the Northern Alliance could make the pursuit and destruction of Al Qaeda a much more manageable task. Second, a strong, non-Taliban government in Kabul would make it much more difficult for residual Al Qaeda to refit and regroup in Afghanistan after U.S. forces departed. Third, the overthrow of the Taliban government—because they supported Al Qaeda—would send a strong and obvious message to other governments that supported or tolerated Al Qaeda and its terrorist operations.

Several military strategies might have achieved the two fundamental military objectives. The easiest strategy would have been to continue the ongoing covert operations seeking to hunt down and capture or kill Al Qaeda's leadership while perhaps increasing the intensity of the effort. Destroying the leadership would probably be a lengthy undertaking and there was no guarantee that picking off Al Qaeda leaders over a considerable period would result in the downfall of the entire organization or that equally effective leaders would not replace their fallen predecessors. Eliminating the leadership quickly would have more of a shock effect on the organization and thus might have a significantly better chance of bringing it down. Unfortunately, purely covert operations tend not to work swiftly in such situations, particularly when the adversary is on guard and taking steps to ensure that it is very difficult to eliminate any of its leaders, let alone all of them.

The biggest drawback to a purely covert strategy was political. The American people had suffered grievous injury at the hands of Al Qaeda in the 9/11 attacks and they expected decisive action. The Bush administration, which had gained office by the narrowest of electoral vote margins, had promised decisive action and needed to deliver on that promise. Further, the United States needed to show the leadership and strength expected of the world's only remaining superpower. Such leadership had the potential to rally friends and convince those who supported or who might consider supporting Al Qaeda that such actions would come with a very high price. Launching a few cruise missiles at Al Qaeda training camps as had been done after the African embassy bombings would not suffice. Nor would purely covert operations that are, by their nature, covert.

At the other end of the strategy spectrum was a conventional war waged in the style for which the U.S. military had been both trained and equipped. Such had been the strategy in Operation Desert Storm, the campaign to drive Iraq out of Kuwait in 1992. However, at least three factors argued against such a strategy. The first factor was the tyranny of political geography. Afghanistan is a land-locked country, the nearest major seaport being hundreds of miles to the south. Bringing in major combat units, particularly armor and artillery, would have been a long process and at the mercy of other states that might not favor a major influx of U.S. military forces into Southwest Asia. The same constraint held true for logistics support of a large mechanized force. Afghanistan is a logistician's nightmare.

A second compelling reason for not bringing in major U.S. conventional forces, particularly ground forces, was the tyranny of topography. The northern two-thirds of Afghanistan, the area that includes most of the major cities including Kabul, is dominated by high mountain ranges cut by deep, narrow valleys. More than 50 percent of all the land area of the country lies above 6,500 feet. Such is not the terrain for high-speed, armor-heavy, maneuver warfare as practiced by the U.S. military. The Soviets learned this lesson well in the 1980s.

A third reason for not bringing in major U.S. conventional forces was that the Taliban army was totally outclassed. It had been unable to secure control over a significant portion of northern Afghanistan, which remained under the Northern Alliance. Much of its success in other parts of the country can be attributed to payoffs and bribes of local commanders rather than military prowess. Much of its equipment was left over from the Soviet occupation and suffered from poor maintenance and questionable reliability. In a sense, going to the enormous effort and paying the enormous costs of bringing a significant conventional force to Afghanistan to fight the Taliban army would be massive overkill.

The military strategy finally settled upon was a creative hybrid that took advantage of the ongoing civil war between the Taliban and the Northern Alliance. Western high-tech prowess, particularly air and space power and precision-guided munitions, would be used in conjunction with Special Operations forces on the ground to help the Northern Alliance defeat the Taliban. As the rout of the Taliban proceeded, a limited number of regular units would then be brought in to spearhead the search for Al Qaeda in the mountains of northeastern Afghanistan. This strategy played to U.S. technological strengths, put the fewest number of American lives at risk, put the least strain on the U.S. military, and created a very small U.S. military "footprint" in Afghanistan. The weakness of the strategy was that it paid scant attention to the probability of post-combat strife that would be caused by a weak central government, rival warlords, the remnants of the defeated but still defiant Taliban, and the prospect that remaining Taliban might find a hospitable sanctuary among fellow Pashtuns in neighboring Pakistan.

In the prewar planning process, postwar military objectives in Afghanistan were, at best, an afterthought. As mentioned in the previous section, there was a political objective of stabilizing post-conflict Afghanistan through humanitarian and rebuilding efforts. The military role in achieving that objective was not spelled out with any precision although it would obviously be difficult, if not impossible, for humanitarian and rebuilding efforts to succeed without sufficient security.

In many respects the lack of prewar thinking about postwar military requirements in Afghanistan is not surprising. Military operations began in Afghanistan less than a month after the 9/11 attacks. During that short interval military commanders and their staffs were focused on detailed military plans and preparations and the movement of forces and equipment into position. Other parts of the administration concentrated their attention on such crucial tasks as obtaining allied support (Great Britain provided the largest amount of ground and air support for the campaign but many others participated as well); securing basing rights as close to Afghanistan as possible, most notably in Uzbekistan and Tajikistan; obtaining overflight rights for aircraft launching from the United States and various overseas locations, and so forth. Unfortunately, lack of sufficient prewar attention to and agreements on postwar objectives and strategies eventually resulted in significant problems after the fall of the Taliban government. The unfortunate result has been a costly, protracted struggle that continues as of this writing (early 2009) with no end in sight. Giving short shrift to post-combat military considerations (now often referred to as Phase IV operations by the U.S. military) would later plague efforts in Iraq, the next stop in the war on terrorism.

Political Considerations

American military involvement in Afghanistan began in October 2001 and has been continuous ever since. In October 2009, that involvement will have exceeded eight years of combat operations, making Afghanistan the longest American military adventure in U.S. history. Because of the length of the American participation in hostilities there, one would normally have expected some negative public reaction both to the duration of the action and to its seeming inconclusiveness. The United States has been in Afghanistan under the guises of the International Security Assistance Force and Operation Enduring Freedom for a long time and has little to show for it. Yet, opposition to the operation remains minimal and support, if not terribly overt, remains steadfast.

Afghanistan, in other words, appears to defy the normal American aversion to long—and especially indeterminate—wars. Much of the reason for this apparent anomaly arises from two unique aspects of the involvement, each of which has been raised previously. One is the identification of the war in Afghanistan with the broader campaign against international religious terrorism. This linkage is reinforced by the proximity of the American entrance into Afghanistan with the events of 11 September 2001, and the connection between the Taliban government and Al Qaeda. Despite the patent lack of success of the campaign to capture or kill bin Laden or to destroy his terrorist network, the goal of eradicating terror remains sufficiently powerful to provide a support base for continued support.

The other aspect is the juxtaposition of the Afghan campaign with the war in Iraq. Although these two conflicts have flimsily been connected through the slender tendril of terrorism, the major connection is that the Iraq War has eclipsed the Afghanistan adventure sufficiently to draw attention away from it. In one sense, this diversion of attention has had the negative effect of drawing resources away from Afghanistan that supporters of the Afghan campaign argue might have allowed for more decisively positive outcomes there than have occurred. In a more important sense, however, preoccupation with Iraq has provided a veil behind which the Afghanistan campaign has gone on without much public scrutiny. As almost all observers would agree, the evolution of the war in Afghanistan and perceptions of it would have been far different had there been no war in Iraq.

The lack of adverse domestic public reaction to Afghanistan has also partly been the result of public confusion over what exactly the United States seeks to accomplish in this conflict. To the extent most Americans can articulate a basis for their ongoing support of U.S. efforts, it is based in the desire to destroy bin Laden and Al Qaeda, the first and most consensually accepted basis for involvement. The desire to create a stable, democratic Afghanistan, a much more controversial and problematical goal, has mostly escaped public consideration, except as a step to create an anti-terrorism environment in postwar Afghanistan. Were Afghan stability the primary declaratory political objective of the United States, it would almost certainly have raised greater criticism than it has. Whatever lesson other potential enemies may have learned about the wrath of defying the United States has presumably already been learned from the fall of the Taliban. The possible success of the Taliban against the U.S.-backed regime in Kabul could, of course, cause future opponents to either modify or even unlearn those lessons.

What this suggests is that the effort in Afghanistan has received a comparative free ride in terms of American public support. The general determinants of support identified in Chapter 1 generally reinforce this conclusion. Until late in the presidential campaign of 2008, there was very little media scrutiny of Afghanistan, and there has been a shortage of critical commentary on the conduct and progress of the war. If the Iraq War winds down while the Afghan campaign continues at its present or even an enhanced pace, the media are almost certain to devote more energy to it.

The expense of the war in Afghanistan has also not been a major matter of concern, because the costs involved for the United States have generally been portrayed as part of the effort in Iraq rather than as an independent source of demand for funding. The Iraq effort, of course, has been much larger and more expensive than Afghanistan, making it the bull's eye for those who would attack U.S. military policy in the region on economic grounds. As costs decline in Iraq, Afghanistan will represent an increased part of the expenses and will become more controversial. The issue will become particular heated if the United States decides to implement pledges to provide massive economic assistance to Afghanistan, as it has promised since 2001 but hardly fulfilled at all. Finally, the Afghanistan campaign's worthiness as a continuing goal has not undergone detailed scrutiny because of the conflation of the different political objectives with the popular goal of eradicating terrorism.

Unlike the war in Iraq, the Afghanistan adventure has received sustained international support, although there has been some erosion as time has gone by and success has remained elusive. The worldwide outrage at 9/11 easily extended to authorization and annual renewal of the International Security Assistance Force (ISAF) mandate by the United Nations. Even though military direction has been passed to the control of NATO, it remains a coalition of nearly 40 members with troop strength of around 50,000. The deployment of additional U.S. forces in 2009 (mostly under the independent guise of Operation Enduring Freedom) changes some of the dynamics of the situation on the ground, with indeterminate political impact. As the purpose of fighting has become more concentrated on repelling the Taliban and supporting the Karzai government (rather than direct assault on Al Qaeda), some of that support has decreased, and individual governments have begun to question their ongoing commitment of forces in the face of questionable progress to this second goal of the overall operation.

Military Technology and Technique

Operation Enduring Freedom was and remains to this writing in 2009 a struggle involving startling technological contrasts. On one side was a coalition led by the United States, the world's only remaining military superpower, whose military might is built around the fruits of modern scientific research, prodigious technological development, and sophisticated industrial production. On the other side was a radical fundamentalist Islamic sect (the Taliban) consisting primarily of members of the Pashtun tribe, whose religious beliefs and philosophical outlook probably would be more at home in the fifteenth century than the twenty-first. On paper, Enduring Freedom seemed to be a total military mismatch until the Taliban, after what was thought to be a crushing defeat in 2001, decided to fight on from foreign sanctuaries using guerrilla tactics.

The Taliban army on the eve of Enduring Freedom in the fall of 2001 consisted of 45,000 troops of which 25 percent were foreigners drawn to the Taliban cause by its radical Islamic fundamentalism. Some of the trappings of a relatively modern military force were available to the Taliban, but in reality they were a facade. In terms of armor, an estimated 100 tanks (T-54/55 and T-62) and perhaps 250 armored fighting vehicles of various types were available, having been left behind by the Soviets when they abandoned their occupation of Afghanistan in 1989. However, the quality of the maintenance work done on those armored vehicles was suspect and thus their utility was, at best, questionable. Also available to the Taliban were about 200 artillery pieces of various calibers that the Taliban had used effectively against their Northern Alliance foes. Some former Soviet aircraft were also available but, again, their maintenance and thus their airworthiness were very questionable as was the skill of those who might fly them. There was some speculation the aircraft might be most useful to the Taliban as flying bombs, a Taliban version of the Kamikaze attacks in the Pacific during World War II. Finally, the Taliban did not have any sort of integrated air defense system, although estimates indicated they had about 300 operational air defense guns of various calibers, an unknown number of man-portable SA-7 surface-to-air missiles, and 20 surface-to-air missile launchers. As it turned out, the lack of an effective air defense system was their Achilles heel. The coalition strategy would capitalize on U.S. and allied airpower to assist the Northern Alliance forces in their struggle to defeat the Taliban army and throw the Taliban government out of Kabul.

Although the technology available to the Taliban was unimpressive, their human resources demanded respect. Many members of the Taliban had formerly been part of the *mujahadin* force that had given the occupying Soviet military a decade-long "bloody nose," leading eventually to the Soviet withdrawal from Afghanistan in 1989. By 1992, the *mujahadin* had overthrown the Afghan communist government that the Soviets had left behind. Many of those same *mujahadin* later became the Taliban fighters who swept aside the forces of other Afghan warlords and seized power in Kabul in 1996. They controlled all of Afghanistan, save a portion in the northeast that remained under the control of a loose confederation of non-Pashtun warlords generally referred to as the Northern Alliance.

The Taliban fighters were not a technologically sophisticated force but they had proved to be tough and resourceful, although there was some evidence that their victories in the southern part of Afghanistan against their warlord adversaries were at times the result of bribes paid rather than military prowess. However, there was no doubt that they could effectively use various kinds of artillery, and, as they had demonstrated against the Soviets, they could be very effective using the shoulder-fired anti-aircraft missiles that were available to them.

The coalition forces, dominated by U.S. forces, could hardly have presented a starker contrast in terms of technological sophistication. The contrast was particularly acute in air and space power, upon which the coalition would rely to defeat the Taliban in conjunction with Northern Alliance ground forces. It was a strategy that played to the strengths of the coalition and against the weaknesses of the Taliban.

Most of the types of aircraft used in Afghanistan had seen previous combat over Iraq during the Gulf War, Operations Northern and Southern Watch, and/or Operation Desert Fox. Of equal importance was the significant combat experience of coalition personnel in operating, controlling, and commanding aerial weapon systems. However, the emergence of one new aircraft type is worth special mention. Unmanned aerial vehicles (UAVs) had been used initially for persistent surveillance and target designation for strikes by conventional aircraft. Their considerable success in this role led to their adaptation for the strike role using two Hellfire missiles. Their small size and small engine made them virtually unseen and unheard on the ground when flying at typical operational altitudes, and their slow speed and long loiter time made them ideal for persistent surveillance of an area looking for adversaries, particularly those who did not want to be found.

Precision-guided weapons such as the Hellfire missile (mentioned before) dominated the aerial campaign supporting the Northern Alliance ground forces. A decade earlier in Desert Storm less than 10 percent of the aerial munitions used had precision guidance. In the first months of Enduring Freedom (October 2001–March 2002) 70 percent of the air-launched weapons had this capacity. The result of the rise of UAVs equipped with precision-guided weapons was a significant improvement in the ability to identify and destroy enemy targets in remote areas that were difficult or impossible to attack with more conventional forces. Another major reason for the increase in precision air strikes was the development and fielding of the Joint Direct Attack Munition (JDAM), which was essentially a guidance tail kit (containing a guidance control unit using an inertial navigation system and the global positioning system) that converted existing "dumb" bombs into extremely accurate "smart" bombs of various sizes. In Afghanistan the JDAM quickly became the aerial weapon of choice dropped in large numbers by many different types of coalition aircraft.

Space operations, which had played an increasingly important role in military affairs for several decades arguably reached their maturity in Enduring Freedom. The space-based global positioning system was crucial to the accuracy of many aerial weapons as well as crucial for the movement of forces on the ground. Perhaps the greatest impact of space operations was the ability to communicate around the globe in "real time." Satellite-based global communications provided the capability to communicate instantly between Central Command headquarters in Tampa, Florida, the Combined Air Operations Center (CAOC) in Saudi Arabia, and forces directly involved in combat operations in Afghanistan. This produced a "common operating picture" that often had considerable benefits in terms of efficiency and effectiveness. However, these capabilities were not without drawbacks.

The most obvious problem with real-time global communications is the natural tendency for decision-making authority to elevate their importance, particularly with forces involved in the politically charged atmosphere of the Middle East and Southwest Asia. Thus, first in Afghanistan and later in Iraq, the most senior American military leadership took advantage of the space-based global communication system and involved themselves directly in the minutia of employing forces eight time zones away. There are potential dangers in overcentralizing decision-making authority in terms of dealing effectively with unanticipated problems or taking advantage of fleeting opportunities.

The military technique developed to use the sophisticated coalition airpower was both effective and picturesque. The coalition would covertly insert special operations forces (SOF) teams, including combat air controllers, into northern Afghanistan. They would attach themselves to the forces of the rival Northern Alliance warlords. Their initial task would be to help coordinate the actions of the disparate Northern Alliance forces and then to call in coalition airpower to support their operations. The objective, of course, was to defeat Taliban forces quickly and topple the Taliban government in Kabul. As this fighting progressed, some conventional coalition ground forces would be brought into Afghanistan to help search for and trap Al Qaeda. As events played out (discussed in the next section) the military techniques used were quite effective. Further, they presented some very interesting images of American SOF forces on horseback carrying twenty-first century communications gear calling in massive amounts of modern airpower in support of Northern Alliance cavalry charges reminiscent of the American Wild West in the nineteenth century.

Somehow, it all worked, and worked quickly to inflict the initial defeat on the Taliban forces, which quickly fled to sanctuaries in the tribal areas of western Pakistan. In the following years when the Taliban regrouped and began to foray back into Afghanistan using guerrilla tactics against the new Afghan government and its supporters, UAVs became the lead element in efforts to find and then strike the Taliban camps in Pakistan, with special emphasis on destroying the Taliban leadership. Needless to say, bombing strikes, even if made by UAVs, have caused political difficulties between the coalition forces in Afghanistan and the government of Pakistan. Additionally, it has caused trouble for the Pakistan government within its borders as coalition attacks demonstrated that the Pakistan government was not able to fully control its own territory either on the ground in the tribal areas or in the skies above the tribal areas.

Military Conduct

Although military combat operations began on 7 October 2001, much in the way of preparations had been underway for some weeks. Most important were negotiations to secure overflight rights allowing efficient routing of combat aircraft, which would launch from bases in the Middle East, the island of Diego Garcia in the Indian Ocean, aircraft carriers in the North Arabian Sea, several bases in Europe, and even direct flights from the United States. Military-to-military cooperation and detailed coordination with several allies was also needed for operations to be effectively coordinated. All of this was accomplished with remarkable celerity and combat operations against the Taliban government of Afghanistan commenced less than a month after the devastating attacks on the World Trade Center.

The opening blows, conducted under the cover of darkness, were air strikes by B-1B and B-52 bombers flying from Diego Garcia, B-2 bombers flying nonstop from the United States, Navy fighter-bombers launching from carriers operating in the North Arabian Sea, and cruise missiles launched from U.S. and British surface ships and submarines. The targets of the early strikes were primarily military with emphasis on Taliban air-defense capabilities, but they also included strikes on Al Qaeda training facilities, suspected places where Osama bin Laden might be hiding, and direct attacks on Taliban leaders including the destruction of Mullah Omar's residence, which he reportedly vacated only minutes before it was destroyed. It is also worth noting that shortly after the bombing began, U.S. air force transports dropped over 30,000 food packets and medical supplies intended for expected refugees fleeing Afghan cities that were under attack.

The raids continued day after day, surprising some with their intensity. Even more surprising to uninformed observers during those early days was the apparent lack of bombing attacks coordinated with the operations of Northern Alliance forces against Taliban fielded forces. The problem was that the insertion of Special Operations Forces (SOF) teams with combat air controllers into the mountainous areas controlled by the Alliance had been badly delayed by weather problems. Finally, on 19 October the first SOF team was successfully inserted and was quickly followed by three more teams all of whom attached themselves to the forces of the various Northern Alliance warlords.

As SOF teams began coordinating with Northern Alliance forces in the mountains of the north, much farther to the south on the night of 19–20 October Army SOF Rangers and Air Force SOF aircraft staged a daring airborne raid on a Taliban airfield near Kandahar, the "home" of the Taliban movement. Although described as an intelligence gathering mission, it appears the real significance of putting the raiders on the ground was psychological—a demonstration to the Taliban that coalition ground forces could go anywhere at any time, even deep into the heart of "Taliban country."

By early November 2001 90 percent of all coalition air strikes were in direct support of Northern Alliance offensive operations in what could best be called close air support directed by the SOF team air controllers. As a result, the previous near-stalemate on the ground began to show progress that would quickly gain momentum. The first major victory was the fall of Mazar-i-Sharif in the north-central part of Afghanistan near the border with Uzbekistan on 10 November. In addition to a great deal of SOF directed bombing of Taliban front line targets, the Northern Alliance victory also featured the use of both infantry and armored vehicles (including tanks) and what can only be described as a cavalry charge. The fall of Mazar-i-Sharif began a rapid domino effect in other Taliban-controlled cities in the north. Taloqan, to the east of Mazar-i-Sharif fell with little resistance. After eleven days of bombing Taliban defenses, Khanabad was captured and Kunduz surrendered to the Northern Alliance forces. Thus by 23 November, there were no longer any Taliban strongholds in northern Afghanistan.

Meanwhile, the next target for the now proven combination of coalition airpower and Northern Alliance ground forces was the Afghan capital, Kabul. After capturing the air base at Bagram to the north of Kabul, heavy air strikes hit the Taliban defenses deployed on the Shamali Plain, the key route to Kabul, and upon targets in Kabul itself. Pounded relentlessly from the air and pressured by the emboldened Northern Alliance forces, the Taliban and their Al Qaeda allies abandoned their defensive positions and Kabul itself on the night of 12 November and fled the city in two directions. Many went south toward Kandahar, home base for the Taliban movement. Others headed east toward the formidable Tora Bora mountains. Northern Alliance forces entered Kabul on the afternoon of 13 November.

Kandahar was now squarely in the crosshairs, although U.S. bombing efforts were also stepped up in the Tora Bora mountains where Osama bin Laden and many of his Al Qaeda loyalists had taken refuge in cave complexes. In the battle for Kandahar, a Pashtun tribal force led by Hamid Karzai (who would later become Afghanistan's president) put pressure on the city from the east while the Northern Alliance closed in from the north. Intensive bombing of the city continued, and after a futile attempt by Taliban leader Mullah Omar to negotiate amnesty for himself and the other top Taliban leaders, he and some of his loyalists slipped away toward the mountains in the northwest. Other Taliban leaders fled east to Pakistan. On 9 December the anti-Taliban forces occupied the city.

Although Afghanistan was now under the control of the Northern Alliance, the United States was still in pursuit of Al Qaeda and particularly its leadership with special emphasis on Osama bin Laden. As previously mentioned, many Al Qaeda and some Taliban had taken refuge in the mountainous terrain and cave complexes in the Tora Bora mountains. This led to an intensive three-week bombing campaign followed by ground forces. Although there were massive secondary explosions caused by the bombing and fierce fighting by the ground forces, the results were disappointing. It was clear that many of the enemy had been killed, a few had been captured, but many had slipped away, and there was no sign of Osama bin Laden. Some of the Al Qaeda survivors escaped to yet another huge mountain cave complex at Zhawar Kili, by the Pakistan border near the town of Khost. The result of coalition attacks there was much the same as at Tora Bora—heavy bombing, many secondary explosions but no sign of bin Laden. Finally, U.S. and Afghan forces launched a complex offensive action involving U.S. and international SOF, regular U.S. army forces, and Afghan militia against Al Qaeda forces who had taken refuge in the Shah-i-Kot valley area near the Pakistan border. Operation Anaconda, as it was called, was fought between 2 and 19 March 2002, and suffered from unfortunate lapses in planning and coordination which only added to the problems of fighting in a very difficult physical environment against a well entrenched and fanatical enemy. Although many Al Qaeda forces were killed and captured, hundreds escaped across the border into Pakistan, apparently including bin Laden.

It seemed to many observers that the first major operation in the global "War on Terror" had been won. The Taliban, which had provided a home and protection

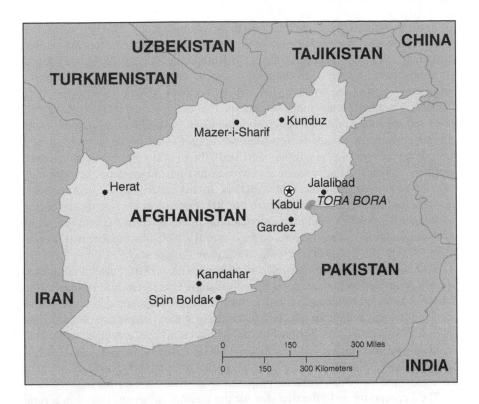

for Al Qaeda, had been thrown out of power. Al Qaeda forces had been attacked and defeated. The remnants of both the Taliban and Al Qaeda had gone into hiding or had fled into the tribal areas of Pakistan, and a new government, led by Hamid Karzai, was in power in Kabul. NATO would now lead the coalition of forces assisting the United States with the security and rebuilding of Afghanistan, including the training of a new Afghan military and a national police force. However, appearances were deceiving. The first phase of the war had resulted in a Coalition victory, but the war had not been won.

The Taliban and Al Qaeda had taken refuge in the mountainous tribal territories of western Pakistan (the Federally Administered Tribal Areas or FATA), an area heavily populated by Pashtuns and only formally controlled by Pakistan. From there the Taliban and Al Qaeda combined with another similarly extreme religious group, Hizb-i-Islam, and began a recruiting drive and set up a training operation. They were aided and abetted in their activities by Pakistanis (some reportedly with close ties to the government and to the military) who were sympathetic to their radical political-religious cause. In 2002 they began offensive operations back into Afghanistan staging armed attacks, including suicide operations and remotely detonated bombings. The attacks grew every year in number and severity including over 5,000 in 2005. It was clear at the beginning of 2006 that the new Afghan government faced a major insurgency.

Several factors led to the rapid growth of the insurgency. The key, of course, was a secure base in the tribal areas of western Pakistan. Other factors also played important roles. The new government in Kabul controlled little beyond Kabul itself and could not provide essential services—including protection from the insurgents—across the country. The former Northern Alliance warlords, whose "alliance" had been much more of a marriage of convenience than anything else, had been appointed provincial governors and often worked at cross-purposes with the government in Kabul. Government corruption at all levels, somewhat of a historical tradition in Afghanistan, remained vibrant under the new national government. The new Afghan national army was just forming and by the beginning of 2006 numbered less than 30,000 soldiers, totally inadequate to meet the scale and scope of insurgent operations seen in 2005. Further, as the insurgency began, only about 10,000 U.S. and other international forces were in Afghanistan, some of them with political restrictions on their use. By 2005 this number had risen to about 30,000, but the insurgency was already well under way.

The insurgency has continued to grow, with many areas in the Pashtun-dominated regions controlled by the insurgents. Equally disturbing is that radical Islamists from other countries are volunteering to fight for the insurgents, and the insurgents have received training from Iraqi militants, particularly in the construction of bombs and the use of suicide attacks. Meanwhile, the Afghan army has continued to expand, nominally numbering over 80,000 at the end of 2008. Likewise the NATO-led International Security Assistance Force (ISAF) has grown to over 60,000, while U.S. forces have grown to over 30,000 and are projected to increase even further in 2009.

The appropriate and effective size for the counter-insurgent forces in a rural insurgency is usually quite high because the critical function of government forces is first and foremost to protect the population from the insurgents rather than to attack the insurgents. This requires a great deal of manpower out in the field, living with those they are protecting, rather than living in cantonment areas and venturing forth on occasion to conduct "search and destroy" operations—to coin a term from the Vietnam War, which is an unfortunate but probably appropriate reference.

Unfortunately for all concerned, including your authors, there is no closure on the struggle in Afghanistan at this writing. How it will conclude depends on a host of factors not the least of which is the concurrent struggle in Iraq, the subject of the next chapter. Other important factors include the performance of the Afghan government and the perception of the government's performance by the Afghan people; the performance of the new Afghan national army; the fallout from the global financial crisis that began in the fall of 2008 which might curtail U.S. and NATO military efforts in Afghanistan; and the policies of the new Obama administration in Washington D.C. as a result of the national elections in November 2008.

Better State of the Peace

As already noted, one of the ways in which this chapter and Chapter 10 are distinguished from earlier chapters is that they deal with ongoing rather than completed conflicts. Since the war in Afghanistan is not over, there is no postwar political environment in which to assess the achievement of one or another set of political objectives. To make matters worse, the outcome itself is likely to be somewhat ambiguous when it is reached, with no clarion event ending the hostilities or grand "peace" agreement setting out definitive conditions against which to measure which side accomplished what political objectives. Finally, the end of this war will quite possibly be more of an interlude than a definite conclusion of the struggle for power in Afghanistan that is part of the political objective of all sides.

The present concern, of course, is with American objectives and whether they will have been achieved. Will the postwar condition in Afghanistan be one that the United States prefers? That, of course, depends on what those objectives were and whether they can be achieved in such a manner that they are embraced rather than simply being forced upon former adversaries. If the latter is the case, Afghan history suggests that what the United States would prefer as an outcome will be viewed as less than that by the contending forces in Afghanistan.

Beyond "sending a message" to future potential adversaries about the consequences of defying the United States, there are two political objectives the achievement of which would constitute an American better state of the peace. Those objectives, of course, are the eradication of Al Qaeda (including its leadership) and the institution of a stable government in Kabul. In original conception, these two goals were related and sequential: the Taliban government had to be overthrown and replaced in order to create an environment that would allow for the suppression of Al Qaeda and the prevention of its reemergence. In the early going—primarily in 2001—the second goal was partially realized with the overthrow of the Taliban. The extension of that accomplishment to a stable, preferably democratic regime in Kabul remains incomplete and increasingly problematical given Taliban resurgence. The goal of destroying Al Qaeda remains unfulfilled.

The destruction of Al Qaeda was, of course, the preeminent goal of engaging in Afghanistan, without which it would have been impossible to justify or sustain an American military effort there. The problem is that operationally, the ongoing war has been primarily directed at achieving the secondary goal of a stable Afghanistan, while the operational dictates of suppressing Al Qaeda have moved across the Durand Line to Pakistan, where most Al Qaeda leaders reside and operations occur. In a very real sense, stabilizing Afghanistan is essentially irrelevant to the United States unless Al Qaeda is eliminated. Since Al Qaeda can continue to exist in Pakistan as long as that country is unwilling or unable to take decisive action in the Federally Administered Tribal Areas (FATA) to force the Pashtuns to send Al Qaeda packing, the realization of the two goals may be independent of one another.

There are clearly operational disconnections between efforts to achieve the two goals as well. Elimination of Al Qaeda has proven impossible to date using standard conventional military means, and the application of conventional military force through air and ground raids into the sanctuaries in Pakistan only makes the political situation worse, enraging the Pakistanis by violating their sovereignty and antagonizing the Pashtun tribesmen when those raids result in "collateral damage" against them. Covert military operations, which presumably are occurring anyway, provide an alternative military approach to the Al Qaeda problem, but, as noted, such an approach is slow, agonizing, and uncertain, particularly as long as the Pashtun code of hospitality inhibits the collection of actionable intelligence about the whereabouts of Al Qaeda targets.

Conventional military solutions by ISAF and Operation Enduring Freedom forces seem less inappropriate to the second political objective's military dictate of providing a shield behind which the Afghan government can stabilize. Having said that, the tyrannies of political geography and topography make such operations difficult, especially in the Pashtun areas which are the major sources of contention and where the opposition has the advantages of centuries of experience in repelling outsiders and a hostile terrain with which they are much more familiar than the intruders.

It is possible that the strategies that have been adopted to achieve either or both political objectives are simply inadequate or misdirected to reaching a better state of the peace. Although the efforts to achieve the goals are virtually independent of one another at the operational level, there is a common element that runs through both and which represents the most exploitable way to achieve one or both goals. That common element is the Pashtuns.

Although largely unacknowledged in U.S. statements of purpose or action, Pashtuns are at the heart of opposition to achieving either or both of the U.S. objectives that would comprise a better state of the peace. The problem with destroying Al Qaeda has been one of penetrating the tribal regions on both sides of the Durand Line in order to find and attack the enemy. The entire area under contest is, of course, part of Pashtunistan. As long as the Pashtuns conceal bin Laden and Al Qaeda, that task will prove, as it has in the past, to be very difficult, even impossible. Solving the problem requires gaining the assistance of those who now hide the opponent. In Afghanistan, the primary source of opposition to the Karzai government is the Taliban, who are also Pashtuns. As noted, not all Pashtuns are Taliban, but virtually all Taliban are Pashtuns. To a degree not always acknowledged, the civil war/ insurgency is a Pashtun/anti-Pashtun contest, in which the United States finds itself opposed to the largest ethnic group in the country, the Pashtuns.

Finding a *modus vivendi* with the Pashtuns would seem to be a way to facilitate achieving both American objectives—eliminating Al Qaeda and bringing stable governance to Afghanistan. Afghan history does not commend the use of military force as a way to "convert" the Pashtuns, which means that a purely military solution is problematical. Late in 2008, the United States announced it would enter into discussions with amenable elements of the Taliban (Pashtuns), aimed at splitting off moderate Pashtuns from support of the Taliban. Whether such an approach will succeed or not remains to be seen, but reaching out to the Pashtuns would seem to be a necessary, if not necessarily sufficient way to try to achieve a better state of the peace in Afghanistan.

10

IRAQ WAR

The American invasion and conquest of Iraq in 2003 was the second major military action by the United States that arose from the dynamics of the post-9/11 experience. Like the war in Afghanistan, it was largely justified as a response to the international terrorism associated with Al Qaeda, although the connections between the Saddam Hussein regime (whose overthrow was the major purpose of the war) and Al Qaeda have proven to be tenuous or nonexistent. In planning for the war, the presumption of American military hegemony influenced willing decision makers to believe the war would be a quick, decisive, even easy, mission, a calculation that proved to be tragically wrong.

Unlike the case of Afghanistan, there was a thread of continuity to the U.S. interest in Iraq. After the 1990–91 Persian Gulf War, there was a widespread view that the United States did not really complete the victory in Desert Storm because it did not physically remove Hussein from power, a task that was physically possible given the overwhelming rout of the Iraqis by the U.S.-led coalition in Kuwait. President George H. W. Bush had resisted calls to push on to Baghdad, reasoning that such a move would shatter the anti-Saddam coalition and ensnarl the United States in a long and costly struggle within Iraq. The passions against Hussein receded during the 1990s but were not extinguished altogether. When George W. Bush achieved office in 2001, he brought with him a sizable number of advisors who were deeply unhappy about the failure to smash the Iraqi regime a decade earlier and who were determined to rectify that situation. In 2003, they got their opportunity.

The result was the Iraq War. By the time the last American combat forces are removed in 2011 as part of the Status of Forces Agreement signed in 2008 by the United States and Iraq, the war will rank with Vietnam as the country's longest and most controversial war—although Afghanistan could eclipse both those distinctions. Because the war has not yet been completed and its outcomes determined and evaluated, it is difficult to make definitive judgments, just as is the case in Afghanistan. The best that can be done is tentative analysis based on the record to date.

The Iraq War has been controversial in at least three basic ways. The first and most basic question is whether it should have occurred at all. As noted in Chapter 8, one way to look at the war that began in 2003 is as a continuation—even a culmination—of the Persian Gulf War of 1990–91. The basis of that reasoning, of course, is the assertion that the failure to overthrow Saddam Hussein in 1991 left the first war uncompleted, a process that reached fruition with the invasion in 2003. In this view the Iraq War is a delayed operational holdover from 1991. Certainly, there is a thread of continuity involved since, as will be shown in the next section, many of those who argued in 1991 that the United States should have "sealed the deal" against Hussein then were prominent in the decision-making process that led to the 2003 invasion.

The controversy over whether there should have been a war, however, exists quite apart from the question of continuity, and has two aspects. One of these has to do with the veracity of the claims that were made at the time justifying the need for war, particularly questions about Saddam Hussein's alleged possession of weapons of mass destruction (WMD) and his supposed ties to terrorism. This criticism has ranged from whether the claims were true or false to whether those who made the claims knew they were false at the time they made them. The other aspect is whether the political objective of the war—which changed across time— was important enough to impel the United States to war. Did, for instance the overthrow of Saddam Hussein warrant the effort the United States has expended in its behalf? Another area of controversy, however, concerns the way the war has been conducted, especially in the first three years of its execution.

The situation can be stated simply. When war against Iraq was being contemplated by the United States, planners within the Bush administration expected a short, decisive, and relatively bloodless (in terms of American casualties) campaign. Secretary of Defense Donald Rumsfeld, for instance, was fond of predicting that the American part in the war would be substantially over within 129 days of the original invasion, and that by the end of the summer of 2003, there would be no more than a small residual U.S. force in Iraq. Such a scenario was extrapolated into a low-cost, high-yield effort that could be completed before the worth of the enterprise was seriously questioned, and that the debate over worth would be influenced by the low costs involved. Had the projections been correct, the war probably would not have been controversial in any significant way; it would have had more or less of the impact of the Spanish-American War (Chapter 11) and might well be remembered in a similar light. Unfortunately, those projections were not realized. Part of the reason was probably faulty assumptions in the prewar planning, as well as a lack of prewar planning for post-combat operations in Iraq.

Controversies regarding the conduct of the war were sequential. They began as a debate among intellectuals, over the legality of the invasion. The physical action that began the war was a cross-border invasion of Iraq by U.S. forces on 20 March

2003. Such an action is, in international legal terms, an act of aggression, which is forbidden under the United Nations Charter, of which the United States is, of course, a signatory (as is Iraq). In fact, the U.S. invasion and conquest of Iraq was only the second instance in the history of the United Nations where one member attacked, conquered, and occupied another member of the world body (Iraq's invasion of Kuwait is the other). On the surface, the action appeared illegal and was widely denounced outside the United States.

The United States countered such accusations on two related grounds, *preemption* and *prevention*. Preemptive acts have international legal precedent, justifying an initial (preemptive) action by one state if it has reason to feel an attack against it is imminent (the standard was first articulated by U.S. Secretary of State Daniel Webster in 1837). Imminence means there is good physical evidence that an attack is about to occur unless action is taken to prevent it (attacking an enemy army massing on one's borders is the most obvious example). Hardly anyone makes, or accepts, the case that such a condition existed before the U.S. invasion of Iraq. Instead, the invasion is justified under the Bush doctrine's provision of prevention, a looser standard that asserts the "right" to initiate action to prevent a future action against the United States. This declaration is a unilateral statement of U.S. policy, not a legal justification, and it is almost universally rejected on international legal grounds. Most international legal authorities have concluded that the U.S. action was, in strictly legal terms, illegal.

There have been more concrete criticisms of the war's conduct, and in particular the planning for the invasion and especially its aftermath. The initial invasion, of course, was a military success, as discussed later in this chapter. The planning problems, however, arose from projections (or lack thereof) about what would transpire after the invasion was completed. Prewar planning emphasized that the war would be swift and relatively painless, as already noted, but it erred about the post-invasion necessities facing the triumphant American-led coalition. So-called Phase IV operations presupposed that the United States would rapidly stabilize the situation after overthrowing the Iraqi government and be able to leave in a short period of time. This explains the absence of a full-scale occupation plan, including plans both for maintaining a large American force in Iraq for an extended period of stabilization and for engaging in massive development of the war-torn country. These assumptions proved utterly—and, many critics inside and outside the government argued, entirely predictably—false, as the ensuing events amply demonstrate.

This lack of preparation meant, among other things, that estimates of the cost of the war in blood and treasure were wrong by orders of magnitude: the administration initially projected a cost of around $10–12 billion for military operations, and recovery costs for Iraq that would largely be covered by Iraqi oil revenues. Whether the administration actually believed its own estimates or whether they were purposely minimized to undercut opposition to the war's authorization remains contentious. Moreover, a serious question remains about whether Congress would have authorized the war in its 2002 Resolution had there been realistic estimates of the actual costs of the war.

The absence of planning for the occupation (Phase IV) has been well documented in numerous books on the war, some of which are listed in the bibliography of this book. Its major consequence has been an open-ended occupation of Iraq, which has become increasingly unpopular in both the United States and Iraq itself. Official costs have exceeded $600 billion for the United States as of this writing in early 2009, and many expect the total accounting to top $1 trillion. In addition, the occupation has been marked by embarrassing episodes such as accusations of torture at Abu Ghraib prison, a multi-headed insurgency that has flared up and died down periodically, and the emergence of an internal terrorist movement in the country (Al Qaeda in Iraq). Presumably this whole process is moving toward some form of resolution under the Status of Forces Agreement (SOFA) signed by the United States and Iraq in December 2008. Under the terms of that agreement, the United States must remove all combat forces from Iraq by the end of 2011. Whether those provisions are met according to the timetable and whether doing so brings meaningful closure to the Iraq War experience remains to be seen.

The Iraq War is clearly a unique part of the American experience, but in ways that are not entirely apparent because the war has still not been concluded and its ultimate outcomes and impacts have yet to be reached and interpreted. As it becomes the second longest war in American history (surpassing Vietnam), it has become a very unpopular war and one that casts a pall on the record of the Bush administration that started and presided over it. Some of the sources of public disapproval cannot be fully assessed until the war is concluded and there is sufficient temporal perspective to assess its total impact and meaning.

It is possible, however, to offer some interim judgments about the reasons for going to war and how the war has been carried out. What makes both Iraq and Afghanistan unique in this volume, of course, is that certain parts of the framework that have been applied cannot fully be assessed for these wars. Most prominently, the ultimate test of a war and its success or failure, and the better state of the peace, cannot be described and measured since assessing the postwar world requires an end to the war. With that rejoinder in mind, the analysis proceeds.

Issues and Events

One of the problems of dealing with an unfinished war is that it is difficult to determine exactly what underlying issues led decision makers to decide to initiate war. In some wars, one can start from some dramatic precipitating event—Pearl Harbor, or the secession of South Carolina, for instance—and move backward toward the disagreements that led to that precipitant. In the case of the Iraq War, there was no obviously compelling action by either side that one can use as a "peg" from which to move backward toward underlying issues or forward through a clear series of proximate events. While there were forebodings, a grim sense of inevitability that war was looming, they were not precise or overwhelming enough to make the decision obvious. The situation is made all the more difficult because the major actors in the decision process, and especially the inner circle within the Bush White House, have yet to produce official explanations in the form of memoirs and the like. Thus, the indeterminacy of assessing the better state of the peace must be shared with the question of issues and events—with ramifications for determining political and military objectives.

There are three different possible explanations of underlying causes leading to the decision to attack Iraq. Each is plausible and has some evidence to support it, and it is possible that all three were contributory at one level or another. Since the record remains unclear at this point, all we can do is describe each and allow the reader to reach a decision about which one(s) were most important.

The first possible explanation might be described as the *neo-conservative* motivation. It consists of two related strands of argumentation. One, to which allusion has already been made, is the notion that the failure to overthrow Saddam Hussein in 1991 was a mistake and that the 2003 war was a way to finish the business begun in Kuwait over a decade earlier. Certainly, many of the decision makers who were prominent and presumably influential in the George W. Bush administration when the decision to go to war was made had been calling for "regime change" in Iraq for many years. Vice President Cheney—who had been secretary of defense during the Persian Gulf War—comes immediately to mind in this regard. President Bush himself apparently came to share this view, expressed in his belief that Saddam had authorized an apparent assassination attempt against his father, President H.W. Bush, in Kuwait in 1993.

The neo-conservative case contained a second, and related, thread: the promotion of democracy in the Middle East. The idea was that a stable Middle East, among other things, would help insure the physical security of Israel, and could best be achieved by spreading democracy among the nondemocratic countries of the region. Iraq, which the neo-conservatives considered the single most destabilizing force in the area under the dictatorship of Saddam Hussein, was viewed as a linchpin for beginning the process of democratization. If Iraq could be transformed into a democratic oasis, they reasoned, democracy might spread to other states that would emulate the improved Iraqi condition. Since Hussein was clearly unwilling to allow his power to be challenged through a peaceful evolution to democracy, then "conversion by the sword" in the form of the overthrow of his regime seemed a way to bring about so-called "regime change." In the process, closure would also be brought to the conflict left unresolved in 1991.

A second possible explanation is *Saddam Hussein's intransigence.* The regime of Saddam Hussein was not popular anywhere in the world, because of the intransigence of Saddam Hussein in his attitude toward the outside world, and especially in the level of his noncompliance with United Nations resolutions after the end of the 1991 war. Even after his country was defeated, he turned his remaining armed forces against Kurds and Shiites who had tried to overthrow him in a bloody reprisal, and he resisted attempts to monitor his compliance with the requirement to disarm, destroy weapons stockpiles, and shut down WMD programs. In the process, he undermined his own standing in the international community, meaning there was little sympathy for him as the United States moved inexorably toward war, and his actions provided easy examples for those seeking to make a case for his overthrow. Saddam Hussein was in many ways, his own worst enemy.

This explanation, of course, is closely tied to the neo-conservative brief. The monstrosity of the Hussein regime was, after all, a major reason that advocates argued for getting rid of him in 1991, and his continued snubbing of international norms only added to the case for his overthrow. Clearly, there were few more tyrannical regimes anywhere in the world, making Iraq seem an ideal candidate for the policy of democracy promotion.

The third, and least publicly acknowledged possible explanation is influence over or control of Iraqi oil reserves. From the time of the invasion until the end of its term in January 2009, the Bush administration dismissed any suggestion that this overtly geopolitical motive was part of their reasoning for going to war, and as such, it thus would appear not to be a reason worth exploring. It is raised in this context because it is less implausible than official denials would suggest, and because it may offer a more defensible geopolitical rationale than the official reasons for the war.

The United States had been excluded from access to Iraqi oil for over thirty years prior to the 2003 invasion, one of the results of the nationalization of those oil fields and the consequent expulsion of private American and British oil companies from them. Known Iraqi oil reserves are the third largest of any country in the region (after Saudi Arabia and Iran), are relatively unexploited, and are of particularly high quality. At the time of the decision to invade Iraq, the Saddam Hussein government was conducting negotiations for future access to these riches, discussions from which the United States was once again excluded. In the context of the times, there was also a growing realization of the need for U.S. energy security (a sufficient guaranteed supply of oil), and as it is well known, the Bush administration had deep interests and ties to the petroleum industry. The denial of any interest in Iraq's oil was questionable.

The pursuit of oil is an "attractive" motivation in a traditional, geopolitical sense. One of the difficulties that plagues attempts to divine American reasons for the war lies in the question, explored more fully later in this chapter, of whether those reasons were worthy of war. This last formulation is classically realist in orientation, tying the decision to go to war to threats to vital national interests. For traditional realists, the stated reasons for the Iraq War have never seemed compelling, especially considering the extensive war that Iraq has turned out to be. Access to petroleum, however, is indeed a vital interest of the United States because of the country's dependence on foreign oil. Following this line of reasoning, oil may have been a sufficient reason to go to war, even if it is not acknowledged as such.

The underlying issues surrounding the American decision are open to discussion, and so are the actual events leading to war. As already noted, there was no dramatic event—no sinking of the USS *Maine*, for instance—that provided the precipitant, nor were there other forms of provocation to which one can tie the decision. The major, and unanswered questions are: what made the United States decide to attack Iraq, and when was the decision made?

There is considerable disagreement on both counts. Official versions about proximate events tend to center on issues such as Iraqi possession of WMD and resistance to UN inspections of places where such WMD might be fabricated or stored, and links to global terrorism. Neither explanation has fared well under scrutiny; no WMD were ever found, and links to terrorism have never been firmly established. The remaining question for historians is not whether these were valid reasons for war, but rather whether those who made the decisions *knew* them to be false. Put a slightly different way, the remaining question is whether the stated reasons for war were actually motivating proximate events or convenient excuses to enable implementation of a decision made earlier. That question, in turn, leads to the additional question of when the decision was made to go to war.

As memory of the Bush administration fades, this has become a lively, contentious debate, particularly over the role and function of the 9/11 attacks in the decision to go to war in Iraq. It is a debate that will begin to be settled as the people involved in the decision process publish and otherwise reveal their accounts of those decisions, although the clarity and veracity of some accounts will almost certainly be questioned. Thus, the final judgment will await the verdict of historians and other analysts sometime in the future.

There are three general possibilities. One is that the decision was reached before and independently of the 9/11 attacks. Support for this conclusion comes in the public advocacies of many of those who dissented from the 1991 decision not to unseat Hussein at the time (notably the neo-conservatives). As early as 1999 (at the Citadel in South Carolina and the Reagan Library in California) then candidate Bush gave speeches that suggested overthrowing Hussein was a course he favored (one of the authors here makes the case for this alternative in *What After Iraq?*). Certainly there was support within the administration for this option before 9/11; whether that advocacy resulted in a more-or-less firm commitment to invade before 9/11 is a matter of debate. If it is correct, however, then the terrorist attacks were enabling events rather than causal elements in the decision.

The second possibility is that 9/11 was the catalyst for the decision. Some accounts have suggested that one of the first concerns raised by President Bush after 9/11 was possible complicity by the Iraqis. Certainly, the 9/11 attacks precipitated a rush of military responses that formed the "war on terror," and Iraq became a focus of that effort. Whether the attacks provided a real turning point or simply created an atmosphere of willingness for military activism that made war with Iraq acceptable is, once again, a question for which there is no definitive answer at this point.

The third possibility is that the administration was genuine in its statements that the proximate events leading to the war were indeed suspicions about weapons of mass destruction possessed or being developed by Iraq, the unwillingness of the Hussein regime to allow adequate inspection of possible WMD sites, and the perceived connection to terrorism. In the post-Bush era, there is a growing attempt by those implicated in those interpretations of Iraqi actions and motives (the intelligence community in particular) to distance themselves from the assertions that were made at the time. If the administration "cooked the books" to justify what they wanted to do anyway by setting up the conditions for war, then these explanations again lead to the suspicion that they were manufactured: enabling conditions to carry out the decision made some time in advance of the actual public pronouncement of intent.

Regardless of when President Bush and his inner circle decided to invade Iraq, that decision was publicly evident in late 2002. In October 2002, the Congress passed the Authorization to Use Force Against Iraq resolution, thereby creating legitimacy for the invasion. Parallel attempts to legitimate actions were pressed at the United Nations, although the most support the Security Council would approve was UNSCR (United Nations Security Council Resolution) 1441, which threatened but did not endorse the use of force against Iraq. President Bush reiterated his concerns with a litany of alleged Iraqi violations regarding WMD in his State of the Union address in January 2003. On 20 March 2003 American forces crossed the border into Iraq.

Political Objective

The political objective in Iraq has been a matter of contention, and that has added to the unease many Americans have felt about the war. At the time of the invasion itself, it was largely defined as the total objective of overthrowing the tyranny of Saddam Hussein. For long-time supporters of completing the job left unfinished in 1991, Hussein and his despotic regime represented such a reprehensible situation that this objective was justification enough for war. Many of the neo-conservatives, for instance, long had favored what amounted to a personal vendetta against Hussein; President Bush may have shared this sentiment.

The overthrow of Hussein, however, left two questions unanswered in terms of the adequacy of "regime change" as a goal per se. One was whether Hussein's tyranny was so outstanding as to justify the actions taken. There is, unfortunately, no shortage of cruel, despotic, and reprehensible regimes in the world, which the United States tolerates and, historically, has even supported. Why single out Saddam Hussein? The other and related question was whether regime change in Iraq was important enough to justify the actions being contemplated and eventually implemented. What vital U.S. interests were served by war? In order to meet the criteria for a good political objective, the rationale for war against Iraq had to be framed in terms that would answer these questions.

There were explanations that addressed both questions and which formed an argument adequately persuasive to create initial support for the war. The basic reason that Saddam Hussein was singled out from the world's dictators was because of the issues of WMD, the war on terrorism, and his intransigence in the face of international demands to open his country to inspections (which he agreed to do in the weeks before the invasion but apparently too late to dissuade the United States). His use of WMD against Iran and the Iraqi Kurdish town of Halabja in the latter stages of the 1980s Iran-Iraq War gave greater salience to this line of argumentation. Saddam Hussein was a particularly dangerous despot who posed a greater threat to the United States and the world than, say, a Robert Mugabe in Zimbabwe.

The answer to the broader geopolitical question of the necessity of the action was its potential impact on the critical Middle East region. Here the neo-conservative goal of democracy promotion provided the underpinning for regime change. The basic argument was that the replacement of Hussein with a democratic regime would stabilize the region. Starting from the observation that Israel (whose security was integral, even crucial, to the argument) is the only democratic country in the Middle East, the idea was that if Iraq were to become a democratic beacon, it could lead to transformation of the region. The neo-conservatives argued that a democratic Iraqi state would be like Ronald Reagan's "shining city on a hill" for nondemocratic countries in the region, whose people would demand democratization as well. If the region could be democratized, it would also become less radical and unstable, because at least in theory, democratic societies do not make war on one another. This desirable condition would be especially beneficial to Israel, since free and democratic Muslim states would have common values with democratic Israel.

This statement of the political objective proved adequate to be persuasive as preparations for war and the rapid, highly successful conduct of the actual invasion unfolded. The rationale for war was premised on a swift and painless overthrow of the Hussein regime followed by a quick withdrawal and turnover of responsibility to a new, democratic Iraqi regime that had, among other things, eschewed WMD and ties to terrorism: a rapid "mission accomplished," in the unfortunate words of the banner on the U.S. aircraft carrier where President Bush announced the end of major combat operations in Iraq on 3 May 2003. Had Rumsfeld's prediction that the troops would be out of Iraq by the end of summer 2003 come to pass, the Iraq War might well be remembered as a great American victory with few questions asked. There had been no real debate on the merits prior to the Congressional authorization of force, and the question might have remained moot had the projections been accurate and the war over before a backlash could mount. It did not, of course, work out that way.

The rationales for war in Iraq have not proven durable for large segments of the American public. While they might have been adequate to support a short and glorious kind of military campaign of the kind that almost never occurs, they did not hold up as well for a protracted and difficult involvement. Not all the criticism of the war has been on principle; much has been about the competence of the efforts to achieve the objectives. Nonetheless, the objectives and the premises underlying them have been the subjects of mounting debate that can be described in terms of assaults on the political objectives.

infrastructure, and supporting efforts to set the conditions for long-term political and economic stability.

Overthrowing Saddam's government either by a quick decapitating air strike or through the methodical defeat of his military forces and occupation of the country were missions the coalition military forces were well-trained and equipped to execute. As planning progressed there was considerable controversy over the details of how to inflict quick defeat on the Iraqi forces, particularly concerning how many forces would be required to do so. But there was never any serious doubt that a relatively quick and decisive military victory was likely, provided the war remained "conventional." The biggest fear was that Saddam would use chemical and biological weapons against coalition forces just as he used them in Iraq's very bloody war with Iran and against rebellious factions of the Iraqi population. The development and reported possession of such weapons had been a major justification for the war and would play a significant role in developing the coalition's military strategy.

Unfortunately, the coalition military forces were not nearly so well-trained and equipped to occupy and pacify a defeated Iraq. They were ill-prepared to deal with the sectarian violence that erupted between Islamic factions vying for political power in the war's aftermath. Nor were the coalition forces prepared physically or mentally to deal with an insurgency waged against them by people they had just "liberated" from Saddam's despotic regime.

The American and the coalition military strategy for the war in Iraq was heavily influenced by the secretary of defense, Donald Rumsfeld, who had also served as secretary of defense during the Ford administration in the mid-1970s. Rumsfeld was a man of strong opinions who knew what he wanted and how to make things happen. Nor was he intimidated by the professional military. He had been a Navy aviator in the 1950s and retired from the Naval Reserve with the rank of captain (colonel in the land and air services).

Rumsfeld was determined to change the American military, to not only equip it with the latest technology, but also to change its way of thinking in order to fully capitalize on the U.S. technological advantage. Rumsfeld believed that the unfolding drama in Afghanistan (see Chapter 9) illustrated his point. Unable to quickly insert massive ground forces into landlocked Afghanistan, U.S. Special Forces personnel, armed with laptop computers and high-tech communications gear and often riding on horseback, called in precision air strikes on Taliban forces, which allowed the forces of the so-called Northern Alliance to quickly rout the Taliban and take control of the Afghan government in Kabul. The Afghan model seemed to confirm the wisdom of Rumsfeld's drive to change the thinking of the U.S. military. At the time the strategy for Iraq was under development, of course, the defeated remnants of the Taliban had not yet reconstituted and had not yet begun their insurgency in Afghanistan that continues to bedevil the United States and its allies as of this writing (2009).

Secretary Rumsfeld was dissatisfied with OPLAN 1003-98, the plan that the

U.S. Central Command (CENTCOM) had "on the shelf" for a potential invasion of Iraq. The plan called for a force of more than 400,000 troops, which would seriously constrain U.S. options because of the deployment time required. Rumsfeld also believed the size of the planned deployment was wildly excessive in relation to Iraqi capabilities. The Iraqi military had proved woefully inadequate and inept when facing modern western military forces in 1991, and their condition had not improved in the ensuing decade. The collapse of the Soviet Union, Iraq's former military mentor and supplier, posed significant problems for the Iraqi military in terms of both training and equipment. The almost daily beating inflicted during a decade of "no-fly zone" enforcement had also taken a significant toll, as had Operation Desert Fox.

The result was a yearlong planning exercise within CENTCOM to produce a plan acceptable to the strong-willed secretary of defense. With each iteration, Rumsfeld demanded ever-smaller forces, faster deployment, and greater speed in combat operations. CENTCOM commander General Tommy Franks pressed his boss's case by preaching to his planners that "speed kills." Speed became the dominant theme for the war strategy: deployment speed to the theater; speed in seizing and securing the oil fields, which many feared Saddam might sabotage as he had a decade earlier in Kuwait; and speed getting to Baghdad. The Iraqi Army/ Republican Guard could not be allowed to flow into Baghdad for a climactic and bloody battle of attrition on the city's streets. To achieve deployment speed, the forces would be only about one-third the size called for in the original OPLAN 1003-98. There would not be a month-long preliminary aerial bombardment as there had been a decade earlier in Desert Storm. Rather a very short preliminary "shock and awe" bombing concentrated on Baghdad would suffice as a prelude and would hopefully make it very difficult for Saddam's government and military to take effective action against the coalition invasion.

On the ground, the push to Baghdad from Kuwait would be through two corridors, which would prevent the Iraqis from effectively concentrating their defenders. Marines would move through the more heavily populated area east of the Euphrates River after assisting the British contingent in securing Basra and the nearby southern oil fields. The Marines would approach Baghdad from the south and east. The Army forces would advance toward Baghdad through the desert areas west of the Euphrates River and thus approach the capital city from the south and west. The strategy also called for a third force invading Iraq from the north out of Turkey to secure Kurdistan and the northern oil fields, but Turkey, which had its own interests in Kurdish issues both in Iraq and in eastern Turkey, refused to allow the coalition to launch the northern wing of the invasion from Turkish territory.

The development of the military strategy for the war with Iraq lasted for over a year. Unfortunately, planning for "Phase IV," the war's aftermath, was far less thorough, just as it had been in the case of Afghanistan. Even in late 2002 military leaders had not received definitive Phase IV policy guidance. Some Phase IV planning did get under way at the Combined Forces Land Component Command

(CFLCC) headquarters. Those efforts produced an outline plan called "Eclipse II," which envisioned a rolling transition from combat to stability operations, from south to north, to begin before the fall of Baghdad. However, little of this information got to field commanders responsible for implementing the transition.

Two important assumptions apparently colored thinking about post-combat operations. The most widespread assumption was that authority would be turned over to the Iraqis very quickly after the fall of Saddam's government and that the coalition would administer Iraq through the existing government ministries. Second, and a corollary to the first assumption, it was assumed that although Saddam's military and political control apparatus would be destroyed by the conflict, somehow the government ministries and control infrastructures would remain intact and able to function. The same assumption was made about the Iraqi police.

In January 2003, National Security Presidential Directive 24 created the Organization for Reconstruction and Humanitarian Assistance (ORHA), which would be the first postwar planning office and would report to the Department of State. Retired Army Lt. General Jay Garner was tapped to lead the organization that would be responsible for civil and humanitarian affairs and reconstruction in postwar Iraq. Garner had considerable experience in humanitarian operations, having led Operation Provide Comfort, the humanitarian relief effort in northern Iraq in 1991.

One month later, and less than a month before combat began in Iraq, ORHA had its founding conference. But little time was left for planning how to deal with a defeated and disorganized postwar Iraq. Combat operations would soon commence and would conclude far faster than most had expected. Once again, as in Afghanistan, the development of a strategy to win the postwar peace had been an afterthought. Once again, winning the peace was overwhelmed by efforts to develop a strategy to win the war. And once again, the results would be unnecessarily tragic.

Political Considerations

Although neither as widespread nor as emotionally furtive, public opposition to the Iraq War has served much the same function as it did in Vietnam, forcing the United States to withdraw from a theater of operations without completing all the military missions necessary to achieve the political objective. As in Vietnam, there will be a lingering debate over whether this politically forced withdrawal was the reason the United States was less than totally successful, or whether the failure was the result of unrealistic and unattainable political or military goals. That debate has not been entirely engaged over Iraq and will not be until the withdrawal is completed and the situation in that country congeals. Quite obviously, if things go well—by U.S. standards—in Iraq, the ensuing debate will be a good bit less fractious than if events unfold in direct contradiction to American desires, as was the case in Vietnam.

As in all wars, purely political concerns have affected the conduct and outcome in Iraq. Domestically, opposition to the war helped elect the American president who will preside over ending the involvement. Lingering questions raised in opposition focus on the meaningfulness of the objectives in the first place (should the United States have invaded?) and the conduct of the war (the competence of the effort, particularly Phase IV operations). Internationally, foreign opposition to the war has been a factor, as has been the intractability of the situation itself in Iraq.

Public reaction to the Iraq War has been a roller coaster ride. Initially, the public fairly uncritically accepted the Bush administration's explanation of the need for action (largely deriving from the extension of the "war on terror"), and the physical invasion proceeded without obvious difficulties, enhancing public support. That support probably reached its zenith with the enigmatic "mission accomplished" pronouncement on 3 May 2003. Support gradually turned into opposition as the war proceeded. The inability to find WMD or plausible ties to terrorism played a part, as did a torrent of reporting that suggested the incompetence of the occupation. As the war dragged on, opposition hardened as Americans increasingly questioned both the worth of the effort and the prospects for completing the task successfully. From late 2004 forward, roughly two-thirds of the American population voiced their consistent opposition to the war. The 2006 election provided a referendum of sorts on the war, and the Republicans were swept from control of both houses of Congress by a Democratic majority promising an end to the conflict.

The "surge" of 2007 appeared to reverse the negative military situation on the ground and revive some support for continuing. As the 2008 presidential campaign began to take form, the election promised to be another statement of support for or opposition to the war, with Republican candidate John McCain supporting "staying the course" in Iraq and Democratic nominee Barack Obama, a consistent opponent of the war, promising withdrawal. The purity of impact was, of course, diluted by the economic recession that grabbed attention in the last two months of the campaign and dominated voter concerns. Nonetheless, antiwar candidate Obama prevailed, and among his first acts after the inauguration was to reiterate his determination to withdraw all American combat forces from Iraq within 16 months, an assignment he ordered his military staff to prepare to implement.

Obama's determination to end the Iraq War also spoke to the other political concern surrounding the war: international opposition. With the exception of Great Britain (whose oil companies, possibly coincidentally, had also been excluded from Iraq), virtually all of America's friends and allies opposed the U.S. invasion and especially the prolonged occupation of the country. The United States was able to cobble together a coalition of sorts to make the war appear to be an "allied" effort, but beyond American and British involvement, the contributions of other countries were mostly symbolic and arguably motivated more by a desire to curry American favor than by concern about Iraq.

The situation in Iraq itself has made the international situation even more difficult and has strained further support for the American political objectives. It has done so in at least two ways, both of which are the direct result of the enormous media scrutiny that has accompanied the occupation phase of the operation. One problem has been that coverage of the American presence has bred a public opposition to the American objectives. It is difficult for an occupying power, for instance, to maintain order, and at the same time to argue it is "helping" the occupied people achieve either stability or freedom, when that very presence seems to contradict either development. When Iraqis are asked for their free and open opinion on their country's situation, they most frequently responded that they would be much happier if the occupation ended and the Americans left. Occupations are best administered outside of detailed public scrutiny, and this occupation—including all its attendant problems—has been very widely covered.

Further complicating the situation is the argument that the situation in Iraq would likely deteriorate back into some form of chaotic violence if the United States departed "prematurely." The structure of Iraqi society, divided fundamentally along religious (Sunni versus Shia Islam) and ethnic/national lines (Arabs versus Kurds) is sufficiently delicate and volatile; it is not clear what happens once the United States departs. There has been a built-in bias to delay whatever explosion might occur by hoping that a continued presence will somehow make things "better," which translates basically as more stable. No one associated with the occupation has successfully defined at what point that goal of a sustainable peace will be achieved or been able to argue entirely persuasively that a continued U.S. presence is aiding the process of Iraqi national self-determination. History may well conclude that given the nature of Iraqi society, it was not possible for the United States to achieve its goals under any circumstances, meaning that the pursuit—and especially the continuation of that process—may simply be futile.

That political considerations ultimately have served to undermine the U.S. effort in Iraq should not be a surprise. It is an ingrained part of the American democratic tradition that the country does not like or support long wars unless it can be demonstrated that the protracted sacrifices such wars require are clearly worthwhile. These problems would not have arisen had the United States been able to conduct the swift, painless campaign the administration described and apparently expected in 2003. There was ample opposition within the expert community that suggested the Iraq situation would become essentially what it has turned out to be, but that advice was ignored as the United States rushed into war. The public will support military efforts of lesser urgency and consequence if they are over quickly and easily, but when the interests involved are marginal or questionable and the effort drags on for an extended period—as it has in Iraq—a negative public reaction is essentially inevitable. The ubiquitous coverage provided by international electronic media only amplifies the scrutiny under which such actions evolve, and it is likely one of the "lessons" of Iraq (discussed more fully in the last chapter) will be to take that possibility into account before embarking on future military adventures.

Technology and Technique

In terms of the technology available and the techniques employed, the 2003 war was one of the greatest mismatches in modern military history. Unfortunately for the coalition, and more specifically for the United States, what happened in the years after Iraq's crushing military defeat cast a pall over the entire venture and raised the specter of losing the war after having won the battles.

On the eve of hostilities in 2003, the Iraqis were in a desperate situation. Their military had been gutted twelve years earlier by Operation Desert Storm and since that time had suffered serious attrition due to air attacks that were part of the "no fly zone" enforcement and Operation Desert Fox. The Iraqi predicament was compounded by their isolation from former suppliers, most notably the defunct Soviet Union. The technology on hand was at least a dozen years out of date (particularly important in electronic technology) and most of it was much older than that. The common Iraqi tank, a Soviet T-55, for instance, was based on a fifty-year-old design. John Keegan, a noted military historian, dubbed the T-55 a "death trap" when facing modern armor. Although having outdated technology is a serious problem in modern warfare, when that outdated equipment is badly serviced and short of spare parts the situation can, and did, become catastrophic for the Iraqis.

The pride of the Iraqi military in 1991 had been its quite sophisticated and fully integrated air-defense system. Naturally it was a prime target in the weeks-long bombing that preceded the Desert Storm ground invasion. It continued to take a terrible pounding throughout the 1990s and into the new century as part of the "no fly zone" enforcement. Further damage occurred when it was pounded again during Operation Desert Fox. As a result, during Operation Iraqi Freedom the Iraqi ground-based air defense system was totally ineffective—the coalition lost only one fixed-wing aircraft (an A-10 close support aircraft) to enemy ground fire. Further, what was left of the Iraqi air force did not launch a single sortie against the coalition. Rather, Iraqi airmen were ordered by Saddam to hide their aircraft—many were found completely buried in the sand—to avoid their destruction and thus be available for some unspecified future use.

The U.S.-led coalition's biggest fear was that the Iraqis would use their technological trump card—chemical and biological weapons. There seemed to be good reason to fear these weapons as Saddam had used them against the Iranians in the Iran-Iraq War and had on several occasions turned them on dissident Iraqis as well. Senior U.S. political leaders raised the specter of Saddam turning his chemical and biological arsenal on his neighbors. Secretary of State Colin Powell had gone before the United Nations Security Council with what was presumed to be hard evidence that the Iraqis had both facilities to manufacture such weapons and stockpiles of the weapons, posing an unacceptable threat to everyone in the region. As it turned out, the evidence was faulty. Saddam had ordered his stocks of chemical and biological weapons destroyed.

The Iraqi technological shortcomings were compounded by leadership and organizational problems. Success as an Iraqi military officer depended at least as much on political and personal loyalty to Saddam Hussein as on professional competence, a common pattern in many third-world militaries. This problem was compounded by a dysfunctional three-tiered organizational structure designed much more to ensure internal security for Saddam's regime than to meet and defeat external threats to Iraq.

The largest tier was the regular army, which consisted of eleven infantry, three mechanized, and three armored divisions totaling 150,000–200,000 personnel. Although the army was the largest element of the Iraqi military, in Saddam's eyes it was the least trustworthy politically. As a result their equipment was the worst of the lot—old, badly serviced, and lacking sufficient spare parts. In all respects it was not nearly the army that had marched into Kuwait in 1991 (forty divisions at that time).

The elite (relatively speaking) Iraqi force was the Republican Guard which had originally been an army brigade designed to act as a kind of praetorian guard to protect then president Arif from coup attempts in the 1960s. Saddam had expanded it to six divisions (three infantry, two mechanized, and one armored), put only the most politically trustworthy officers in charge, and provided the best of the available equipment. But even Saddam's patronage could not overcome Iraq's political and economic isolation in the 1990s and thus the Republican Guards' equipment was also badly serviced and short of spare parts. There was also a "Special Republican Guard," a three-brigade security force rather than a force organized and equipped for high intensity combat operations. Altogether, the Republican Guards could field about 20 percent of the total Iraqi force, approximately 40,000 to 50,000 troops.

The third element of the Iraqi armed forces was the irregular force collectively known as the "Fedayeen" (martyrs), the most prominent of which was the "Fedayeen Saddam." These were anti-western, political and/or religious zealots (many from other Muslim countries), who were expected to fight fanatically behind the coalition's front lines to disrupt operations and attack supply lines. Many early reports from coalition forces noted how stubbornly the Fedayeen fought and how much difficulty they caused for the coalition forces, particularly along their extended and ultimately vulnerable supply lines.

The coalition forces could not have presented a more startling contrast to the Iraqi military. Dominated by the U.S. military, the coalition forces were modern, well tested, and professionally led. Following the collapse of the Soviet Union, the breakup of the Warsaw Pact, and the consequent end of the cold war, the United States became the sole military superpower. As a consequence the Coalition had well-tested and proven professional leadership from the highest levels of command down to the noncommissioned officers in the field—the operational backbone of any military organization.

The leaders and those they led were equipped with the most modern military equipment. U.S. forces benefited significantly from heavy military spending during the 1980s and 1990s for high-technology weapons, many of which received a significant baptism by fire in Operation Desert Storm. Of equal importance, new training methods emphasizing modern, high-intensity conflict became a significant focus in the U.S. military, particularly joint training involving ground, air, and naval forces. The end result was a high-tech ground, air, and naval force with superb training. Further, it had significant combat experience throughout the 1990s that would serve it well in the initial combat against Iraq in 2003. However, it did not prepare the U.S. forces well for the low-tech, low-intensity insurgent warfare that would follow in the wake of the initial coalition military success.

In terms of technology, the trend toward higher and higher percentages of precision guided munitions usage continued. During the heavy fighting from 14 March 2003 until 18 April 2003, 68 percent of the aerial munitions used were precision guided. As the conflict unfolded, the importance of finding Iraqi forces using overhead sources—manned and unmanned aerial vehicles and spaced-based systems—was reemphasized from earlier conflicts. This was especially obvious during a three-day sandstorm (24–27 March) that blinded the coalition ground forces and stopped them in their tracks. Overhead-sensing devices were able to detect Iraqi ground forces despite the storm and strike them with deadly precision.

The coalition's basic strategy, as noted earlier, was based on speed created by the coalition's huge technological advantage that would offset what some perceived as the disadvantage of a force that was far too small. Speed would be aided and abetted by several techniques. First, the long aerial bombardment in Desert Storm would be replaced by a very short "shock and awe" bombing aimed at the senior Iraqi leadership. Ground forces would quickly move forward fighting only when they encountered resistance rather than looking for enemy forces. Overhead assets could look for the enemy forces and engage them while ground forces raced to Baghdad to seal it off, to prevent Iraqi forces from retreating into the city and its environs and staging a climactic and bloody street-by-street battle.

One of the key techniques to make the strategy successful with a minimum ground force was to find a way to protect the coalition's vulnerable left (western) flank as their forces advanced. The coalition's right flank was Iran, no friend of Saddam's regime. The solution to this problem was to insert U.S. special operations forces (SOF) along with British and Australian commandos into Iraq's western desert. In addition to screening the coalition's left flank while the main body of the coalition forces raced north, the SOF and the commandos would also look for Iraqi WMD storage and production sites.

In the north the Turks, after much negotiation, refused the coalition request to let a full division of U.S. forces land and transit Turkey into northern Iraq. To present some sort of threat to Baghdad from the north, the alternative was to insert SOF forces into Iraqi Kurdistan to work with the Kurds. Hopefully this northern threat would keep Iraqi forces in the north from falling back on Baghdad.

It appeared to many observers (including some senior ranking military officers) that the Iraqi campaign was being planned "on the cheap" and had a far smaller force than required for a quick, smashing military victory. In truth, the coalition had more than enough force deployed to achieve an overwhelming military victory but unfortunately that was not to be the end of the struggle. As in Afghanistan, winning the peace would be much more difficult and much more costly in blood and treasure than winning on the battlefield, and would require far more manpower.

Military Conduct

In a nationally televised address on 17 March 2003, President George W. Bush delivered the coalition's final ultimatum telling Saddam Hussein that he and his sons had to leave Iraq within forty-eight hours. The year-plus long military planning effort and national and international political argument concerning the justification for war with Iraq were coming to an end. The invasion was scheduled to begin just four days hence on 21 March. However, even in the best-planned wars unanticipated events and opportunities can change the timetables.

"Predator" unmanned aerial vehicles had spotted several oil wells burning in the southern Iraqi oil fields, providing a stark reminder of Saddam's notorious burning of the Kuwaiti oil facilities as his forces retreated from that country in 1991. Fearing the worst, General Franks asked his commanders whether the invasion could be moved forward to thwart Saddam's presumed plan. His commanders responded that the ground attack could be safely moved forward twenty-four hours. However, the accompanying aerial attack could not be rescheduled, because it had too many aircraft on tight timetables arriving from throughout the region, from the United Kingdom, and from the United States for the initial bombardment. Franks decided to move the ground invasion ahead of the air attack by twenty-four hours, a very unorthodox move that would totally surprise the Iraqis. Thus the "overt" invasion would begin on 20 March. However, a "covert" invasion of northern and western Iraq by U.S. special operations forces and British commandos would begin on 19 March.

Coalition ground forces began their advance into Iraq from Kuwait on 20 March with 20,000 British troops quickly turning east toward the port of Basra and its nearby oil fields. Meanwhile, 60,000 Marines headed north, advancing on the east side of the Euphrates River toward Nasiriya and then on to Baghdad. West of the Euphrates 65,000 Army troops headed north bound for Najaf and then through the so-called "Karbala Gap" and on to Baghdad. All would be assisted by overwhelming coalition airpower capable of delivering precision close support and locating major Iraqi forces. The following day, coalition airmen, some of whom launched from Europe, the United States, and the Indian Ocean island of Diego Garcia, conducted a so-called "shock and awe" bombing of Baghdad's government and military targets, which had originally been scheduled to coincide with the ground invasion.

As the bombers attacked Baghdad on 21 March, the British arrived at Basra and sealed off the city. With an eye toward postwar reconciliation, the British sought to minimize casualties, particularly among civilians. Accordingly, rather than launching a frontal assault, the British waited patiently while their agents inside the city attempted to incite a popular uprising against Saddam's forces, an effort that eventually would prove fruitless.

By 22 March, the U.S. Army's lead units were already 150 miles inside Iraq and halfway to Baghdad. The next day the Marines, who were already in Nasiriya, became involved in heavy fighting with both regular Iraqi military and paramilitary

Fedayeen forces. Three days later, Army forces west of the Euphrates began what would be a fierce battle at Najaf. Thus far there had been some serious fighting but the forces both east and west of the Euphrates were making very rapid progress toward Baghdad.

Weather interrupted the campaign on 24 March with a very large sand- and rainstorm that would seriously slow progress for three days. Visibility on the ground was reduced to near zero for both Iraqi and coalition forces by what one senior officer, General David Petraeus, called "a tornado of mud." From high above, however, many sensor systems had no problem looking down through the "muddy" atmosphere and locating Iraqi units. Iraqi locations were then the subject of an aircraft attack with precision weapons.

By 26 March, despite the effects of the sandstorm, Army forces had encircled Najaf, which was pounded by heavy air and artillery bombardment of Iraqi forces for several days before mop-up operations began. That same day, paratroopers landed in northern Iraq (Iraqi Kurdistan) to work with special operations forces previously inserted in the region and with the Kurds. This operation was also intended to provide a credible threat to Baghdad and Saddam's regime from the north. It was this action that convinced Qusay Hussein, Saddam's eldest son, that the "real" invasion of Iraq was from the north, and he issued orders for the Republican Guard divisions to turn and move to the north to protect Baghdad.

Meanwhile, on 2 April, the leading Army units moved into and through the "Karbala Gap," a narrow strip of land between the Euphrates River and Lake Rezzaza that had been identified as perhaps the key piece of real estate for the Army's dash to Baghdad. U.S. commanders expected a tough fight for this passageway to the Euphrates and on to Baghdad and they received all-too-believable intelligence warnings of probable Iraqi attacks using chemical or biological weapons. But the Republican Guard divisions assigned to defend the Karbala Gap had been moved toward the new threat from the north, and the biological weapons warnings proved unwarranted. By that evening, Army units controlled the narrow gap between the Euphrates and Lake Rezzaza and had seized a bridgehead across the river. Meanwhile, the Iraqi commander in the area, General Raad Al-Hamdani, who had been ordered to turn and move his forces to the north, was suddenly ordered to reverse his actions and counterattack the American forces pouring through the gap. Amid this confusion Al-Hamdani was able to rally a brigade from the Republican Guard Medina division for the counterattack, but it was too little and too late. The brigade was slaughtered. On 3 April, Army troops from the 3rd Infantry Division reached the western outskirts of Baghdad at Saddam International Airport.

After turning back a counterattack by the Baghdad Fedayeen and engaging and destroying the remainder of the Republican Guard Medina division, the Army units at the Baghdad airport made the first of their "thunder runs" sending a column of tanks and other armored vehicles from the airport into Bahgdad's southwest suburbs and then back to the airport. They were met with heavy fire and strong resistance but in turn took a heavy toll of Iraqi defenders, leaving a very bloody

was immediately disputed by the secretary of defense and his civilian minions who derisively noted that to think the postwar occupation force would need to be larger than the invasion force was preposterous.

The army began relearning the lessons of Vietnam that it had forgotten. Doctrine was rewritten and new commanders came to the field with a new understanding of what needed to be done. A great deal of retraining was conducted in the field. In early 2007 a "surge" of additional U.S. forces began flowing into Iraq, forces that included five Army brigade combat teams, a Marine expeditionary unit, two Marine infantry battalions, and other assorted support and headquarters personnel. The concept was to put the forces to work protecting the population, particularly in Baghdad and in the areas surrounding Baghdad. The result was that from July through December 2007, insurgent attacks declined by more than 50 percent. Meanwhile, better-trained and better-equipped Iraqi forces were finishing their training and deploying to the field.

As of this writing in 2009, it remains unclear whether the effort of the U.S.-led coalition in Iraq will ultimately be successful. There also remains the question as to whether the new Iraqi military can mold itself into an effective and efficient organization that both understands how to combat an internal insurgency and stand ready and able to protect Iraq from its mercurial neighbors.

Better State of the Peace

The Iraq War has not been completed as of this writing in 2009, and although the American phase of that conflict will formally end when the last American combat troops leave in 2010 or 2011, the evolution of the processes within Iraq set in motion by the U.S. invasion and overthrow of the Iraqi government will not be completed until sometime after Iraq itself reaches some kind of new status quo. All the various rationales for the U.S. instigation of this war are ultimately defined by the disposition of the Iraqi state. The American better state of the peace requires that Iraq emerge as a country whose policies are compatible with the objectives for which the United States went to war. The jury is, of course, still out and will likely remain out for some time to come.

Will American objectives be achieved in the long run? If they are, it will be arguable that the war was a success—a "victory"—although there will likely always be residual disagreement on whether the achievement of those goals was worth the effort and sacrifice required to attain them. If they are not, then the war effort will likely be judged as feckless and that judgment will be harsher. It is too early to tell.

Why did the United States go to war with Iraq? The initial stated reasons, of course, derived from the global war on terror (GWOT) and Iraq's purported connection to it. Thus, Iraqi's alleged possession of biological and chemical weapons and its suspected program to obtain nuclear weapons were threats to the United States primarily in a terrorism context: Saddam Hussein could be in cahoots with terrorists who might use those weapons against the United States. This rationale was always somewhat vapid, but gained credibility within the context of a national obsession with terrorism; outside that context, this argument would have almost certainly been insufficient to justify war. If that was the objective, it was akin to the United States going to war in 1812 over British impressments of American sailors. In 1812, the British had renounced the policy of impressments before the United States declared war; in 2003, Saddam Hussein had not been demonstrated to have either WMD or ties to terrorists. In both cases, war was a response to unsubstantiated grievances.

A stronger case can be made for the proposition that the United States went to war to transform an Iraqi political system needing "regime change." It is undeniable that the Saddam Hussein government was despotic, reprehensible, and a barrier to U.S. interests in stability (and access to oil) in the Persian Gulf region and the Middle East more generally. There are, however, two sources of controversy about regime change as a political objective. The first is whether it was worth it to the United States. Admitting that Saddam Hussein was an embarrassment and a nuisance to the United States, what vital American interests were violated by his continuation in office? And what vital interests did it become possible to realize after his removal? In terms used throughout this book, this is the question of worthiness of the objective.

The second source of controversy was whether regime change producing a desirable outcome was possible. The neo-conservative argument that the replacement of Saddam Hussein would have other positive regional effects was, after all, premised on the notion that a subsequent regime would be a noticeable improvement over its predecessor. Whether that is true, of course, is an imponderable at this point in time, but the indicators are not entirely positive. The goal of a westernized, democratic Iraq serving as a regional model may be impossible given the sharply fractious nature of Iraqi society. If Iraq emerges as a stable, fully participatory political democracy, then this second objection will be overcome, but by 2008 the United States had admitted this was not really the likely outcome by scaling back its expectation from a democratic to a stable Iraq. If Iraq disintegrates into communal chaos and fighting, it will be strong evidence the goal was unattainable.

This question of attainability particularly extends to the international relations of the post-American situation in Iraq. For many of those who consistently argued in favor of war against Iraq over the years, the ultimate reason for regime change in Iraq was the positive impact it would have by stabilizing the region and making it less a source of international contention. For this to occur, Iraq must serve as a beacon for other regional states, an exemplar of the benefits of democratization that causes its neighbors to demand and achieve emulation of the Iraqi example. In turn, regional stabilization will be cemented by an improved Arab-Israeli relationship, since democratic Arab states and a democratic Israel will realize they have more in common as unions of free peoples than can justify continuing warfare. Were those conditions to ensue, the Middle East would cease to be a source of turmoil and conflict with which the United States must contend in pursuing its combined, and often contradictory, goals of a free and secure Israel and secure access to Middle Eastern energy reserves. If the U.S. effort in Iraq leads to all these outcomes, then it will likely be adjudged a success worthy of the sacrifice it entailed.

These long-term, laudatory outcomes are sequential and beyond the immediate scope of the Iraq War itself. Iraqi regime change and transformation have been the immediate goals of the war, and their achievement is somewhere in the future, after the war itself is over in the sense of a visible American military presence in Iraq. If that goal is attained, the other sequential steps become at least possible; if it is not, the preconditions for the other results will not be present, and the war effort cannot be blamed for not producing them. Whether democracy will take root or regional stability will occur is conjectural, of course, and is a criterion of success for the U.S. action in Iraq that will take even longer to unfold.

It is thus impossible to assess a better state of the peace from the Iraq War now, and that limitation will almost certainly remain for some time into the future. The United States was successful in overthrowing the regime of Saddam Hussein and with replacing it with something else that meets at least minimal definitions of political democracy, and that can be viewed as a positive part of the better peace. Whether it was an adequate reason for a long, costly war is a different, but not irrelevant question. The regional goal for the United States is a stabilized Middle East that will, among other things, reduce or eliminate international religious (Muslim) terrorism, provide secure access to energy resources, and help assure the continued security of Israel. The action in Iraq was the linchpin of a strategy to achieve these ends. It is, according to its proponents, the necessary first step, but whether it is sufficient remains to be seen. Given that those ends are somewhere in a murky future, assessing whether a better state of the peace will have been achieved must await the future.

11

AMERICA'S MINOR WARS

In addition to its major military conflicts, the United States has fought a number of smaller wars. Some, like subduing the Barbary pirates, many of the Indian campaigns, and more recently the action in Grenada, were of such minor extent as not to warrant individual attention here. Three historical cases, however, stand out as being of sufficient significance to merit some detailed discussion: the War of 1812, the Mexican War, and the Spanish-American War.

Although each was a unique event, these wars shared some common characteristics. The first and most overarching commonality was that they lacked the polarity of moral crusades, and this made them limited, both in terms of the political purposes for which they were fought and the means available and/or necessary to prosecute them. Because they were not moral crusades, each had within it the seeds of potential unpopularity, and the War of 1812 and the Mexican War were, at best, limited in their popularity. The apparent exception was the Spanish-American War. The public was solidly behind the war when it began and that support never wavered because it was over too quickly and successfully for opinion to turn against it.

The second and third commonalities, which are related, represent common themes in American military history through World War II. These threads are that the United States was unprepared to fight any of these conflicts and, as a result, had to raise and put in the field amateur armies that succeeded almost despite themselves. The militia tradition was alive and, if not well, at least prominent in each case. Of all of America's wars, the country was least prepared for that theater of the Napoleonic Wars we call the War of 1812. The period between the end of the American Revolution and the outbreak of war in 1812 was marked by an almost immediate dismantling of the Continental Army (total active duty strength in 1784 was 80 soldiers protecting military stores). A national security debate dominated by Jeffersonians suspicious of the military had resulted in virtually no expenditure for military preparedness. Similarly, an army had to be raised almost from scratch to fight the Mexican War, although the young West Point system had at least provided the country with a professional officer corps. To fight Spain, an armed force of 275,000 was mustered into service, but the war was over so quickly that only about 35,000 ever saw combat.

Because the United States had no real standing force with which to prosecute any of these conflicts, it had to rely on the traditional recruitment methods of appealing to the state militias and making exhortations to individuals to volunteer (conscription was unthinkable in any of these wars). Because at least some of the political purposes for which the United States entered combat were served in each case, the myth persisted that this was an effective and efficient way to field an armed force capable of prosecuting the full range of military operations. The myth is just that and does not hold up under scrutiny. In fact, the military forces raised in this manner, although competent at some defensive tasks, were generally ineffective and succeeded, when they did, thanks to overwhelming odds or the incompetence of the adversary. Since Americans did not critically scrutinize this myth, it persisted well into the twentieth century, guaranteeing the United States would enter the world wars unprepared as well.

The fourth and final source of commonality was that American territorial expansion was an underlying cause of each. In the War of 1812, a major issue that united the "War Hawks" was the annexation of Canada. The Mexican War was fought for and succeeded in fulfilling America's "manifest destiny" to control the continent from ocean to ocean. The Spanish-American War had as part of its motivation and as a large part of its outcome the creation of an American Empire.

If there were common themes in these three wars, there were unique aspects to each as well. The War of 1812 has the distinction, if that is the proper term, of being the closest thing to a decisive military defeat the United States has ever suffered. In that conflict, the most notable American victory in a land battle (and one of very few victories), the Battle of New Orleans, was fought three weeks after the peace treaty ending hostilities had been signed and was the only occasion when the British attacked strong fortifications. As the result of unpreparedness and an amateur military leadership, the conduct of the land portion of the war was almost a national disgrace. The reason the United States avoided a decisive military defeat and achieved at least part of its objectives was that the British were too tied down and war weary from their campaigns against Napoléon to administer the whipping that was physically theirs to give.

The Mexican War has the distinction of being the first U.S. war that was overtly political in the sense that purely political considerations intruded into the way military operations were conducted. President James K. Polk intruded into military action to an extent previously unheard of. The reason was that the two major commanders in the field, Generals Winfield Scott and Zachary Taylor, were considered the leading contenders for the Whig presidential nomination in 1848 and were hence potential opponents of the incumbent president. Polk is said to have directed his generals so as to minimize their headlines rather than maximize military advantage.

The war with Spain had at least two distinguishing aspects. First, it was the shortest and least bloody of America's historical wars. The invasion of Cuba was a walkover hardly more difficult than the 1983 assault on Grenada, and the major naval engagement at Manila Bay resulted in one American death, a coaler who died of natural causes. Second and more important, the war was the most openly imperialistic American military adventure. Many American citizens were moved to support the war out of humanitarian concern for the plight of the citizens of Cuba, but there was more. That additional motivation was imperialistic and manifested itself in the belief that American manifest destiny could be fully served only through the acquisition of empire. To some, empire was the symbol of great power status. To others, such as Alfred Thayer Mahan, the United States needed an island empire in the Pacific Ocean to secure American access to Asian markets. In the end, of course, America acquired its island empire by stripping Spain of its possessions.

WAR OF 1812

In many ways, the largely unnecessary War of 1812 was the final episode of America's fight for independence and recognition within the family of nations. At the same time, the fact that the United States found reason to declare war on Great Britain provided evidence, if unrecognized at the time, of the de facto bond between the fate of the United States and the affairs of Europe. The two themes came together because it was British harassment of American attempts to trade with both sides in the Napoleonic Wars that created the passions that resulted in an American theater of that conflict.

Issues and Events

Although the United States had won formal independence from England in 1783, it was a freedom that did not carry with it much British respect. Apparent British disrespect took two major forms. The first and most important was British treatment of a burgeoning American maritime enterprise. The United States, after independence had been won, rapidly blossomed as a major trading power with a large merchant marine but without a navy of any note. The decision not to build a "blue-water" navy reflected traditional American distrust of military force in peacetime, but also was based on an assessment that we could not compete with the Royal Navy under any circumstances. As a result, American trade was reliant on the goodwill of the Royal Navy. When the Napoleonic Wars broke out and the United States attempted to engage in trade with both sides (a phenomenon that would be repeated a century later), the British understandably were much more enthusiastic about promoting and protecting trade with themselves than with their enemies, a sentiment the French reciprocated in regard to American trade with Britain. Lacking the means for self-protection, the United States was in a position of intolerable vulnerability.

The other form of disrespect was supposed British activities along the American Frontier. From its territories in Canada, the British dispatched traders and agents along the western boundaries of the United States. Their purported purpose was to engage in trade with Indian tribes who lived along those western reaches, but Americans believed they did more than that, notably fomenting harassment and attacks by the Indians against the settlers as a way to retard American westward expansion.

The issue that provided the proximate cause of war was the British policy and practice of impressment of sailors from American ships into the Royal Navy. This practice was part of the general British harassment of American shipping. It took the form of halting and boarding American merchant vessels and impressing deserters from the Royal Navy who had been granted protection by the United States. Although the problem of desertion was vexatious for the British, impressment was clearly in violation of international law. Having been granted protection and, in some cases, citizenship, these sailors were Americans. Their forcible removal was thus an act of war, and British defense of the action based in the idea of "indefeasible nationality" (that the sailors could not renounce their British nationality as a way to avoid service) was at best a shaky argument. What the practice really demonstrated was the disdain with which the British regarded American independence and sovereignty. Between 1809 and 1812, the annual rate of impressment ran between 750 and 1,000, and nearly 6,000 sailors had been impressed by the time war was declared.

Political Objective

At heart, the war came about because of U.S. frustration over its inability to resolve satisfactorily those matters that irritated British-American relations. From the point at which commercial desires had come into conflict over trade with France, the United States had tried a number of political methods to deal with the problem, but none of them worked. Of all those measures, the Embargo Act of 1807 probably best demonstrated American frustration. This act forbade any American ship to sail from any American port to any foreign port, and thereby guaranteed that the price for American security would be the loss of trade on which much of American prosperity rested.

Growing frustration created a mood in at least part of the American public that led to the election in 1810 of a group in Congress whom the Federalists dubbed the "War Hawks" because of their determination to build up and possibly to employ military force. By 1812 parts of the United States were simply itching for a fight.

That fight had to have a political purpose, and the stated objective was the ending of impressment as a symbol of the British practice of naval harassment. As an objective, ending impressment could hardly have been more ironic: crop failures during the winter of 1811–12 had forced the British government to move to reinstate trade with the United States, and on 16 June 1812 the orders allowing impressment were formally rescinded. The U.S. declaration of war passed Congress two days later. Because of the slow speed of contemporary communications, Congress was unaware of the rescission when it voted, but Congress did not reverse the declaration when it learned that stated objective was no longer at issue.

There was a secondary goal as well. That goal was the conquest of Canada, although there was disagreement on why that purpose was to be undertaken. In the view of the more radical elements within the country, the purpose of conquest was annexation. The North American continent was to be rid of the British once and for all, removing a major barrier to American expansion. To less ambitious individuals, the purpose of the invasion and occupation was to create leverage with the British, a "bargaining chip" to gain concessions from Great Britain on the more fundamental issue of naval harassment.

Military Objectives and Strategy

Put in the simplest terms, the American military objective was to seize Canada. How the seizure of Canada would protect neutral rights, guarantee freedom of the seas, or secure the other objectives for which the United States went to war was never clearly established. Canada was the nearest British target that the Americans could strike, and its capture became the rallying cry for the War Hawks.

The fundamental problem was to determine how to defeat the British forces and drive them from Canada. The seat of British power north of the border was the city of Quebec, which was the most obvious American target. However, as long as the British navy controlled the Saint Lawrence River, the city was virtually impregnable.

The second choice was Montreal, a city that could easily be reached by the river-lakes invasion route used by both sides during the Revolution. In retrospect, it appears that a strong, concentrated thrust against Montreal by the small American Army might well have succeeded. However, American planners cast their eyes on several other targets as well.

On the northern shore of Lake Ontario, Kingston and York presented tempting targets as did the British forts between Lake Ontario and Lake Erie in the area known as the Niagara Frontier. Finally, Fort Malden opposite Detroit seemed to be a fine target. Attacking these places could have provided excellent diversions to complicate British defense problems. Unfortunately, as planning progressed, these targets became less diversionary and were transformed into major military objectives that seriously diluted the American effort. Given the poor state of American forces, a concentrated thrust against Montreal was a reasonable strategy, although far from a guaranteed success. Dividing the small American forces over several objectives virtually guaranteed failure on all fronts.

The final campaign plan called for three simultaneous thrusts into Canada. The first, composed of the main regular Army, was to advance from Plattsburg, New York, by the river-lakes route against Montreal. The other two thrusts, composed mostly of militia, were to cross the Niagara River to attack British forts on the other side and to advance from Detroit on Fort Malden. The limited American skill in military planning and staff work was revealed in the lack of a coordinated start time for the three offensives.

The British plan was simply to hold on. The difficulties with the United States were a minor affair compared with the struggle with Europe. The British gave almost complete discretion to their generals on the scene, even though they did not represent the cream of the British army. By 1814 all of this would change as the British were able to concentrate far greater forces in North America and undertake an offensive strategy.

At sea the American navy wrestled with the problem of confronting the mighty British fleet. Commodore John Rodgers experimented with battle squadrons formed around the few heavy American frigates, but had little success. American captains preferred single-ship sorties to prey on English commerce and engage single English warships. They achieved some success with this tactic, but their impact was of little consequence to the outcome of the war. The navy was unable to lift the British blockade of American ports and could not prevent the destructive and embarrassing British raids in Chesapeake Bay.

Political Considerations

The War of 1812 was, until the Vietnam experience, America's least popular war, and the major political consideration during its conduct was how to gain enough support to continue it. Lack of popular support was largely sectional, but grew because of the incompetent manner in which the war was conducted and the debilitating economic consequences of its dragging on.

Such support as the war had at its onset was regional. The War Hawks, by and large, came from the South and the Western Frontier, and these were the areas where support was greatest. Their interests were in subduing the Indian "allies" of the British and opening up the continent for expansion. Sentiment for invading Canada was, consequently, greatest in these regions as well. Ironically, the part of the country that would have benefited most from ending naval harassment was commercial, maritime New England, where opposition was greatest. In New England, there was widespread belief that the United States had no real military chance against Great Britain and that the only possible outcome was a military defeat that would destroy what little was left of American commerce with Europe.

If public opinion was a problem at the beginning of the war, it increased as the war continued. Part of the growing opposition was fueled by the incompetence of the military effort and the obvious fact that no military progress was being made. The dreary outcome of the Canadian campaigns reinforced New Englanders' skepticism of the enterprise, and the fact that the United States had few victories in land battles until New Orleans meant the government had few glories with which to rally support. Moreover, paying for the war threatened to bankrupt the treasury by 1814,

American overseas trade was reduced to a trickle by the Royal Navy's blockade, and the combination of these economic factors produced runaway inflation. Lack of public support was a bad problem that got worse, and only the American victory at New Orleans served to assuage the bitter memories and allow some positive reconstruction of support for the government after the war ended.

Internationally, the war has to be considered within the context of the overall struggle in Europe. European issues created the conditions that brought the war about, and the conclusion of the titanic struggle against Napoleon set the stage for its conclusion. As a part of that great struggle, the North American theater was militarily little more than a minor irritant to Great Britain that warranted a minimum military holding action (fortunately for the Americans). In the end, Britain agreed to a cessation of hostilities not because of any negative assessment of the military situation, but because the British people were war weary at the end of the Napoleonic Wars and wanted peace.

Military Technology and Technique

The War of 1812 was fought only three decades after the final battles of the American Revolution. The military technology available to the antagonists was nearly identical to that available in the Revolution. The standard infantry weapon remained the smoothbore flintlock musket. Its slow rate of fire and its limited range and accuracy put a premium on linear formations using massed volley fire. Such tactics, as discussed in the chapter on the Revolution, put a premium on continuous drill and training. Iron discipline was required if opposing lines of infantry were to march toward one another properly aligned for volley fire and to get within effective range of the smoothbore muskets.

Discipline and training were the weak points of the American army. Virtually disbanded after the Revolution, the regular army had a total strength of only about 11,000 men by 1812, and nearly half of them were recent recruits brought into service for the conflict. These forces were augmented by militiamen, many of whom served only short tours of duty lasting from one to six months. Although their intentions were good, these forces appeared as rabble compared to their well-trained and disciplined British opponents. The saving grace was that there were few British professionals available to defend Canada because the British were occupied in Europe with the Napoleonic Wars. Only 6,000 British regulars were stationed in Canada in 1812, augmented by several thousand militia and Indians.

American troops were inexperienced and so were the officers who led them. Nearly 30 years had passed since the Revolution, and only the most senior American officers had any experience fighting a modern European army. At lower levels, officers were almost wholly deficient in the technical skills of warfare. Few realized the benefits of constant drill, and those who did had to rely on copies of European drill manuals. The only bright spot in this otherwise dismal picture was the presence of some graduates of the United States Military Academy at West Point, which had been founded in 1802. However, only 89 officers had graduated by 1812 and all were junior in grade.

At sea the Americans faced the world's greatest naval power with a fleet of between 15 and 20 vessels. (Fortunately, the British fleet had considerable obligations throughout the world in opposing Napoleon's forces.) At the heart of the American naval forces were seven frigates, three of which were classified as heavy frigates. These three ships, each of which mounted 44 guns, were the most powerful of their type in the world. Larger and more heavily gunned than comparable British frigates, the American ships were constructed of fir rather than oak. Unlike the army, the navy provided excellent training, with particular emphasis on gunnery. American naval gunners were among the most proficient anywhere.

One major change from the days of the Revolution was in the availability of arms. The American government had encouraged a significant arms industry and had established two government arsenals, one at Harper's Ferry, Virginia, and the other at Springfield, Massachusetts. Although arms were available in plentiful numbers, getting them to the troops in the field along with the other wherewithal of war was a difficult task. The logistical capabilities of the American military had been all but ignored in the period since the Revolution.

Military Conduct

The ill-conceived American offensive into Canada got off to a poor start when the three prongs of the planned offensive failed to begin at or near the same time. The first prong to start was led by William Hull, governor of the Michigan Territory, who sought to capture Fort Malden across the Detroit River from Fort Detroit. Hull marched from Dayton and had to cut 200 miles of road through the wilderness over which to haul supplies. He crossed the Detroit River on 12 July 1812 with a force of roughly 2,000 (about 1,500 of whom were Ohio militiamen). Meanwhile, Fort Malden had been reinforced by a small British force led by Gen. Isaac Brock. The elderly Hull suddenly became worried about his supply lines, which were vulnerable to raiding parties that could be landed across British-controlled Lake Erie or to attacks by Britain's Indian allies. Giving in to his fears, Hull retreated across the river into Fort Detroit on 7 August. Brock quickly followed, and bluffed Hull into surrendering his entire command on 16 August. Brock's victorious forces numbered only 730 Canadians and about 600 Indians. Thus, the first American offensive ended in total disgrace with hardly a shot fired in anger.

The second prong was on the Niagara Frontier, where Stephen Van Rensselaer, a major general in the New York militia, led a force of more than 3,000 (including 900 regulars) across the Niagara River against Queenston. On 13 October 1812 the regulars crossed the river and quickly captured the heights above Queenston. However, the militia refused to follow the regulars and would not leave American territory. They stood idly by while forces led again by Isaac Brock, who had rushed to Queenston after his victory at Detroit, destroyed the

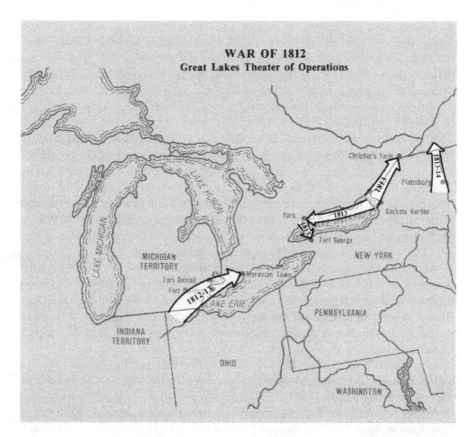

WAR OF 1812
Great Lakes Theater of Operations

American forces on the heights. Thus, the second American offensive also ended in disgrace and defeat.

The third prong had Montreal as its target and was originally planned as the main thrust into Canada. In mid-November, Maj. Gen. Henry Dearborn led his 5,000 troops from Plattsburg to within two miles of the Canadian border, but again the militia portion of his force refused to cross into Canada. As a result, Dearborn returned to Plattsburg and went into winter quarters. Thus, the third American offensive aborted, and the American ground strategy was in total disarray.

At sea the navy had considerably greater success. Facing overwhelming odds, American ships engaged in several brilliant single-ship actions. However, the British blockade was beginning to strangle commerce along the Eastern Seaboard. By the end of 1812, nearly 100 British naval vessels were participating in the blockade, including 11 ships of the line that the American navy could not match.

Early in 1813 Brig. Gen. William Henry Harrison was given command of a large, mostly militia force with orders to recapture Detroit. Moving forward toward the mouth of the Maumee River in January, an advanced unit of his force was destroyed while attempting to relieve the beleaguered residents of Frenchtown. Harrison was subsequently ordered to bide his time until the Americans could seize naval control of Lakes Erie and Ontario. While Harrison waited, Oliver Hazard Perry was building ships at Presque Isle on Lake Erie. By the fall Perry was ready and totally defeated the British squadron at Put-in-Bay. Harrison then pressed his advance; the British evacuated both Detroit and Malden, and were decisively defeated at Moravian Town on the Thames River.

Meanwhile on Lake Ontario, General Dearborn attacked and burned York in April, sailed to Fort Niagara, and in May attacked and captured Fort George. However, the British contingent was allowed to escape, and Dearborn was replaced by Maj. Gen. James Wilkinson. By October Wilkinson had stripped most of the troops from the Niagara area to launch a campaign down the Saint Lawrence River against Montreal. The campaign aborted after a defeat en route at Christler's Farm. Worse, the British quickly captured the almost defenseless Forts George and Niagara. In all, 1813 was not a sterling year for American arms.

Action was renewed in March 1814, when Wilkinson took to the offensive but was defeated by a smaller British force at La Colle Mill on 30 March. Wilkinson fell back on Plattsburg and was replaced by Maj. Gen. Jacob Brown. On 2 July Brown crossed the Niagara River and quickly seized Fort Erie. Pressing on to the north, the Americans defeated the British at Chippawa on 5 July. At Lundy's Lane on 25 July, the Americans fought to a draw with the advance units of British reinforcements from Europe. Brown retired to Fort Erie and the British followed and laid siege to the Americans. On 17 September Brown's successful sortie from the fort broke the siege.

Meanwhile more British reinforcements arrived from Europe after the fall of Napoleon in April 1814. Sir George Prevost led a veteran army of 11,000 south along the river-lakes route to confront a much smaller force of Americans entrenched at Plattsburg. However, a brilliant naval victory by Commodore Thomas Macdonough at Plattsburg Bay defeated the supporting English fleet on the lake and Prevost quickly returned to Canada.

Along the Eastern Seaboard, the British planned raids against both Washington and Baltimore. In August Maj. Gen. Robert Ross's British troops arrived on the Patuxent River supported by a fleet commanded by Adm. Sir Alexander Cochrane. After defeating the Americans at Bladensburg, the British burned most of official Washington. When the British moved on Baltimore, they were heavily engaged at Godly Wood, and then determined that Fort McHenry and the hastily erected defenses around Baltimore presented too difficult a challenge. The entire British force subsequently sailed away.

In the South, the British launched a campaign to capture New Orleans. Maj. Gen. Andrew Jackson rushed to the scene, mobilized all available forces, which were mostly militia, and established a defensive line along the Rodriguez Canal. On 8 January 1815 Sir Edward Pakenham launched an infantry attack against Jackson's well-fortified position. The attack was repulsed and the British suffered 2,100 casualties plus an additional 500 captured. Jackson lost seven men and six were wounded. On 18 January the British withdrew. Neither side realized that the Peace of Ghent, which officially ended the war, had been signed on 24 December 1814.

Better State of the Peace

In the end, something resembling the attainment of American basic political interests occurred, although in small measure as the result of American military activity. The War of 1812 is a war that the United States lost by most military measures one might apply. Unlike Vietnam, however, a better state of the peace for both Great Britain and the United States did come to pass.

If the basic issue was an end to British naval harassment, the objective was achieved before the war began, and the American war effort had nothing to do with a British decision based in domestic considerations. As it worked out, freedom of the high seas and British–American trade served the interest of both countries. Once the Napoleonic Wars were finally concluded in 1815 and the conflict between the two Anglo-Saxon states was resolved, that mutual advantage came to be recognized. The best statement of British recognition is that through the rest of the nineteenth century the same navy that had been the prime tormentor of American commercial maritime activity acted as its primary shield and protector.

The other political objective, the conquest of Canada, quite obviously was not attained, and was in fact abandoned after 1813. Given the resources available to the United States and the sheer size of Canada, it was probably an unattainable objective anyway (as a military matter the objective created problems not unlike those facing the British in the Revolution). Moreover, had the United States pursued that objective with greater rigor and especially if we had had any success, the British might well have been forced to turn full force against the United States after Napoleon was defeated rather than agreeing to the Peace of Ghent.

The primary American objective (the end of impressment) was achieved by indirection, but it was achieved. The question that must be asked is whether anything was learned from the experience. The answer, especially from a military standpoint, is mixed. On one hand, Americans quickly forgot about the ignominy of the war, and thus the method of recruiting armies to fight for America, the militia tradition, was not amended. On the other hand, the extreme deficiencies in military leadership evident throughout the war did lead to the invigoration of the U.S. Military Academy at West Point to train a professional officer corps to fight in America's next war. The fruit of that lesson would first be demonstrated in the Mexican War.

MEXICAN WAR

The war between the United States and its neighbor to the south was a clash between expansionist American nationalism and a protective Mexican nationalism that sought to preserve control over a vast stretch of the western portion of the

North American continent. From the American perspective, the war was a struggle to fulfill the sense of what Americans had come to view as their "manifest destiny" to control the continent from the Atlantic to the Pacific. Equally virulent Mexican nationalists sought continued control of lands that were part of Mexican independence from Spain.

Issues and Events

The issue of manifest destiny had grown in the years after the War of 1812 as Americans forged westward. In that period, the territories of the Louisiana Purchase received their first large-scale settlement, Florida was wrested from Spain, and the location of the northern border of the Oregon Territory was fixed after we had nearly gone to war with Great Britain over the issue ("54–40 or fight"). When that latter dispute was resolved with the border along the Washington-British Columbian boundary, the United States spanned the continent. All that was left were the territories of the Southwest, especially after Texas was annexed in 1845.

These territories were, of course, parts of Mexico that had been colonized originally by the Spanish and that became part of sovereign Mexican land after Mexican independence. The Mexicans had not, however, done much to settle any of the affected lands, which were largely populated by Indians and a few missionaries before the Americans began to arrive in numbers. The largest number of Americans came to California, and the American claim to the land rested largely in the fact that the majority of the population was American and desired to be part of the United States rather than Mexico.

Complicating the entire situation was growing American sectionalism and the resultant tendency to view any attempt at expansion in the context of the balance between slave and non-slave states. The desire to annex southern territories as diverse as Cuba and the Yucatan had been squelched because either would have tilted the balance of power in Congress toward the slaveholding states, and the same issue had been the major factor in delaying the entrance of the Texas Republic into the Union. Ironically, there was also opposition to the war in the southern United States because the lands of the Southwest were poorly suited for the plantation culture.

The proximate event that made war between the two countries inevitable was the admission of Texas into the United States, and it did so in two ways. First, although Mexican authority over Texas had been ended by the military defeat of Mexican forces in 1837, Mexico did not accept the independence of the Texas Republic. Rather, the government of Mexico maintained that it retained sovereign authority and that the Texas Republic was a rump state with no legitimate authority. The recognition of Texas by the United States and the consequent decision to admit Texas into the Union was, in the Mexican view, a simple act of imperialism.

Regardless of the international legal niceties involved, American authority in fact was being exercised in Texas, which raised the second problem. That problem, which ultimately led to the start of the war, was in determining exactly where the boundary between Texas and Mexico was. In the American view, the border was the Rio Grande River. Mexico maintained that the Nueces River formed the boundary. Although the disputed territory was largely unpopulated, American accession to the Mexican claim would have ceded most of west Texas to Mexico. When Gen. Zachary Taylor endeavored to settle the matter by occupation and encountered Mexican forces, war was the result.

Political Objective

The political purposes for which the two nations fought the Mexican War were clear, concise, and symmetrical, as well as limited. For the United States, the objective was the annexation of the territories of New Mexico, Arizona, and California to the United States and the accession of Texas south of the Nueces River. For Mexico, the objective was to keep those same territories as part of Mexico. The objectives were limited in that neither required the physical conquest and overthrow of the opposing government (although that was the ultimate means the United States used to accomplish the task).

The major question was how to go about resolving the issue. Both Mexican and U.S. nationalisms were too strong to allow a diplomatic compromise, and Mexican realization that a majority of the population in the pivotal territory, California, was American meant Mexico could not accede to decision by majority will. At the same time, the lands under question were vast and mostly unpopulated except for indigenous Indian tribes. Hence de facto control in the form of occupation was impractical. The only way left for settling the issue was the sword.

Military Objectives and Strategy

The Mexican War was a new experience for Americans in two major respects. First, it was the first American war in which the United States was the more powerful antagonist. Although the regular Mexican army was larger than the American army, the United States could mobilize and field a far larger force. The American industrial economy was vastly superior to the primitive Mexican economy and far more self-sufficient. In every measurable respect, Mexico was the weaker of the two powers.

The Mexican War was also a new experience because it was America's first offensive war. Although the War Hawks had attempted to turn the War of 1812 into an aggressive seizure of Canada, the Mexican adventure was the first truly offensive war. Offensive wars require offensive military objectives, which caused President James Polk a certain amount of concern.

The problem was how to make the Mexican government give up its claims to the disputed territories. Polk was determined to act as the commander in chief, and he played a significant role in the determination of offensive plans, in the supervision of the military staff, and in the selection of key personnel to lead the war effort. Polk was also concerned that the war be over quickly. He was particularly concerned that a long war might become so unpopular that the Democratic party would suffer at the ballot box in the next election. However, he also believed that the Mexicans could be easily and quickly beaten if attacked with enough vigor.

The Americans' general plan was for an expedition into the northern Mexican provinces to seize and hold them until the Mexican government came to terms. In addition, the Navy's Gulf Squadron was ordered to blockade the eastern coast of Mexico and to seize Tampico as a logistics base. The blockade was particularly significant because the Mexicans had no arms industry and imported nearly all of their armaments. The blockade meant that the Mexicans could rely only on the stock they had on hand. At the same time, the Pacific Squadron was to seize San Francisco and blockade the California coast.

Zachary Taylor was given command of the main effort in northern Mexico (after command had been offered to and refused by Winfield Scott). Taylor achieved considerable success in northern Mexico as he captured Monterrey and Victoria and then soundly defeated the Mexicans at Buena Vista. But despite these American victories and the tightening American blockade, the Mexicans refused to give in. Finally, Polk approved of Scott's plan for landing a major force at Vera Cruz, followed by a direct march on the capital at Mexico City. This campaign, conducted brilliantly under adverse conditions, finally broke the back of the Mexican resistance 16 months after the outbreak of hostilities.

Political Considerations

"Mr. Polk's War," as its opponents labeled it, was not a terribly popular event. As in the War of 1812, its popularity varied considerably by section of the country. At the same time, domestic, partisan politics affected the war's conduct in a way previously unknown in American military affairs.

Because the war's successful completion meant that the extension-of-slavery issue could not be avoided, the war had significant opposition in those parts of the country where the issue produced the most passion. In New England, a large part of the population viewed the whole endeavor, coming as it did on the heels of the admission of slaveholding Texas to the Union, as little more than a Southern plot to add more slave territories to the Union. Southerners, on the other hand, believed the land and climate of the Southwest were inhospitable to the cotton agriculture that underlay plantation society, and they were equally suspicious of Northern intentions. Beyond those Americans in the Southwest, support for the enterprise came from people in the Mississippi Valley and the Northwest who, as pioneers themselves, were most infected by the notion of manifest destiny.

As stated earlier, the Mexican War was, to that time, the most blatantly political war (in terms of its conduct) in American history, and Polk's role in directing the military effort was the most active that a president had ever undertaken. Partly, this phenomenon resulted from the fact that Polk was an activist president. At the same time, the partisan issue of who would be his opponent in an expected reelection bid in 1848 always loomed in the background and influenced military decisions. Had the war been a more difficult military undertaking and had the outcome been in substantial doubt, this intrusion might have created the kinds of howls about political interference that plagued later American military endeavors. As it was, controversy over the conflict was a large factor in Polk's subsequent decision to retire at the end of his first term in office.

Military Technology and Technique

Several improvements in military technology took place between the War of 1812 and the Mexican War. However, most military tactical techniques remained almost unchanged because most of the basic limitations of the principal weapons remained unchanged. The standard infantry weapon was still the smoothbore musket with its limited range and poor accuracy. The standard artillery piece was still the smoothbore muzzle-loading cannon. Both of these weapons, however, had undergone significant improvements that increased their reliability.

The standard flintlock musket was being replaced both in the United States and in Europe by the percussion musket. The development of the percussion cap, although it did not appreciably change infantry tactics, did have a significant impact on war itself. Unlike flintlock weapons, percussion weapons operated reliably regardless of rain or snow. Thus, one of the reasons for seasonal campaigning and standing down in "winter quarters" in past wars was eliminated.

The standard artillery piece had also undergone a significant change. Field artillery had become far more mobile and now could be dragged into battle by horses at a rapid pace and swung into action quickly. In this particular war, American "flying artillery" was superior in both speed of maneuver and in numbers to its Mexican counterpart.

The distances to the Mexican battlefields presented significant logistical problems. Much of the logistic load was carried by sea, and although steamships were available, few were used in the oceangoing transportation effort. However, steam-powered riverboats were used extensively to transport men and supplies to ports of embarkation. River transport was particularly important because few rail lines had extended far enough west and south to be of significant assistance in the logistics effort. Within the theater of operations, supply still depended on horses and wagons, and the American effort was so large that a shortage of wagons hindered some of the early operations.

The Mexican War also witnessed significant American amphibious operations. General Scott's landing at Vera Cruz was exceptionally well done and was the result of considerable planning. Much was learned about amphibious combat techniques that would prove useful during the Civil War. As will be seen later in this chapter, these lessons were, for the most part, long forgotten by 1898.

Although several American general officers acquitted themselves well, Scott came away with a reputation as a great military tactician that remains untarnished today. He won consistently, with limited casualties to both sides. He preferred to defeat the enemy by superior maneuver, thus forcing the enemy to retreat. His campaign from Vera Cruz to Mexico City was a model of eighteenth-century maneuver warfare.

West Point-educated officers also built their initial reputation during this war. None of the American generals had any formal military education. As a group, however, they made good use of the skills of the West Point graduates on their staffs. Scott, in particular, was impressed and claimed that the war might have lasted far longer had it not been for the skilled work of the West Point officers, including many future Union and Confederate generals in the Civil War.

Military Conduct

The Mexican War was not declared until 13 May 1846, but several important battles were fought before that date. Following orders from President Polk, General Taylor with most of the regular American army (about 3,500 men) advanced south from Corpus Christi to the Rio Grande and established a camp opposite Matamoros on 24 March 1846. Additionally, he established a supply depot at Point Isabel some 45 miles to the northeast on the Gulf Coast. The Mexicans reacted by concentrating nearly 6,000 troops at Matamoros.

Both sides glared at each other for a month. Finally on 25 April, the Mexicans mounted a strong cavalry sweep north of the river and overwhelmed a small American reconnaissance force. Taylor reported to the president that hostilities had begun and took part of his command to Point Isabel to protect the depot from the marauding Mexican cavalry. Meanwhile, Mexican Gen. Mariano Arista led his 6,000 troops across the river against Taylor's base camp, which was under the command of Maj. Jacob Brown. Arista laid siege to the camp from 3 to 8 May, but Brown's defense was successful, although he died in the action.

Arista then moved to place his force between the camp and Taylor's force, which was returning from Point Isabel. The two forces met at Palo Alto on 8 May and the Mexicans were forced to retire primarily because of superior American artillery. The following day, 9 May 1846, Taylor attacked a strong Mexican position a few miles to the south at Resaca de la Palma. The Mexican forces broke after a brief but fierce fight, and Arista withdrew across the Rio Grande, retreating nearly 100 miles to Linares. Four days later, the United States declared war on Mexico.

Taylor crossed the Rio Grande and entered Matamoros on 18 May, but was forced to delay further operations while he awaited the transportation (wagons) and reinforcements needed for offensive operations. By August, Taylor had the wagons and men he needed and moved south on Monterrey with a force of 6,000. The city was well fortified and defended by Gen. Pedro de Ampudia with a force of about 10,000. On 24 September, after a fierce three-day fight, Ampudia offered surrender of the city and an eight-week armistice, which Taylor accepted. Polk repudiated the agreement, but the news did not reach Taylor until seven weeks had elapsed. In the

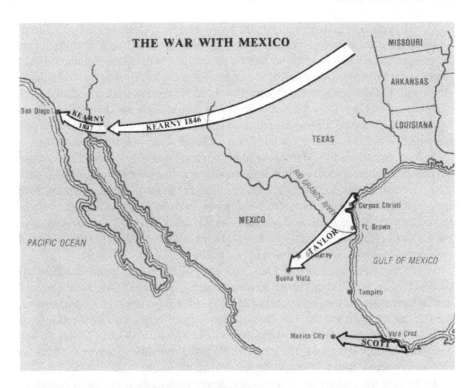

THE WAR WITH MEXICO

meantime, the Mexicans had fallen back to San Luis Potosí where the new Mexican president, Antonio López de Santa Anna, was reorganizing his forces. Taylor, in the meantime, moved slightly south of Monterrey and occupied Saltillo.

To this point, the Mexicans had met with consistent failure. Taylor had marched into northern Mexico and defeated two different Mexican generals, each of whom commanded a larger force than Taylor's. Elsewhere, Brig. Gen. Stephen Kearny had seized New Mexico and advanced into California. Monterey had been occupied by American naval forces in July, and it was clear that Kearny and the naval forces would force a climactic battle with the Mexicans in California.

Despite these grim circumstances, the Mexicans would not accede to American demands. Polk was left with little choice but to follow Scott's plan to attack the seat of Mexican political power, Mexico City. Scott left Washington on 24 November, and after gathering many of Taylor's troops at Point Isabel, sailed to Tampico and established his headquarters.

Meanwhile Santa Anna learned of Scott's plan and resolved to crush the weakened forces of Taylor before Scott could strike. Moving north from San Luis Potosí, Santa Anna arrived at Buena Vista, having lost nearly 20 percent of his force on the grueling march. On 23 February 1847 Santa Anna launched his attack. In a wild and confused battle, he was repulsed, with 500 dead and about 1,000 wounded, and fell back across the desert to San Luis Potosí.

On 9 March Scott arrived with 10,000 troops near Vera Cruz, landed unopposed, and laid siege to the city. After a five-day artillery bombardment, the city's garrison surrendered with little bloodshed on either side. Scott moved inland quickly to get away from the disease-infested coastal plain. On 18 April Scott found Santa Anna with 12,000 men at Cerro Gordo. Thanks to his West Point–trained engineers (including then Capt. Robert E. Lee), Scott was able to flank Santa Anna's position and rout the Mexicans after a sharp fight. The Mexicans loss was 1,000 killed and wounded with 3,000 captured, while Scott suffered about 400 total casualties.

In May, Scott was forced to pause in his advance until those volunteers whose 12-month enlistments had expired were replaced with new enlistees. Finally on 7 August, Scott set out for Mexico City, severing his lines of communication with the coast since he did not have sufficient troops to defend them. Circling south of Lake Chalco and Xochimilco, Scott met Santa Anna's defenders on 20 August in the Battles of Contreras and Churubusco, in which the Mexicans were defeated and forced to retreat within Mexico City's walls.

Peace negotiations ensued as Santa Anna attempted to reorganize his battered forces. Finally, negotiations broke down and on 8 September 1847, Scott attacked and defeated the Mexicans at Molino del Rey. Five days later, on 13 September, the Americans stormed the last bastion outside the city itself in the Battle of Chapultepec. It was here that 100 cadets from the Mexican Military College made a gallant but hopeless stand. The next morning, the remaining garrison in Mexico City surrendered, Santa Anna having left during the night.

Following the capture of Mexico City, peace negotiations began in earnest at Guadalupe Hidalgo and an agreement was reached on 2 February 1848. The last American troops left Mexico City in June and evacuated Vera Cruz in August.

Better State of the Peace

After the U.S. Army had compelled the toppling of a hostile Mexican government and seen it replaced by a regime more amenable to negotiating a settlement on the basis of American objectives, the war ended. The terms of the peace were negotiated in the form of the Treaty of Guadalupe Hidalgo in February 1848. The United States acquired the territories it had desired in the beginning (the disputed areas of Texas, California, New Mexico, and Arizona territories) in return for a $15 million payment to the government of Mexico. After the Gadsden Purchase of 1853 added remaining sections of New Mexico and Arizona to the Union, the territorial boundaries of the 48 contiguous states took on their final shape; manifest destiny was served.

The settlement was certainly imposed. Militarily, the forces of the United States had all but destroyed Mexico's ability to resist the imposition of our policies, and the occupation of Mexico City exceeded Mexico's cost-tolerance, the factor that proved pivotal in the end. At the same time, Mexican hostile will (defined as acceptance and embracing of our policy) was not overcome, and no serious efforts were made to convince the Mexicans of the virtues of manifest destiny. Instead, we sought to buy the Mexicans off in the peace treaty and the Gadsden Purchase. Much of the anti-Americanism that still exists in Mexico surely has its roots in our failure to overcome that aspect of hostile will.

The war had its consequences and its lessons. It provided, among other things, a training ground for the officer corps that would lead the forces (especially Southern) in the Civil War a little over a decade later; virtually all of the major military leaders of the fratricidal conflict received their first major blooding in Mexico. Moreover, the performance of that portion of the officer corps who had attended the academy proved to be a vindication for the West Point system. At the same time, the soldiers who fought under those leaders were, by and large, the same sort of citizen-soldiers who had fought all of America's wars, apparently supporting once again the virtues of the militia system. Manifest destiny was served and, for the time, sated. It would return again a half-century later, and the result would be the Spanish-American War.

SPANISH-AMERICAN WAR

The conflict between the United States and Spain in 1898 was labeled by its most prominent war hero, Theodore Roosevelt, "the splendid little war" and in many ways it was. The war lasted only a little more than three months, and all of America's objectives were achieved at the cost of less than 300 combatants killed in action. In the process, the United States established itself among the world's powers, a force that would have to be reckoned with in the future.

Issues and Events

The underlying, pervasive issue that gave rise to war with Spain was the question of manifest destiny, and it was an issue that had both humanitarian and imperialistic aspects. As a humanitarian concern, there was a rising missionary zeal in the country that reviled repression and sought to share the American political and social experiment. This concern was focused most explicitly on the island of Cuba and the fate of its citizens under Spanish rule. In addition, there had grown in the 1890s the first strong imperialist sentiment in U.S. history (if one does not consider the settlement of the continent an act of imperialism). This sentiment argued that for the United States to achieve the status of a major power, it must have colonies (colonies bestowed great power status in a way not unlike nuclear weapons do today). Since the Afro-Asian world had been thoroughly carved up into European empires, the only way to acquire an empire was to take one away from someone else.

The resurgence of manifest destiny followed the lapse of that concept after the Mexican War. During the interim, of course, the country was convulsed by the Civil War and the process of bitter reconstruction that followed, and national energies not trained on that trauma were focused on the settlement of the American West. It was a time for introspection and not external expansion. By the 1890s, the worst of reconstruction was past, the West was largely settled, and the country had emerged as a commercial and industrial giant. It was time to assert America's place in the community of states and to protect its position in the international economic system. The average citizen might be more moved by America's mission to "save a savage world," but its leaders marched increasingly to the drum of geopolitics.

The proximate events that led Americans into war focused on Cuba. That island so close to the Florida coast had held a special place in the American conscience for half a century. At one time many Americans had considered colonizing Cuba, and its fate was seen as intertwined with ours. Americans had watched with compassion during the "Ten Years' War" Cubans had waged unsuccessfully against Spanish colonial rule between 1868 and 1878, and had watched the Cubans revolt again in 1895. Cuba held a special fascination.

Adding fuel to this fascination and concern was the "yellow journalism" of the New York press. The two giants of the newspaper world, Joseph Pulitzer of the New York *World* and William Randolph Hearst of the rival New York *Journal*, were locked in a titanic circulation war, and coverage of the situation in Cuba became the primary weapon for selling newspapers. To increase circulation, events in Cuba were pictured in especially lurid and sensational terms that undoubtedly magnified and distorted Spanish suppression and acts of terror. Americans who had these sources as their primary basis of knowledge, however, looked on with increasing horror that led to a growing sentiment for war.

The situation in Cuba was also bad for business, and the American business community had a special concern with evolving events. Before the 1895 revolt, Americans had invested more than $50 million in Cuban plantations, transportation projects, and business establishments, and all those investments were threatened by the revolution. Moreover, trade between the island country and the United States was severely hampered.

The movement toward intervention in Cuba grew steadily and inexorably. When William McKinley was elected president in 1896, he sought to avoid war, telling outgoing President Grover Cleveland that he hoped he could avoid American involvement in "this terrible calamity." Events would, however, not allow this to happen. The event that led directly to war was the sinking of the USS *Maine* in Havana Harbor.

The sinking of the *Maine* is shrouded in controversy. The battleship had been summoned to Cuba by American Consul Fitzhugh Lee (who had been given that authority by President McKinley) on a purported courtesy call that was in fact a response to the storming of Havana newspaper offices by Spanish officers in retaliation for negative articles written about the military. The vessel sat at anchor for three weeks under heavy security, but on the night of 15 February 1898 a massive explosion ripped the ship, causing the death of 260 crewmen out of a total crew of 350.

No one knows for sure who sank the *Maine* or why. We will never know because the ship was raised from the bottom of Havana Harbor in 1911, towed to deep sea in the Atlantic, and sunk without detailed inspection, leaving the mystery intact. If the facts were in dispute, however, the apportionment of blame at the time was not. Americans learned of the tragedy in the New York *Journal* on 17 February. The paper's banner proclaimed that "The War Ship *Maine* Was Split in Two by an Enemy's Infernal Machine." That enemy, of course, was Spain, and the incident fanned the flames lit by the revelation of the famous de Lome letter earlier that year (a missive written by the Spanish ambassador in Washington describing the president in especially derisive terms).

The combination of events greatly increased pressure on McKinley to declare the war he sought to avoid. He demanded and received an apology over the de Lome incident. He also received Spanish assurances that the violence in Cuba would end and that they would institute economic reforms. These assurances were too little too late. American war fever could be sated only by fire, and on 11 April 1898, a reluctant President McKinley issued his war message. After 33 years of peace, America was once more at war.

Political Objective

As framed in President McKinley's war message to the Congress, the American political objective in the war with Spain dealt exclusively with alleviating the situation in Cuba. In the process of the war's conduct, however, the United States came into possession of other Spanish territory, creating the empire that was an objective of many Americans but which had not been a stated goal of the administration.

Exactly what was to become of Cuba was not stated in the message. Rather, McKinley said American military intervention was rooted in four concerns: a humanitarian concern over the devastation occurring on the island, protection of American citizens and rights on the island, an end to threats to Cuban-American commerce, and a guarantee that American strategic interests in the area would be honored. In time, this objective translated into Cuban political independence coupled with heavy American economic penetration and control.

The acquisition of empire occurred almost by accident. On 1 May 1898, Commodore George Dewey engaged the Spanish fleet in the "battle" of Manila Bay, sank it in its entirety, and thereby ended Spanish political dominion over the Philippines. The American flag flew over Manila, and it was only after some considerable debate that we decided it should stay there. Likewise, a force was sent to Puerto Rico after the fall of Cuba to overcome the Spanish garrison there, and once that was accomplished, McKinley simply decided to keep the island as a war indemnity.

Military Objectives and Strategy

American military strategy was controlled by President McKinley. Rather than having a well-thought-out plan, strategy and objectives unfolded with events, threats, and opportunities. The essential military problem was that no one knew exactly what McKinley sought as political objectives when the war began. Was it to aid the Cuban rebels or to seize Cuba? Questions remained concerning other Spanish colonies such as Puerto Rico and the Philippine Islands.

The navy had the fewest problems making its plans. As early as 1896, officers at the Naval War College had developed a plan for fighting Spain. The plan called for a blockade of Cuba to starve the Spanish troops followed by the occupation of the island by a small American force aided by the Cuban rebels. Simultaneously, the Americans would attack the Spanish Pacific Squadron at Manila Bay to safeguard American commerce in the Pacific. This general plan was quickly approved.

The commanding general of the army, Nelson A. Miles, proposed a full-scale invasion of Cuba by an 80,000-man regular army to take place in the fall after the rainy reason had passed. McKinley thought such a delay would be intolerable. Miles then suggested that Puerto Rico should be the main focus of American operations.

The first approved and coordinated plan relied on naval action to bring the Spanish to heel. In addition to the Cuban blockade and the attack on the Spanish Pacific Squadron, the plan called for a small army force of 5,000 to land on the Cuban coast and to funnel supplies to the rebels. This plan changed quickly for two reasons. First, on 29 April 1898, news arrived that a Spanish fleet had set sail under the command of Adm. Pascual Cervera. American ships were quickly detached from the blockade to form a "flying squadron" to protect the Atlantic Seaboard and to find the Spanish fleet. Second, a cable confirmed that Commodore Dewey's Asiatic Squadron had smashed the Spanish Pacific Squadron at Manila Bay and asked for 5,000 troops to seize Manila.

McKinley became much more aggressive with the good news from the Pacific. Additionally, the blockade seemed to be having only a limited effect on the Spanish, but it was taking its toll on American ships and men. The plan changed and the target became Havana. Army troops would land near the city and then march on the seat of Spanish power. However, it was soon learned that Cervera's small fleet had arrived and entered Santiago Harbor. The target of the ground attack was quickly changed to Santiago. A force of 17,000 men sailed for Cuba with more to follow as training was completed and shipping became available.

The American force seemed small for the job, as the Spanish army had 150,000 troops in Cuba. However, tropical disease had taken its toll of the Spanish and perhaps only half that number were effective. Worse, the soldiers were scattered throughout the island in an attempt to withhold ground from the rebels. The Spanish army could not quickly concentrate because of the primitive transportation system on the island.

Political Considerations

In some senses, the Spanish-American War was a model event, a prototype for America's wars of the second half of the twentieth century. This may seem an odd statement, since the resemblance between the war with Spain and say, Vietnam or Korea is, to say the least, tenuous. The Spanish conflict is not a model for how the United States has fought wars in the contemporary period, but the analogy has meaning when the conflict is seen as a model for the kinds of limited wars the United States can successfully sustain. The measure of sustenance in America, as in any democracy, is continuing popular support for military action to its conclusion.

Public support for the campaign against Spain not only nurtured the endeavor, it virtually forced it. The war with Spain was the first instance in American history wherein the news media played a crucial initiating role and, as pointed out earlier, the New York circulation war was a critical element in forcing President McKinley to declare war. America's desire for war was strong, and it was sustained throughout the campaign.

The critical question is why this was the case. At least three factors come to mind that distinguish the war with Spain from Vietnam and Korea, but which bear similarity to more recent adventures such as the U.S. invasion of Grenada and the British war with Argentina over the Falkland (Malvinas) Islands. First, the war's stated aim of relieving Cuba was clear, unambiguous, and popular. The liberation of Cuba as the major goal was well known, had overwhelming popular support, and translated readily into military requirements the accomplishment of which were clear and easily measurable. Only when the war spread to empire did the objective become muddy. Second, the war was short and relatively bloodless. The military campaign took only three months and caused modest casualties. Such criticism of the war as did occur was raised after its end and centered on the abysmal medical support conditions that resulted in many needless noncombat deaths among American servicemen. Third, the war was an easily achieved military victory. San Juan Hill and the Battle of Manila Bay were the crowning blows, and they were both walkovers against an overmatched foe. There was plenty of glory and little bloodshed.

Military Technology and Technique

The "splendid little war" was a joint Army-Navy operation, and both services bore little resemblance to their forebears from the Civil War. In many respects, the Spanish-American War was a modern conflict in that most of the weapons and techniques used were much more like those of the twentieth century than those of the Civil War.

The standard American infantry weapon was the .30-caliber Krag-Jorgensen rifle. Unlike its immediate predecessor, the so-called Trap-Door Springfield, the Krag-Jorgensen was a five-shot repeater that used smokeless cartridges. Unfortunately, by 1898, only the regular army had been equipped with the Krag-Jorgensens, and national guardsmen mobilized for the war were forced to use the obsolete single-shot Springfields.

American artillery was plentiful, but technical development had lagged, and the quality of the artillery had fallen far behind that of most European armies. Many pieces still required slow and dangerous muzzle loading, and all fired black powder. The smoke from the black powder instantly gave away artillery positions and made the gunners' situation dangerous against a first-class adversary. Perhaps more important, American gunners still had no method of sighting for indirect fire, which meant that they could engage the enemy only at ranges not much longer than those of the Civil War.

The U.S. navy had made significant technological strides since the Civil War. Spurred on by Mahan and others who argued for a first-class navy and overseas possessions, the navy had embarked on a large building program that had produced, by 1898, the sixth-largest navy in the world. The navy had five battleships, four of which were of the most modern types and listed as "First Class." For example, the USS *Oregon*, which fought in both the Pacific and Atlantic, displaced 10,000 tons, mounted a total of four 13-inch guns on turrets fore and aft as well as eight 8-inch guns, and had a top steaming speed of nearly 17 knots. The navy also had 30 cruisers such as the USS *Olympia*, Dewey's flagship, which displaced 6,000 tons, mounted four 8-inch guns as well as ten 5-inch guns, and could steam at nearly 22 knots.

Although well armed, American forces were not well prepared for a war of any size. The regular army of just over 28,000 was well trained and experienced in the frontier Indian wars. However, it was skilled only in small-unit actions. The largest regular formation was the regiment and few officers had ever seen larger formations. Likewise, the navy was inexperienced in fleet operations.

Perhaps the greatest shortcomings were in joint operations and in amphibious operations. Coordination between the services during the conflict was appalling. The embarkation at Tampa of army forces bound for Cuba was a scene of mass confusion, including the last-minute discovery that there were not enough ships to carry the troops. At the end of the voyage, the landing operations were also chaotic: there were not enough small boats to get the troops and supplies ashore quickly. The landing took four days, in sharp contrast to Scott's landing at Vera Cruz a half-century earlier. That operation, which involved an equally large force, was accomplished in one day.

The Spanish-American War also exhibited the continued growth of modern centralized command and control. President McKinley established a "war room" in the White House complete with detailed maps and markers, and equipped with 25 telegraph lines. These lines connected him with the various military departments and with officers occupying important posts in other cities. McKinley did not hesitate to use his communications capability in directing the efforts of the military staffs.

Military Conduct

The first military action of the war was not an engagement, but the destruction of the American battleship *Maine*, which was officially on a goodwill visit to Havana.

As noted, the circumstances of the explosion that sank the ship on 15 February 1898 are a matter of some debate. Regardless of how it happened and who was responsible, the sinking led more or less directly to the American declaration of war on 25 April. At the same time, the regular establishment of the army was increased from 28,000 to 60,000, and President McKinley called for 125,000 volunteers. A month later McKinley called for an additional 75,000 volunteers.

The navy was ready for immediate action and quickly clamped a blockade on Cuba. On 25 April Commodore Dewey sailed his Asiatic Squadron to Manila Bay. He entered Manila Bay on the night of 30 April, and on 1 May engaged the Spanish squadron commanded by Adm. Patricio Montojo. The engagement was less a battle than an execution. Montojo's fleet was outclassed and outgunned. Dewey's force totally destroyed the Spanish force while losing only one man (via heatstroke) and eight wounded. Dewey then waited for the arrival of sufficient troops to seize Manila.

In the Atlantic, the Americans learned on 29 April that Admiral Cervera had sailed with the main Spanish fleet from the Cape Verde Islands. A small flying squadron was detached from the blockading force to protect the Eastern Seaboard and intercept the Spanish fleet. Surprisingly, Cervera avoided the American forces and slipped into Santiago Harbor on 19 May. Rear Adm. William T. Sampson, who commanded U.S. naval forces in Cuban waters, immediately blockaded the harbor.

By mid-June, despite tremendous logistical snarls and a lack of planning for almost everything needed by a large modern army, Maj. Gen. William R. Shafter was ready to sail to Cuba with 17,000 men. Shafter's forces arrived off the Cuban coast on 22 June and commenced landing at Daiquiri. The landing was unopposed, but confusion reigned, and it was not until 25 June that the full force was ashore. Had the Spanish been able to oppose the landing, the story of the war might have had a far different ending.

After some minor skirmishes, and great difficulties in unloading supplies from the poorly loaded ships, Shafter moved on Santiago. On 1 July he assaulted the San Juan Heights that protected the eastern approaches to Santiago. By nightfall, after confused maneuvering, several sharp setbacks from the Spanish, and the Rough Riders' "charge up San Juan Hill" led by Lt. Col. Theodore Roosevelt, the positions were in American hands. The Spanish fell back to their inner defense line.

On 3 July Cervera led his trapped fleet out of Santiago Harbor in a valiant but doomed attempt to escape the American blockade. As was Dewey's triumph at Manila Bay, the Battle of Santiago Bay was one-sided. Running along the coast, the Spanish ships were overwhelmed with heavy fire and forced aground as burning hulks. All six of the Spanish ships were lost. Incredibly, total American losses were one killed and one wounded.

Faced with insurmountable odds, the Spanish commander in Santiago, Gen. José Toral, surrendered the city on 17 July. The surrender included all Spanish forces in eastern Cuba. Toral was unaware that tropical diseases were taking their first toll on American forces and that Shafter's supply problem remained difficult.

On 25 July General Miles landed on Puerto Rico and after being reinforced, pushed inland. He met almost no opposition as Spanish forces fell back into San Juan. Before Miles attacked San Juan, word arrived that Spain had asked for peace.

Meanwhile, on 25 May 1898, Gen. Wesley Merritt departed San Francisco with the vanguard of troops bound for Manila. He arrived at Manila on 30 June with a force that would eventually total 15,000. The situation was delicate because the Philippine rebel leader, Emilio Aguinaldo, had Manila under siege and had declared a Philippine Republic. While the American political leadership pondered what to do with the Philippines, Merritt and Dewey wanted to take the city as soon as possible. Squadrons from some of the great powers were beginning to appear in Manila Bay, and Merritt and Dewey feared serious problems if American control was not quickly established.

The Spanish commander in Manila, Fermin Juadenes, was willing to surrender but not to the rebels, whom he feared would seek retribution against the Spanish. After secret negotiations, a sham battle was staged on 13 August, and the Americans entered the city. The Spanish were therefore able to surrender to the Americans, and Merritt took control of the city. Neither Juadenes nor Merritt was aware that the war had ended two days earlier.

Better State of the Peace

Because the war was such a military mismatch, achieving the stated political goals forced upon McKinley was relatively easy. Spanish hostile ability was minimal to begin with, and once the United States had sunk the Spanish fleets and thus left the island garrisons isolated, overcoming the vestiges of hostile ability was simple and straightforward. Moreover, the Spanish, fully aware that they stood no reasonable military chance against the Americans, possessed little hostile will, so that both their cost-tolerance and unwillingness to accept our policies were quickly overcome as well.

American objectives were achieved, at least in the short run. Cuba was relieved of the Spanish yoke, but the full independence they expected was only questionably theirs. McKinley's war message had not guaranteed independent status, and although political independence was granted, economic penetration by the Americans left the island in a position of dependency that many Cubans believe was broken only by Fidel Castro's revolution 60 years later. At the same time, the United States acquired an empire in the form of Guam, Puerto Rico, and the Philippines. Those imperialist ambitions were not clearly defined when we entered the war but evolved. Filipino insurgents, for example, first viewed us as liberators and made common cause with us in helping remove the hated Spanish. They began to oppose us when they recognized that the Yankees had no intention of leaving, either. The result was a bloody counterinsurgent campaign and, once that was concluded, an exposed empire that stuck out like a sore thumb in the way of Japanese expansion in the western Pacific.

12

THE AMERICAN EXPERIENCE

The American experience with the use of military force is unique, shaped by history, geography, and a host of other factors, many of which have been discussed in previous chapters. The task now is to sort through that experience to see what patterns may emerge and what may be learned that can enrich our understanding of the present and guide encounters with the future. What, in other words, does examination of America's experience with military force tell about the present and future prospects of employing military force to achieve national ends?

Certain general comments can be made as a preface to a systematic review of the framework that has been employed. The first is the great degree to which understanding American military history is enshrouded in mythology, a result of American ahistoricism. Part of this mythology is the notion of Americans as a pacific people for whom peace is the norm and war is the exception. Rather than seeing peace as an interlude between wars (an attitude held by many Europeans whose history reflects such a view), Americans tend to look at war as a transgression of our normal circumstance.

The record, of course, does not fully support such a view. As the preceding chapters demonstrate, the United States has been involved in six major wars during its existence (including every major European conflict during our history), as well as three categorized in this work as minor. At this writing, the United States remains enmeshed in two wars—Afghanistan and Iraq. Even that listing is not inclusive, as it omits actions such as the taming of the Barbary pirates (Colonel Qadhafi was not the first Libyan with whom we have tangled), the Seminole War, unconventional engagements against Philippine insurgents, the various campaigns against both the eastern and western Indians, the so-called "Banana Wars" in Latin America, and 1990s adventures in places such as Somalia and Bosnia. An inventory of the number of years in which Americans have been at war all or part of the year is a lengthy list that does not comport with any sense of passivity. Yet partially as a result of this myth,

Americans believe that war, when it is thrust upon them, is to be gotten over with as quickly as possible so that they can get back to the more normal business of peace.

There is a second element in the American mythology and that is the myth of military invincibility. Americans may be slow to anger, so goes the myth, but when aroused, Americans are winners. The watchword of the American military is "can do." When called to arms, Americans respond and prevail. The truth is that, although Americans have usually been militarily successful in the long run and have generally achieved those political purposes for which hostilities have been entered, the record is not unblemished. The United States has not always prevailed militarily (for example, the War of 1812) or politically (Vietnam, for instance). Realizing the falsity of this myth would help, among other things, to put the Vietnam experience into proper perspective, along with what will likely be less than totally satisfactory results in Afghanistan and Iraq.

Another part of the myth of invincibility is the belief that the United States has always fought not only successfully, but well. Once again, this assertion does not bear up well under close examination. Rather, at least until the post–World War II period, the United States entered wars unprepared to fight them and either took a long time to prepare to fight (World War I) or fought poorly early in the war while inexperienced troops and their often equally inexperienced commanders gained combat experience (the early North African campaign in World War II). Moreover, the U.S. experience can hardly be described as one of great tactical or strategic brilliance. America has produced, after all, no universally recognized great strategist of land warfare, but has tended to rely instead on superior manpower and weaponry to grind down enemies' ability to resist. Although this approach to warfare has not been pretty or subtle, it has been effective, at least against enemies who played by the same rules. To say these things is not to denigrate the tradition of the American military; it is simply to put it in proper perspective.

There is a second general strand that has run through most of the American experience, and it is a basic, underlying antimilitary bias. This antimilitarism expresses itself in a suspicion about military solutions to problems, a negative attitude toward maintaining military forces during peacetime, and an aversion to military spending when the country is not at war. The signs that one can occasionally find in New England antique shops that read "no dogs or soldiers allowed" are a reminder of this heritage.

Antimilitarism was born in the nation's formative days. As argued earlier, the stationing of British troops on American soil was an important link in the chain that led to the American Revolution, and the Continental Congress watched the army suspiciously throughout the war, fearing it might be used to usurp congressional authority. (This led a frustrated George Washington to remark that he could understand why the Congress might worry about the existence of the army during peace but not during war.) That fear and suspicion led to a quick disarmament immediately following the Revolution and the virtual absence of any military preparedness until the War of 1812.

The result, through most of American history, has been a pattern of a small professional military establishment during peacetime and rapid mobilization of forces when war occurs. Until World War I, this generally meant the calling of the militia, based on the notion that the "citizen-soldier," valued because he was more citizen than soldier, was an effective fighting man. The evidence, of course, does little to support this notion, which should have died conclusively on the battleground of First Manassas in 1861. The net effect has been that we have entered wars with neither the trained personnel nor the equipment to begin hostilities rapidly and successfully.

This particular pattern has, of course, disappeared to some degree since World War II and especially the Korean conflict. Since that time, the United States has maintained a substantial "force in being," prepared to assert itself in a timely fashion when called upon. This state of perpetual preparedness, in addition to being considerably more expensive than the earlier pattern, does represent a discontinuity from America's past.

The period since the Vietnam War has been one of changing attitudes toward the military and older traditions about the military. One victim of the Vietnam experience was universal exposure to potential military service via conscription. Opposition to the war in Vietnam made the draft politically impossible to sustain and resulted in the replacement of the concept of the pure citizen force by a professional force drawn from the citizenry, the all-volunteer force (AVF). Because the AVF is drawn from the citizenry, the notion of the citizen-soldier is not renounced, but it does mean that not all citizens are potential soldiers.

This experience has been mostly positive. The AVF has achieved levels of proficiency and professionalism unattainable in a force partly composed of involuntary members, and the armed forces would be loath to return to a conscription system whereby part of the force was there because of compulsion rather than choice.

There are, however, arguably negative consequences as well. One may be a greater willingness to apply volunteer forces in places and situations where a conscript force could not be used for political reasons. Would the American people have supported the wars in Afghanistan and Iraq if they (or their sons and daughters) might have been drafted for the fight? Also, the AVF is expensive to the point that its viable size is limited by costs as well as by how many Americans will volunteer to serve. The effects of constrained size have created problems in Iraq and Afghanistan and left the United States with few uncommitted manpower resources during their conduct. It is unclear that the AVF concept could be sustained in a larger military contingency.

Issues and Events

In attempting to generalize from the American experience at war, the central question that must be asked is: What kinds of issues have most and least galvanized the American people as they moved down the road to war? Although any generalization runs the risk of oversimplification and hence of raising the ire of the historian, American military history does suggest a criterion that divides the wars that have had strong public support from those that have not. That criterion is the high moral character that could (or could not) be attached to the issues leading the country toward war.

The most popular American military adventures have sprung from issues that were perceived as both absolute and moral in nature. Although there was division within the body politic, the issues of British tyranny and American independence had an absolute and moral character, as did the protection of hearth, home, and way of life for Confederates during the Civil War. The other American wars that were unfailingly popular, the two world wars, were also of this nature. World War I, the "war to end all wars," was fought by Americans to make the world safe for democracy, thus serving the dual interests of ridding the earth of the scourge of war and promoting a morally superior political form. Destroying the total evil of Hitlerian fascism in World War II had a similarly lofty ring that was irresistible once Pearl Harbor had propelled us into that fray.

By contrast, when issues were perceived as less important or of a lower moral content, there have been divisions in the public that have lessened support for wars. America's first unpopular conflict, the War of 1812, had impressment and de facto independence from Britain as underlying issues, and these issues were of such marginal appeal that New England, which would have benefited from the ending of impressment, actually opposed going to war. Territorial expansion in the Mexican War, tied as it was to the extension-of-slavery issue, generated support only in the West, and Lincoln was burdened throughout the Civil War (as well as the period leading to the war) by Northern apathy about union and emancipation. America's major adventures in Korea and Vietnam, where the issue was the containment (but not rollback and eradication) of communism, similarly had limited appeal. The global war on terror provided only temporary support for military action in Iraq and Afghanistan.

This moral sense spills over when it comes to proximate events that bring about war. Although it is an arguably accurate self-depiction, Americans view themselves as a peace-loving, pacific people who cannot, by their very nature, be the initiators of war. The government of Mexico, a number of Indian tribes, and the citizens of Iraq might well take exception to such a characterization, but this perception does create a need to be attacked in order to produce the sense of moral outrage necessary to push the American people to war. In some cases, leaders have recognized and even exploited this fact.

A quick review of the American past reinforces this typification. The British march out of Boston to Lexington and Concord (even if no one knows who shot first) became a rallying cry to begin the Revolution. President Lincoln, anxious to quell the rebellion but unwilling to appear to be engaging in aggressive action that would legitimize secession, was forced to wait patiently for the South Carolina militia to fire on Fort Sumter. Despite the war hysteria whipped up by the yellow press, it took the sinking of the USS *Maine* (by persons unknown) to engender the rallying cry in the war against Spain. The German sinking of the *Lusitania* provided an impetus in 1916, and there is no better example than the Japanese attack on Pearl Harbor. In more contemporary times, the North Korean invasion of South Korea compelled the United States to enter that conflict; President Johnson used the Gulf of Tonkin incident as the excuse to prosecute the Vietnam War; and Iraq's invasion of Kuwait compelled an American response. Widespread atrocities committed against populations by their own governments have been a trigger since the end of the cold war. A clarion call for the eradication of terrorism has been the prevailing rationale since 9/11.

Political Objective

The role of the political objective is absolutely critical in the conduct of war by a democracy, because it provides the rallying cry to generate and sustain the public support without which a democracy cannot long fight. The sweep of the American experience suggests that there are four criteria that define a "good" political objective, that is, an objective Americans will support. To the extent that an objective meets these criteria, it is likely to develop and sustain public support. To the extent that an objective violates one or more criteria, it is likely to suffer erosion of public support. The four criteria are: the objective must be simple, straightforward, and unambiguous; it must be morally and politically lofty; it must be overwhelmingly important; and it must be seen to be in the best interest of most Americans.

The first criterion says that a good political objective can be easily understood by everyone. Ideally, the objective should be reducible to a catchphrase or slogan that both captures the essence of the main objective and serves as a rallying cry (e.g., "Remember the *Maine*" as a way to simplify assumed Spanish perfidy). The second criterion, following on the earlier discussion, suggests that a good objective can be turned into a crusade that appeals to people's moral sense ("Make the world safe for democracy"). The third criterion means that the attainment of the objective must be vital to the United States and failure to attain it disastrous. This criterion would be best exemplified in the situation where losing a war would physically endanger the integrity of American soil, and the only example in U.S. history was the Southern side in the Civil War (the British threat in 1814 was more limited). The fourth criterion, of most importance when there were still great sectional differences between Americans, refers to the need for a majority of groups to view the objective as important. In contemporary times, the third and fourth criteria can be merged into a single criterion of perceived self-interest.

The critical importance of an appropriate objective can be demonstrated by looking at which American wars violated which criteria. The criterion of simplicity is rather clearly violated in four instances: Vietnam, Korea, the War of 1812, and Iraq. As argued earlier, the problem of the objective in Vietnam was, at least partially, that average Americans did not understand why their country was at war, and the various means that were used to try to convince them (the Munich analogy and the domino theory) provided neither clarity nor simplicity to aid understanding. In Korea, the problem was not in understanding objectives but in knowing which was operative at any point in time. What was not so clear was why and when the objectives changed. In the War of 1812, the agendas of impressment (an issue resolved before war was declared) and of annexing Canada (either as a "bargaining chip" to end naval harassment or to expand the United States) were similarly confusing. In Iraq, the objective kept changing as successive reasons given for the war proved untrue or unprovable.

The criterion of moral loftiness has been even more often abused, with at least five instances from the historical record where one could question the moral force of the objective. The first instance was the War of 1812, in which the morality of seizing Canada was particularly questionable. In the Mexican War, manifest destiny was but a thin disguise for naked American aggression and seizure of sovereign Mexican territory. One of Lincoln's recurring difficulties was convincing his countrymen that forceful union, and then union plus abolition, were worthy of their support and blood. In the twentieth century, the principle has arguably been violated twice: in the Korean and Vietnam wars. In Korea, the problem was associated not with the original and eventual objective of containing North Korea but rather with not liberating it from "godless communism." If the evil was sufficiently dire to require the sacrifices of war, then it should be exorcised, not simply contained. In Vietnam, this sort of dilemma was compounded by support for a succession of either venal or incompetent South Vietnamese governments. Destroying Al Qaeda represented such a moral charge in Afghanistan; subsequent support for the Karzai government does not meet it so clearly.

The third criterion reflects the importance (worth) to national security of attaining the objective. This is always somewhat difficult to gauge for a government that fights primarily in an expeditionary manner, when the physical security of hearth and home is not directly in jeopardy and when threats to the homeland are abstract extrapolations from the situation at hand. In fact, the only American war experience in which territorial integrity was really at issue was the Confederate side in the Civil War. Support for the Rebel cause was quite high among Southerners throughout the long conflict.

The question of worth has been raised as a significant concern in six American wars—two in the nineteenth century, two in the twentieth, and two in the twenty-first. In the nineteenth century, the question was raised in both the War of 1812 and about Union conduct of the Civil War. Whatever its feasibility, annexing Canada was not a major priority for most Americans, and the naval harassment issue affected only a small slice of the population, mainly in New England. Much of Lincoln's problem in maintaining support for his war effort arose from the fact that many Northerners did not care whether the South left the union, or at least did not think preventing secession was worth fighting over.

The necessity of achieving the objective was questioned significantly in the twentieth century during both the Korean and Vietnam wars. In the case of Korea, the importance of "merely" ridding South Korea of its North Korean invaders was widely questioned, but only the second time this objective was adopted. At that point, the war had become a static contest of attrition about the 38th parallel, and the war seemed increasingly pointless and worthless. In Vietnam, a principal burden of U.S. officials was trying to convince the American people that there was something about the outcome worth the country's involvement and our sacrifice. Those attempts were ultimately unsuccessful. The emergence of "humanitarian interests" in the 1990s suggests a reassessment of what constitutes a worthy cause, but whether those interests justify much American sacrifice in blood remains an open question. Although they are ongoing and assessments are thus incomplete, the "wars of choice" in Afghanistan and Iraq raise questions of worthiness as well.

The final criterion is the perceived vitality of the objective to most Americans. In the nineteenth century, when this most often was a problem, the divisions of Americans tended to be along sectional lines that have largely disappeared as an important factor in U.S. foreign policy. During the twentieth century Americans have been more divided by political philosophies, demographic classifications, or socioeconomic factors. Extended to cover these newer divisions, the fourth criterion is still valid.

Using this slightly extended definition, three wars immediately come to mind as violating the criterion. The first two were in the nineteenth century and involved sectional differences. In the War of 1812, the war was opposed by New Englanders because they feared the result would be to destroy New England's commerce with Europe. Support was greatest, ironically, in the South and along the frontier. These sections would not have benefited directly from the achievement of the war's objectives, but they nonetheless produced most of the war hawks. Similarly, the Mexican War was popular only in the Midwest and the West. It was opposed in New England on the grounds that it was a ploy by Tennessean James Polk to extend slave territories, and it was opposed in the South on the grounds that the additional territories were unsuited for the plantation system. The Vietnam War similarly divided Americans along political lines (liberals in general opposition, conservatives in support) and along demographic lines (the young, college-aged middle and upper-middle classes in greater opposition to the war than the lower, less-educated classes).

The post-Vietnam experience with the AVF may have moderated the impact of questions regarding vitality of interests in using force. The change may arise from the depersonalization of the war decision for most Americans: since 1972, no American who did not volunteer for armed service has had his or her life placed in direct jeopardy by the decision to go to war. To date, all of America's post-Vietnam adventures at arms have been of sufficiently limited extent that a general mobilization has not been necessary. As a result, support, opposition, or apathy about involvement all lack personal consequences for those who choose to avoid harm's way.

Nonetheless, a pattern of the kinds of political objectives that the American people will and will not support for a sustained period of time emerges, especially if one adds two other considerations—the limited or total nature of the objective and the duration of active fighting (and hence dying) by Americans. The latter distinction is made because the period of combat is a better indication of sacrifice and deprivation than is overall length of a war. World War I, when we were technically at war for over a year and a half but engaged in combat for less than six months, illustrates this point most dramatically.

These two criteria are added to make more explicit a hypothesis about the objective that has until now been somewhat implicit: political democracies support long wars entailing sacrifice better if they are total in purpose rather than limited. Hence, if a limited purpose is to be pursued, the war should be relatively short and painless (which, in contemporary terms, appears to mean the absence of significant American casualties or the meaningful possibility of involuntary service).

The reasons for this assertion follow from the criteria for a good objective. The total objective of removing an evil regime (a Hitler) easily meets the criteria of simplicity, moral loftiness, worth, and general interest. Limited wars, on the other hand, do not seek to exorcise some overwhelming evil (and hence are questionably lofty). They are generally fought for some geopolitical reason that is questionably important or straightforward (hence violating worth and simplicity) or that no more affects the average American than the dynastic wars of the eighteenth century affected the average French peasant (violating general interest).

These concerns suggest some observations about the role of the political objective. The first thing one notices is the sharp contrast between the wars of total and limited purpose. With the exception of the Union side in the Civil War, all of the violations of the criteria occur in the justification of limited wars. Moreover, in general there appears to be no particular relationship between total and limited wars in terms of duration. In terms of actual combat, the total wars average out to a little over three years in duration, but so do the limited wars (these calculations do not include the ongoing wars in Afghanistan and Iraq, the total duration of which remains to be seen but in any event will increase the average length of America's limited wars). Where duration is of interest is within the category of limited wars; the only one that violated none of the criteria (or at least none became salient) was Teddy Roosevelt's splendid little war against Spain: active military conduct spanned barely three months and resulted in minor American casualties.

What is particularly striking about grouping the wars this way is the relationship of the criteria to the public's support for a war. The most popular of America's wars have been total, with the exception of the war with Spain. The world wars were the most widely supported of all, followed closely by the Confederate side in the Civil War. One can argue that the majority of Americans were not steadfast in their support of the Revolution, but revolutions rarely have anything like majority support. There was opposition in the Union to the Civil War, but that may have been at least partially a response to shock at the consequences of modern total war and the ineptitude of early Union military leadership.

By contrast, America's least popular wars have also been limited in purpose. The two most unpopular conflicts were Vietnam and 1812, which also share the distinction of violating all the criteria for a successful objective. The Korean conflict comes in third on the list of violations, but it may also best illustrate the point about what kinds of wars the American public will and will not sustain. Because the question of popularity is easier to assess after a war is completed and passions about it have subsided, judgments about Afghanistan and Iraq remain to be seen.

It seems clear that Americans prefer to fight either wars of total purpose or limited wars that are bounded both in time and in the level of sacrifice associated with their conduct. If this is the case, and it remains difficult to sustain support for other kinds of wars, then there are important implications for the future application of American military force.

Military Objectives and Strategy

The military objectives pursued by the armed forces have varied widely during the American experience, as have the strategies they adopted. It was not until the Civil War that U.S. forces fought for unlimited military objectives, best described as destroying the enemy's ability to resist. Military objectives during the American Revolution were time dependent and limited, ranging from throwing the British out, to preventing the British return, to achieving a decisive battlefield victory, and finally to preserving the Continental army until the peace treaty was signed. The objectives in both the War of 1812 and the Mexican War were limited and circumscribed, as were the military and political objectives during the Spanish-American War.

Even in the Civil War, one set of Americans in that awful struggle was pursuing limited military and political objectives: the Confederates sought to convince the federal government that the price of keeping the Union intact would be too high in blood and treasure. The political objective was not to destroy the government in Washington, but rather to end its authority in one part of the former Union. The Union side's political and military objectives, on the other hand, were unlimited: the North had to destroy the Confederate ability to resist and the Confederacy itself if the Union was to endure.

It was out of this savage struggle to preserve the Union that an enduring American military tradition evolved. Since 1864 the favored American military strategy has followed the tradition of Ulysses S. Grant during his campaign in northern Virginia. Grant's method was simply to take advantage of the North's superior resources and overwhelm the Army of Northern Virginia. The strategy was to attack relentlessly, giving Robert E. Lee and his forces no rest. Grant could replace losses from his vast manpower pool for which the South had no equivalent. The Grant tradition was somewhat apparent in World War I but became predominant in World War II, when the United States figuratively drowned its enemies in a sea of men, firepower, superior technology, and logistics.

The total, brutal style of warfare begun by Grant and practiced in the world wars became the American way of war. Thus, when limited war made a return during the cold war, Americans seemed to have forgotten the much larger part of their military heritage—limited wars fought for limited purposes. The traditions of U.S. Grant did not apply well in Korea and Vietnam because of the potential for direct superpower confrontation, the specter of a nuclear holocaust, and, in the Vietnam case, because much of that struggle involved combating enemy forces that were using guerrilla tactics. By 1991 the cold war was over and the threat of escalation to nuclear confrontation had receded. The stage was set for a return to the Grant tradition in the Gulf War. The trip back to that tradition seemed complete when the Iraqi forces were totally overwhelmed by U.S. and coalition numbers, firepower, technology, and logistics in the first Persian Gulf War (Desert Storm). The only thing that saved the Iraqi regime was that the war was fought for limited political objectives—a fact that many Americans came to rue in the years that followed.

The traditions of U.S. Grant were not well suited for the war in Afghanistan in 2001. Although overwhelming airpower played a significant role, U.S. ground troops were primarily advisors with most of the fighting on the ground done by Afghan troops of the Northern Alliance. The aftermath of the Northern Alliance victory was and remains an insurgency, not the kind of struggle that lends itself to the total war model.

The early part of the war against Iraq that began in 2003 was a circumstance made for the Grant tradition—overwhelming power applied without interruption in a race to the heart of the enemy nation. The result was a very rapid and total military defeat of the Iraqi military, the ouster of Saddam Hussein, and, it was assumed, victory. But as in Afghanistan, the war in Iraq morphed into a multifaceted insurgency and religious civil war in which much of the sophisticated arsenal and training of U.S. troops was of very limited use, and in which the Grant approach and tradition were counterproductive. What success the U.S. forces have had in countering the insurgency has come from relearning the lessons of Vietnam, not from applying the approach of Ulysses S. Grant.

Political Considerations

Domestic and international concerns, which are primarily political in character, inevitably affect and are affected by the conduct of war. Although this statement is unexceptional and even self-evident, American military and civilian officials have often acted as if this Clausewitzian dictum did not exist. The problem is much more severe in limited-war situations than it is in instances of total war.

Total war tends to lessen the friction between the military and civilian authorities. This is so to some measure because a total war calls for maximum military effort, thereby lessening the politically defined shackles on the conduct of hostilities associated with limited war. That said, political considerations enter even into total-war situations. Grant, during the climactic 1864–65 campaign against Lee's Army of Northern Virginia, tried to obscure the reportage of the large number of casualties he was incurring for fear that public opinion would be adversely affected. In World War II, a central disagreement between the Americans and the British was whether the war against Germany should be fought with the earliest possible destruction of the Nazi armies as the top priority or whether the shape of the postwar political map should be the overriding concern.

These kinds of considerations are mild compared to military-civilian frictions during limited wars, and especially the limited wars of the second half of the twentieth century. As Clausewitz pointed out, war tends to feed on itself and thus expand and intensify. When military objectives, strategy, and resulting combat are constrained because political objectives are limited, there is a basis for friction. A limited objective almost by definition means that the military will fight with less than its total capacity, with "one arm tied behind its back." This is especially true as technology has increased the lethality of war machines. Political constraints against the military using all its resources are generally motivated by a fear of widening the war and transforming its objectives in the process. In a world where the ultimate transformation could be to total war with thermonuclear weapons, constraints are likely to be tight indeed.

Aspects of both the Korea and Vietnam cases illustrate this problem and the frustration and friction that inevitably emerge. In retrospect the problem has at times been almost comical. An extreme instance was the instruction to U.S. pilots during the Korean War to bomb only the southern half of bridges spanning the Yalu River, in order to avoid attacking Chinese territory and consequently altering fundamentally the nature of the war. The instruction created an impossible task for the military, which became frustrated and angry because of this political "interference."

A more general and vexing problem that arose in both Korea and Vietnam was sanctuary. Sanctuaries were areas to which enemy forces could retire without danger of attack because the areas were off-limits to American military efforts. In the Korean War, the People's Republic of China was the sanctuary. It was not attacked for fear of starting a general Asiatic war that military wisdom said should be avoided and which would have tied down so many American forces as to invite Soviet activity in Europe. In the Vietnam War, the most important sanctuaries were those parts of North Vietnam on the proscribed list for American bombers during most of the struggle, especially Hanoi and Haiphong. The reason for making sanctuaries of Haiphong and Hanoi was to avoid sinking Soviet or Chinese ships and thus to avoid drawing either or both into the war. At the same time, placing these areas off-limits made it impossible to attack the conduit of war materiel that substituted for an industrial web in North Vietnam.

What these examples show is that political restrictions placed upon the military in limited wars make the effective prosecution of hostilities more difficult. Politically motivated constraints probably make such wars last longer and reduce the likelihood of winning by some degree, but one of the lessons of the American contemporary experience is that such dynamics are a fact of life. Sanctuaries in Pakistan for Taliban and Al Qaeda fighters pose a similar problem for the continuing struggle in Afghanistan.

Political interference (if that is the proper term) is absolutely predictable in such circumstances. Anyone surprised that political authorities imposed limitations on the military in Vietnam was ignoring the Korean experience—which the military as an institution largely did, a process generally replicated over Vietnam as well. The sanctuary example may best make the point. Some analysts of the Vietnam case have concluded that the American military should not allow itself to be placed in another combat situation where the enemy is unilaterally granted a sanctuary, because the military task is too greatly compromised. From a military viewpoint, such a conclusion makes perfect sense. At the same time, a look at a world map does not reveal many places where American military assertion would not raise the possibility that sanctuaries would be granted for fear of broadening the war.

The other largely domestic political consideration, to which allusion was made in a previous section, is the question of public support for hostilities. Public support is absolutely necessary if a democracy is to conduct war. The problem is that there are influences that almost automatically erode the support for a war. Two of these are worth raising and considering.

The first has to do with the distinction between wars of total and limited purpose. Traditional total wars (total for both sides) have disappeared from major power politics because the extreme possibility is a general nuclear war, which is unacceptable to all. Limited wars, on the other hand, are often ambiguous and debatable in merit and tend to lose support as they lengthen or as they exact increasing sacrifices, especially in American blood. Leaving the fighting and dying to volunteer professionals moderates this concern somewhat, but when wars drag out and become very costly (as in Iraq), they also become more unpopular.

The second influence is the news media, especially the electronic medium (television). The major effect of television was first felt in the Vietnam War, which was transformed into a "living-room war" that simply could not be avoided because it covered the screen every evening. Television, by bringing pictures into our living rooms, personalizes war in a way the printed word cannot. It is one thing, for instance, to read about how many casualties there were in a firefight; it is quite another to have the maimed bodies flashed on your television screen. Moreover media coverage is relentless, especially in an age of global, 24-hours-a-day news networks with insatiable appetites for news. The result is an extensiveness and pervasiveness of coverage that makes even quite limited military events ubiquitous and unavoidable.

The impact of all this, and especially global television, on the conduct of war is not well understood. At one level, at least, the effect is to deglamorize war by showing its most raw and most destructive manifestations. The result is to make electronic coverage of war at least implicitly pacifistic. War has always been most glamorous to those who have never seen it, and the television camera guarantees that everyone—not just the participants—sees what war looks like. The almost inevitable result is that television coverage erodes support for physical hostilities to the extent it is allowed unfettered coverage of combat. At the same time, publicity about great suffering in places such as Somalia, Bosnia, and Kosovo has made military action more appealing. It may also be that relentless coverage of carnage eventually desensitizes us to it and increases our tolerance for atrocity. The jury is still out on the influence of the electronic media.

Two international considerations emerge from the American experience. The first of these is the interdependence of the United States with the rest of the world. Although Americans have thought of themselves as independent and aloof from the affairs of Europe in particular, such clearly has not been the case. A critical element in the birth of this country was the direct alliance with France and indirect alliances with Spain, the Netherlands, and tsarist Russia (the latter two through the League of Armed Neutrality). In the case of each "ally," the motivation to aid the revolutionary cause was solely geopolitical in nature, an attempt to get even with Britain because of the outcome of the Seven Years' War. For that matter, the issue of taxation, which had its roots in the stationing of British troops in the colonies after the French and Indian wars, was but an outgrowth of the war in Europe that had strained the British economy and forced Lord Richard Grenville to look for alternate sources of revenue.

American interdependence was also demonstrated during the Civil War. Until the battle of Antietam, the Confederacy conducted a lively courtship of Britain and France that was opposed by the U.S. government, and the failure of the Confederate States of America to gain European recognition and succor may have been critical to the eventual Southern failure. Likewise, the government of the United States, despite its best efforts at neutrality, became involved in the European wars that punctuated the nineteenth and twentieth centuries. Our War of 1812 was nothing more than the New World theater of the Napoleonic Wars, and U.S. involvement was made necessary by our commercial interests. Similarly, recognition that a German-dominated Europe would be intolerable for American commercial efforts made attempts to remain neutral in World War I futile. When the second great conflagration came about, the American instinct toward aloofness was rapidly drowned in the necessity of avoiding a permanent German victory on the continent.

After World War II, the United States recognized this interdependence by making the longest and most extensive and expensive peacetime military commitment in its history, the North Atlantic Treaty Organization (NATO). NATO remains the keystone of American defense posture, but its future beyond European borders seems problematical. Maintaining the security of Europe is clearly an American vital defense interest, not withstanding the lack of threat to that interest since the end of the cold war.

Problems of the alliance point to a second international political concern, the problem of coalition warfare. The United States has fought two real coalition wars—the world wars—in which there were significant contributions by several states, each of which had somewhat different political objectives. Vietnam, Korea, and Desert Storm were technically coalition wars, but the United States dominated to such an extent that in each case, our objectives were those of the coalition. In Afghanistan and Iraq, both efforts are also officially coalition actions. This designation is somewhat easier to sustain for NATO involvement in Afghanistan, but except for Great Britain, none of the other allies in Iraq have made more than a token military contribution in what is overwhelmingly an American effort. In World War I, Wilson's vision of the peace lost to Georges Clemenceau's punitive vision, with the detrimental effect of paving the way to the second global war. In World War II, the Americans overrode Churchill's greater concern for the postwar map, which allowed the Soviets to capture more territory than they might have otherwise. Both cases point to the historical deficiencies of Americans as geopoliticians.

Military Technology and Technique

A constant trend throughout military history has been the increasing power and destructive efficiency of weapons. The change over 200 years of American history is stark, ranging from inaccurate and unreliable smoothbore muskets to highly accurate nuclear-tipped intercontinental ballistic missiles and precision-guided "smart bombs." Although these weapons represent extremes of the spectrum, advances in firepower, accuracy, and range can be found in every category of weapons.

The impact of increased lethality has been both dramatic and multifaceted. The most obvious and dramatic impact has been in the general destructiveness of war. In the American Revolution, damage was limited to the immediate battlefield, and total casualties to both sides were minimal. In World War II, casualties numbered in the tens of millions, many of whom were noncombatants killed and injured as whole cities were laid waste.

The fact that cities were destroyed in wholesale lots brings home the fact that traditional "home fronts" are now on the front lines because of the range and destructiveness of modern weapons. Worse, perhaps, the home front has been rationalized as a legitimate military target because of the importance of industrial production to the prosecution of modern warfare. Although naval blockades have indirectly attacked the home front for centuries as a method of producing strife and inconvenience, modern strategists can directly target industrial capacity as a central part of the overall war effort.

Bombs falling from the sky, whether they fall on industrial plants or troops in the field, typify the impersonal nature of modern war due to the increased range and firepower of weapons. Modern warfare is a far cry from the battles of mounted knights in the Middle Ages. It is also far different from eighteenth-century warfare, in which linear formations often fired one volley and then pressed the battle home with the bayonet. Hand-to-hand fighting still takes place in modern war, but the range and accuracy of modern weapons place most of the killing at a distance.

Technology has eliminated "waiting until you see the whites of their eyes"; so also it has eliminated some of the perceived glory of combat. This trend reached an epitome of sorts in the electronic "shooting gallery" during the Persian Gulf War and the intercontinental separation of unmanned Predator aircraft and their controllers in the Afghanistan and Iraqi wars.

Although further increases in the range and accuracy of weapons can be expected, we may be near the end of useful increases in firepower, at least in terms of the largest and most dangerous of man's weapons. Many would argue that the firepower of thermonuclear weapons is so great that they are, in effect, unusable. Man's inventive genius has turned its attention to weapons whose energies are far more adaptable to precise and concentrated application, thus reducing unwanted "collateral damage."

Whatever the future of weapons development, the increased lethality of today's weapons has been accompanied by several technological trends. The first is that increases in range, accuracy, and firepower have been paid for by increased complexity. Even today's most common weapon is an engineering marvel compared with earlier versions of the same type of weapon. As examples, compare the Revolutionary War musket with the modern infantry assault carbine or the World War I–era biplane with the modern supersonic jet fighter.

Second, complex weapons are generally much more costly to produce than simple weapons. They are more expensive both in terms of absolute cost and in terms of the time required to design, perfect, and produce them. A third major technological trend in American military history has been the steady mechanization of warfare. Until the Civil War, the American military was powered by muscle and wind. In that great struggle, steam power began to replace both muscle and wind, and the trend in mechanization accelerated from that time to the present. The impact of mechanization has been staggering. Perhaps the greatest single effect has been to make warfare three-dimensional. The development of the internal combustion engine and its subsequent application to aircraft added a new dimension to warfare. The airplane added mobility and flexibility to armed forces, shrank distances, brought the home front to the front lines, and vastly complicated the problems of and increased the possibilities for military commanders.

Mechanization has also speeded the pace and scope of war. Armies on wheels and tracks move many miles a day and arrive ready for combat. Often the secret to survival in battle, whether on land, on the sea, or in the air, is speed of movement. Thus, modern tanks can travel and shoot at 30 to 40 miles per hour, modern submarines travel submerged at speeds as fast as World War II surface combatants, and modern aircraft fly and fight at supersonic speeds. Mechanization has meant that military power can be projected within minutes over great ranges.

The fourth major technological trend is that the electronic age has come to the battlefield, a trend that gathered speed in the Vietnam War and has reached the point where weapon systems now rely on miniaturized solid-state electronics. "Smart" bombs, laser designators, infrared homing devices, beam-riding missiles, computerized fire-control systems, and other sophisticated electronic devices have become commonplace in the arsenals of modern military powers. To some extent, the unconventional, asymmetrical opponents the United States has encountered in places like Afghanistan and Iraq have sought to negate this impact by using tactics designed to evade detection (not massing fighters, for instance) or operating in territory where it is difficult to bring technologically sophisticated force to bear (the mountainous terrain along the Afghan-Pakistan border, for example).

Finally, as was seen in both Afghanistan and Iraq (on both occasions), instantaneous global communications, a product of the space age, have had an enormous impact on the ways in which wars are prosecuted and how they are perceived. Local commanders have to deal with higher headquarters half a world (and up to twelve time zones) away. Rather than letting the results of local command and control speak for themselves, global communications provide distant higher headquarters the opportunity and the means to micromanage events on a "real time" basis. Further, space-based global communications allow newsmen the opportunity to broadcast their interpretation of events "live" as they unfold—interpretations that may be biased, misinformed, or incomplete. Even in the best of circumstances, the ability of the press to influence public opinion on a nearly instantaneous basis has made it imperative for commanders to be much more media "savvy."

What conclusions can be drawn about the role of advances in the technology of war in the American experience? Perhaps the most obvious conclusion is that modern warfare has become a very complex and expensive undertaking. War now emphasizes the importance of the economic base to national defense and, within the economy, the importance of the heavy industrial and technological bases. However, in spite of the importance of technological innovation in the conduct of America's wars, superior technology has never been the decisive factor in any American war. The struggle to use technology effectively and to cope with enemy technology has been much more important.

Until recently, superior technology on the battlefield has rarely been the decisive factor in war for at least four reasons. First, technological advantage is incremental, affecting the margins of combat capability. The technological gap between contending forces tends to be small and, although the gap can be important in a battle, it is generally not large enough to be the decisive factor in the outcome of a war. The gaps are also often quickly filled. Technological advances are based on known physical properties that can often be duplicated quickly by the enemy. As a result, any given technological advantage tends to be short-lived.

Second, how technology is used is as important as the technology itself. Possession of superior technology does not mean that its advantages will be exploited effectively. The story of military technique is the story of attempting to use technology in the most effective manner. The rifled weapons of the Civil War caused significant changes in infantry tactics, but not until bloody lessons had been learned at such places as Fredericksburg. In World War I, it took more than two years to determine how to use the tank effectively. In World War II, the effectiveness of strategic bombing was hindered by our search for the vital industrial targets whose destruction would cripple Germany and by diversions of strategic bombing forces to nonstrategic roles.

Third, the side with the less-developed technology seeks methods to negate the advantage of superior technology. This can be done by obtaining weapons of equal power and by constructing effective defenses. North Vietnam established an integrated antiaircraft missile system to defend themselves from U.S. air attacks. Or, superior technology can be offset by superior strategy and tactics. This was well demonstrated by the Viet Cong in South Vietnam and by the insurgents in Afghanistan and Iraq. When insurgents use guerrilla tactics they tend to negate the effects of superior American firepower by eliminating lucrative targets for that firepower. These asymmetrical impediments are now part of the standard opposition the U.S. confronts.

The fourth reason that superior technology is often not decisive is that for political reasons it is not used to its full extent. This has become evident in the post–World War II era as American political leadership has attempted to prevent the spread and escalation of American wars. For example, in Korea and Vietnam the United States did not use all the weapons at its disposal, nuclear weapons being the most obvious example. For the same sorts of political reasons, the use of weapons has been circumscribed whenever the enemy was permitted to take sanctuary in areas immune from the application of our superior technology. In the Korean case, for example, China was a sanctuary. In the Vietnam conflict, Laos and Cambodia were "off limits" to U.S. forces for much of the war, and China was a sanctuary for all of the war.

The ability to produce and field sophisticated weapons that provide great firepower combined with the tradition of overwhelming our enemies has produced a significant trend in American military technique. Modern American strategists and tacticians have sought to substitute fire and steel for American blood. Strategic bombing in World War II was an attempt to find a way to victory that would minimize American bloodshed. Harry Truman's rationale for using nuclear weapons against Japan was based, at least in part, on the desire to save lives that would have been lost had American forces invaded the Japanese home islands. In Korea, and particularly in Vietnam, fire delivered by artillery and aircraft was used to reduce American casualties in infantry operations.

Military Conduct

Numerous factors, many of which have already been discussed, affect the conduct of war. However, there are several subjects that should be addressed in terms of synthesizing the American experience in military conduct. Korea and Vietnam brought to light two major misconceptions about the American military experience that have led to considerable frustration and debate and are being tested in contemporary terms in Afghanistan and Iraq. The first of these misconceptions concerns the types of wars Americans have fought.

In both Korea and Vietnam, heated debate arose over the political limitations placed on military conduct. There was much hue and cry over forcing the military to "fight with one hand tied behind its back" and there were demands to "turn the military loose." Critics pointed to World War II as the traditional and proper model for American military conduct. However, a significant part of the American tradition is one of war fought with limited means. Of the wars covered in this volume, only three—the Civil War and the two world wars—were conflicts in which the nation was mobilized and the military was "turned loose."

The second misconception centers on the involvement of civilian authorities in conducting warfare. Many critics have decried the tendency of the American civilian leadership to become directly involved in the details of military campaigns, rather than leaving the conduct of war to the military professionals. The most famous incident of this type was the confrontation between General MacArthur and President Truman during the Korean conflict. The debate reached its peak during the Vietnam War and centered around control by the White House of the aerial bombardment campaign over North Vietnam.

To a large extent, the protests over civilian control are made in ignorance of American military tradition. As has been pointed out in this volume, civilian control over military campaigns has been the rule rather than the exception in American military history. In the Mexican War, President Polk ordered field commanders to start and stop their campaigns based on his personal political motives. In the Civil War, President Lincoln played musical chairs with his generals and even took responsibility for the defense of Washington, D.C., during the Peninsular Campaign. In 1898 President William McKinley took a personal hand in directing preparations for war from the newly installed White House "war room." Only the two world wars seem to offer exceptions to the general trend. But even in World War II, political considerations determined the overall shape of the military effort and often overruled the desires of military leaders.

The trend toward close civilian control accelerated during the cold war. This acceleration was a product not only of the American tradition but also of the fear of escalation to nuclear war as well as the increased capacity for close control provided by instantaneous worldwide communications. Korea and the MacArthur–Truman confrontation provided a foretaste of the close political control exercised in Vietnam. The selection of bombing targets by the "Tuesday Lunch Group" in Washington, D.C., gradual escalation of the bombing campaign, and bombing pauses were typical examples of the close civilian control exercised during the Vietnam War. Desert Storm appeared, at first glance, to reverse the trend, because it appeared that the civilian leadership stayed out of the planning and conduct of the war. In reality, there was an enormous amount of interplay between the senior civilian and military leadership in Washington and the American forces in the Gulf region. Plans were presented, amended, and approved at the highest policy levels before they were implemented. During the actual campaign, overbearing civilian "interference" did not hamper Gulf War commanders, because potential problems were addressed and resolved before the fact. This trend has continued in Afghanistan and Iraq.

The "overwhelming" approach of U.S. Grant combined with a love of high technology defines the preferred American method for conducting military operations. An overwhelmingly high-tech military force is the American ideal. The combination of overwhelming force and technology produces a view of war somewhat akin to an engineering project and a resource management problem. The secret to success, in this view, is to apply overwhelming resources (numbers and technology) efficiently. This engineering approach to war served the United States well when the enemy was a modern military power that fought in the same general style. But the Vietnam experience provided a harbinger that clever, asymmetrical strategies can negate the advantage of overwhelming resources and high technology. Part of that lesson included the corollary that victory on the battlefield does not necessarily translate into victory in war. Although American arms were nearly always victorious in battle and inflicted far more damage than the enemy was able to inflict, the outcome of the war was determined by a myriad of other factors, notably including the impact of the war on home front morale. War is a complex struggle between rival societies and battle is only its most obvious and deadly manifestation.

Better State of the Peace

The period since the end of World War II has been wrenching for Americans in terms of their relationship to armed violence. The United States entered the second world conflict as a geopolitical babe in arms and emerged as the world's most powerful state, with international responsibilities that simply could not be shirked. The United States was a superpower and had to learn to act like one. Part of the responsibility was learning to employ the military instrument of power in a way that would serve American national interests. Conceptual struggles over the applications of military force since 9/11 demonstrate that this challenge continues to the present day.

Part of the reason for this difficulty, of course, is America's relative inexperience in the geopolitical game. Historically buffered from the realities of international power politics and having the luxury of never having to develop a tradition wherein the military played a prominent part, Americans have tended to regard military force as something of a novelty. It has not been viewed as just another instrument of political power. War has been an abnormality that occasionally intruded on the more normal condition of peace.

The learning process has been made more wrenching and difficult because of at least two other factors. On one hand, there is the radical transformation of the political power map of the world since 1945. First, the traditional European world powers were relegated to regional status, and in their place emerged only two military superpowers—the United States and the Soviet Union. Then, with the demise of the Soviet Union, only one superpower remained. The postwar introversion of

Europe accelerated the processes of nationalism in former colonies in Africa and Asia. The international system of 1945 that consisted of about 60 states grew to encompass more than 190, and many of the new states have been sources of instability and competition. The role of the United States in this evolving environment remains in transition, as does the system itself.

On the other hand, technology—which has steadily expanded the range, accuracy, and firepower of weapons since the early nineteenth century—has transformed the means of war to the point that the ends may have been changed fundamentally as well. Nuclear weapons are the obvious apex of this process and raise the basic question of whether any political objective can be achieved by military means when nuclear weapons might be used.

The result of all this may be to transform the kinds of political objectives for which military force can be employed, forcing us back to an age of limited warfare, with better states of the peace defined on a different model than that of the world wars. Warfare fought for total political purposes and employing total means may simply have gotten too dangerous and too expensive to wage.

There is a certain irony in all this. In earlier eras, political purposes had to be limited because the means available for conducting war were too modest for anything else. It may be that now the tables are turned. War may have to be limited in its purposes because the means are too great to be sensibly employed for anything but circumscribed ends. There will probably never be another amphibious assault of the nature and grandeur of Normandy, partly because the firepower available to the defenders would make the attempt suicidal but also partially because the objectives that would give rise to such an effort would probably result in escalation beyond any controllable and politically meaningful level. In addition, such an assault or its defense may be too expensive to be rationally entertained.

There, of course, is the rub. Over the decades since the end of World War II, the United States has done a splendid job of preparing to fight the next total war, but preparations have been so thorough and complete that the United States can no longer contemplate fighting that for which it is most ready. Countless billions of dollars have gone into the preparation, but conducting the war would probably mean the end of civilization as we know it.

Since it is difficult to envisage a better state of the peace after any future total war, one is faced with two possibilities. The first is that the United States will not involve itself in future wars because those wars are unaffordable. Certainly there was, in the immediate wake of Vietnam, sentiment to this effect, but subsequent experience clearly shows that the end of armed conflict is not about to occur. And that leaves the second possibility: the military instrument of power will continue to be employed, but it will serve limited political objectives and involve the controlled application of violence at levels well below total war. Limited war, in other words, is the only likely possibility for gaining a better state of the peace militarily.

Such a prospect affects the way we think of better states of the peace and their attainment. Wars fought for limited purposes are not, by and large, wars of societies pitting their total resources against one another. In those circumstances, the questions of hostile ability and will are somewhat different than in previous times. In the case of combating an opponent seeking to achieve total political objectives, it may be necessary to overcome hostile ability. If both sides are fighting for limited objectives, destroying the enemy's armed forces (hostile ability) may not only be inappropriate, it might cause the nature of the war to change and expand, possibly beyond tolerable limits. Rather, the limited wars of the future may be fought with the purpose of overcoming hostile willingness to continue (cost-tolerance).

Future wars are most likely to occur in the developing world, including Asia, the Middle East, Africa, and Latin America. They are likely to be limited in scope and purpose, at least as viewed from the outside. Many of these clashes have been and will be quite passionate, even brutal, for the principals, but the efforts by which the United States and others to impose peace will be quite measured.

The eventual outcomes in Afghanistan and Iraq will impact how and why America goes to war, at least for a time. Should the better states of the peace (democratization and stabilization, for instance) prove impossible to achieve, there may be another phase of introversion as after Vietnam, to evaluate the feasibility of some of the political objectives and the ability to craft and implement military objectives and strategies to achieve different goals. The outcomes in Afghanistan and Iraq almost certainly will temper the American sense of omnipotence in a world where it may be the remaining superpower but not the universal hegemon. The degree of chastening, of course, will depend on the extent to which better states of the peace are realized in the two wars.

If this characterization is accurate, it will also be somewhat discomforting, because wars so defined do not fit well into the American military self-image. Potential uses of military force may bear more resemblance to Vietnam than to World War II, but on a much more constrained scale. Rather than envisioning great crusades against some monstrous Hitler figure, Americans are looking at carefully measured applications of force to accomplish a limited goal, moreover, in situations where the adversaries almost certainly view the conflict as considerably more important than do Americans. These observations are particularly relevant to the situations today in Afghanistan and Iraq.

Selected Bibliography

Alexander, Bevin. *Korea: The First War We Lost.* Revised edition. New York: Hippocrene Books, 1998.

Ambrose, Stephen E. *Americans at War.* Jackson: University Press of Mississippi, 1997.

———. *Citizen Soldiers: The U.S. Army from the Normandy Beaches to the Bulge to the Surrender of Germany June 7, 1944–May 7, 1945.* New York: Simon and Schuster, 1997.

———. *D-Day, June 6, 1944.* New York: Simon and Schuster, 1994.

Atkinson, Rick. *Crusade: The Untold Story of the Persian Gulf War.* Boston: Houghton Mifflin, 1993.

Baldwin, Hanson. *Battles Lost and Won.* New York: Harper and Row, 1966.

Barnett, Coffelli. *The Swordbearers. Supreme Command in the First World War.* New York: William Morrow, 1975.

Bemis, Samuel Flagg. *The Diplomacy of the American Revolution.* Bloomington: Indiana University Press, 1957.

Blair, Clay Jr. *The Forgotten War: America in Korea, 1950–1953.* New York: Random House, 1987.

Bowman, John S. *The Vietnam War: An Almanac.* New York: World Almanac Publications, 1985.

Brands, H.W. *T.R.: A Life.* New York: HarperCollins, 1998.

Brodie, Bernard. *A Guide to Naval Strategy.* New York: Praeger, 1965.

———. *War and Politics.* New York: Macmillan, 1973.

Brodie, Bernard, and Fawn M. Brodie. *From Crossbow to H-Bomb: The Evolution of Weapons and Tactics of Warfare.* Revised and enlarged edition. Bloomington: Indiana University Press, 1973.

Brown, Harold. *Thinking About National Security: Defense and Foreign Policy in a Dangerous World.* Boulder, CO: Westview Press, 1983.

Burg, David F., and L. Edward Purcell. *Almanac of World War I.* Lexington: University Press of Kentucky, 1998.

Burg, Steven L., and Paul S. Shoup. *The War in Bosnia-Herzegovina: Ethnic Conflict and International Intervention.* Armonk, NY: M.E. Sharpe, 1999.

Calvocoressi, Peter, and Guy Wint. *Total War.* New York: Pantheon Books, 1972.

Campbell, Kenneth J. *A Tale of Two Quagmires: Iraq, Vietnam, and the Hard Lessons of War.* New York: Paradigm, 2007.

Carr, E. H. *The Twenty Years Crisis: 1919–1939.* London: Macmillan, 1939.

Carter, Samuel, III. *The Final Fortress: The Campaign for Vicksburg* 1862–1863. New York: St. Martin's Press, 1980.

Catton, Bruce. *The Civil War.* New York: Fairfax Press, 1980.

———. *Reflections on the Civil War.* Edited by John Leekly. Garden City, NY: Doubleday, 1981.

Chandler, David G. *Atlas of Military Strategy.* New York: Free Press, 1980.

Chandrasekaran, Rajiv. *Imperial Life in the Emerald City: Inside Iraq's Green Zone.* New York: Alfred A. Knopf, 2007.

Churchill, Winston S. *The Second World War.* Six volumes. Boston: Houghton Mifflin, 1948–1953.

Clausewitz, Carl von. *On War.* Revised edition translated and edited by Michael Howard and Peter Paret. Princeton: Princeton University Press, 1984.

Coddington, Edwin B. *The Gettysburg Campaign.* New York: Charles Scribner's Sons, 1979.

Coffman, Edward M. *The War to End All Wars: The American Military Experience in World War I.* New edition. Lexington: University Press of Kentucky, 1998.

Coll, Steve. *Ghost Wars: The Secret History of the CIA, Afghanistan, and Bin Laden, from the Soviet Invasion to September 11, 2001.* New York: Penguin Books, 2004.

Collins, Joseph J. *Choosing War: The Decision to Invade Iraq and Its Aftermath.* Washington, DC: Institute for National Strategic Studies, National Defense University, April 2008.

Commager, Henry Steele. *The Blue and the Gray.* New York: Fairfax Press, 1982.

Cordesman, Anthony H. *The Lessons and Non-Lessons of the Air and Missile War in Kosovo.* Washington, DC: Center for Strategic and International Studies, July 1999.

———. *The Iraq War: Strategy, Tactics, and Military Lessons.* Washington, DC: Center for Strategic and International Studies, 2003.

———. *The Iraq War: Progress in the Fighting and Security.* Washington, DC: Center for Strategic and International Studies, 2008.

Corry, John A. *Prelude to a Century.* New York: Fordham University Press, 1998.

Craven, Wesley F., and James L. Cate, eds. *The Army Air Forces in World War II.* Reprinted edition. Washington, DC: Office of Air Force History, 1983.

Crews, Robert D., and Amin Tarzi, eds. *The Taliban and the Crisis in Afghanistan.* Cambridge, MA: Harvard University Press, 2008.

Dale, Catherine. *Operation Iraqi Freedom: Strategies, Approaches, Results, and Issues for Congress.* Washington, DC: Congressional Research Service, March 2008.

Davis, Burke. *Gray Fox: Robert E. Lee and the Civil War.* New York: Fairfax Press, 1981.

Diamond, Larry. *Squandered Victory: The American Occupation and the Bungled Effort to Bring Democracy to Iraq.* New York: Times Books, 2005.

Donovan, Timothy H., Jr., Roy K. Flint, Arthur V. Grant, Jr., and Gerland P. Stadler. *The American Civil War.* West Point, NY: Department of History, United States Military Academy, 1980.

Dupuy, R. Ernest, and Trevor N. Depuy. *The Encyclopedia of Military History.* New York: Harper and Row, 1977.

Eisenhower, David. *Eisenhower: At War, 1943–1945.* New York: Random House, 1986.

Eisenhower, Dwight D. *Crusade in Europe.* New York: Doubleday, Inc., 1948.

Eisenhower, John S.D. *Agent of Destiny: The Life and Times of General Winfield Scott.* New York: Free Press, 1997.

———. *Intervention!: The United States and the Mexican Revolution, 1913–1917.* New York: Norton, 1993.

———. *So Far from God: The U.S. War with Mexico, 1846–1848.* New York: Random House, 1989.

Esposito, Vincent J. *The West Point Atlas of American Wars.* New York: Praeger, 1978.

Ewans, Martin. *Afghanistan: A Short History of Its People and Politics*. New York: Harper Collins, 2002.

Falls, Cyril. *The Great War.* New York: G.P. Putnam's Sons, 1959.

Farrell, Robert H. *American Diplomacy: A History.* New York: W.W. Norton and Sons, 1959.

Farwell, Byron. *Over There: The United States in the Great War, 1917–1918.* New York and London: Norton, 1999.

Ferguson, Niall. *The Pity of War: Explaining World War I.* New York: Basic Books, 1999.

Foote, Shelby. *The Civil War: A Narrative.* New York: Random House, 1958.

Frank, Richard B. *Downfall: The End of the Imperial Japanese Empire.* New York: Random House, 1999.

Franks, Tommy, with Malcolm McConnell. *American Soldier.* New York: Regan Books, 2004.

Freedman, Lawrence, and Efraim Karsh. *The Gulf Conflict 1990–1991.* Princeton: Princeton University Press, 1993.

Freeman, Douglas S. *Lee's Lieutenants.* New York: Charles Scribner's Sons, 1942.

Fromkin, David. *The Independence of Nations.* New York: Praeger Publishers, 1981.

Fuller, J.F.C. *The Conduct of War.* New Brunswick, NJ: Rutgers University Press, 1961.

Futrell, Robert F. *Ideas, Concepts, Doctrine: A History of Basic Thinking in the United States Air Force, 1907–1964.* Maxwell Air Force Base, AL: Air University, 1972.

———. *The United States Air Force in Korea, 1950–1953.* Revised edition. Washington, DC: Office of Air Force History, 1983.

Gaddis, John Lewis. *Strategies of Containment: A Critical Appraisal of Postwar American National Security Policy.* Oxford, UK: Oxford University Press, 1982.

Gelb, Leslie, and Richard K. Betts. *The Irony of Vietnam: The System Worked.* Washington, DC: Brookings Institution Press, 1979.

Gilbert, Martin. *The First World War: A Complete History.* New York: Henry Holt, 1994.

Glantz, Aaron. *How America Lost Iraq.* New York: Jeremy Tarcher/Penguin, 2005.

Gold, Gerald et al., eds. *The Pentagon Papers. New York Times Edition.* New York: Bantam Books, 1971.

Gordon, Michael R., and General Bernard E. Trainor. *The General's War.* Boston: Little Brown, 1995.

———. *Cobra II: The Inside Story of the Invasion and Conquest of Iraq.* New York: Pantheon Books, 2006.

Gray, Edwyn A. *The Killing Time.* New York: Charles Scribner's Sons, 1972.

Guernsey, Alfred H., and Henry M. Alden. *Harper's Pictorial History of the Civil War.* New York: Fairfax Press, 1987 (originally published in 1866).

Gurr, Ted Robert. *Why Men Rebel.* Princeton: Princeton University Press, 1970.

Haass, Richard N. *Intervention: The Uses of American Military Force in the Post-Cold War World.* Washington, DC: Carnegie Endowment for International Peace, 1994.

Hallion, Richard P. *Storm Over Iraq.* Washington, DC: Smithsonian Institution Press, 1992.

Hammes, Thomas X. *The Sling and The Stone: On War in the 21st Century.* St. Paul, MN: Zenith Press, 2004.

Hansell, Haywood S., Jr. *The Air Plan That Defeated Hitler.* Atlanta: Higgins-McArthur/ Longido and Porter, 1972.

Hartmann, Frederick H. *The Relations of Nations.* 6th edition. New York: Macmillan, 1983.

Hashim, Ahmed S. *Insurgency and Counter-Insurgency in Iraq.* Ithaca, NY: Cornell University Press, 2006.

Hassler, Warren W., Jr. *With Shield and Sword: American Military Affairs, Colonial Times to the Present.* Ames: Iowa State University Press, 1984.

Hastings, Max. *D-Day and the Battle for Normandy.* New York: Simon and Schuster, 1984.

Herring, George. *Americas's Longest War: The United States and Vietnam, 1950–1975.* New York: McGraw-Hill, 1995.

Hersh, Seymour. *Chain of Command: The Road from 9/11 to Abu Ghraib.* New York: HarperCollins, 2004.

Hickey, Donald R. *The War of 1812: A Forgotten Conflict.* Urbana and Chicago: University of Illinois Press, 1989.

Hillen, John. *Blue Helmets: The Strategy of UN Military Operations.* Washington, DC: Brassey's, 1998.

Hilsman, Roger. *American Guerrilla: My War Behind Japanese Lines.* Washington, DC: Brassey's, 1990.

Holbrook, Richard. *To End a War.* New York: The Modern Library, 1999.

Huntington, Samuel P. *The Clash of Civilizations and the Remaking of World Order.* New York: Simon & Schuster, 1996.

Hynes, Samuel. *The Soldiers' Tale: Bearing Witness to Modern War.* New York: Penguin, 1997.

Isikoff, Michael, and David Corn. *Hubris: The Inside Story of Spin, Scandal, and the Selling of the Iraq War.* New York: Three Rivers Press, 2007.

Kahin, G.M., and John W. Lewis. *The United States in Vietnam.* New York: Dell, 1969.

Karnow, Stanley. *Vietnam: A History.* New York: Viking Press, 1983.

Kattenburg, Paul M. *The Vietnam Trauma in American Foreign Policy, 1945–1975.* New Brunswick, NJ: Transaction Books, 1980.

Keegan, John. *Fields of Battle: The Wars for North America.* New York: Knopf, 1996.

———. *A History of Warfare.* London: Hutchison, 1993.

———. *Six Armies in Normandy: From D-Day to the Liberation of Paris, June 6th–August 25th, 1944.* New York: Viking, 1982.

———. *The Iraq War.* New York: Alfred A. Knopf, 2004.

Keegan, John, and Andrew Wheatcroft. *Who's Who in Military History.* New York: William Morrow, 1976.

Knox, Donald. *The Korean War, Pusan to Chosin: An Oral History.* San Diego: Harcourt Brace, 1985.

Koenig, William J. *Americans at War: From the Colonial Wars to Vietnam.* New York: G.P. Putnam's Sons, 1980.

Kristol, William, and Lawrence F. Kaplan. *The War over Iraq: Saddam's Tyranny and America's Mission.* New York: Encounter Books, 2003.

Krepinevich, Andrew F., Jr. *The Army and Vietnam.* Baltimore: Johns Hopkins University Press, 1986.

Lambeth, Benjamin S. *The Transformation of American Airpower.* Ithaca, NY: Cornell University Press, 2000.

Leckie, Robert. *The Wars of America.* Revised and updated edition. New York: Harper and Row, 1981.

Lewy, Guenter. *America in Vietnam.* New York: Oxford University Press, 1978.

Liddell Hart, B.H. *The Real War 1914–1918.* Boston: Little, Brown & Co., 1964.

Linderman, Gerald F. *The World Within War: America's Combat Experience in World War II.* New York: Free Press, 1997.

Long, E.B. *Civil War Day by Day: An Almanac 1861–1865.* Garden City, NY: Doubleday, 1971.

Lyons, Terrence, and Ahmed I. Samatar. *Somalia: State Collapse, Multilateral Intervention, and Strategies for Political Reconstruction.* Washington, DC: The Brookings Institution, 1995.

MacIntyre, Donald. *The Naval War Against Hitler.* New York: Charles Scribner's Sons, 1971.

Maclear, Michael. *The Ten Thousand Day War.* New York: Avon Books, 1982.

Mann, James. *Rise of the Vulcans: The History of Bush's War Cabinet.* New York: Viking, 2004.

Mao Tse-tung. *Mao Tse-tung on Guerrilla Warfare.* Translated by Samuel B. Griffith. New York: Praeger, 1961.

McCaffrey, James M. *Army of Manifest Destiny: The American Soldier in the Mexican War, 1846–1848.* New York: New York University Press, 1994,

McMaster, H.R. *Dereliction of Duty: Lyndon Johnson, Robert McNamara, The Joint Chiefs of Staff, and the Lies That Led to Vietnam.* New York: Harper Perennials, 1994.

Miller, David. *The Cold War: A Military History.* New York: St. Martin's Press, 1999.

Millett, Allan R., and Peter Maslowski. *For the Common Defense: A Military History of the United States of America.* New York: Free Press, 1984.

Millis, Walter. *Arms and Men.* New York: Capricorn Books, 1967.

Moore, Harold G. (Lt. Gen. Ret.), and Joseph L. Galloway. *We Were Soldiers Once . . . and Young: Ia Drang—The Battle That Changed the War in Vietnam.* New York: Harper Perennials, 1993.

Morgenthau, Hans J., and Kenneth J. Thompson. *Politics Among Nations: The Struggle for Power and Peace.* 6th edition. New York: Alfred A. Knopf, 1985.

Morrow, John H. Jr. *The Great War in the Air.* Washington, DC: Smithsonian Institution Press, 1993.

Moyar, Mark. *Triumph Forsaken: The Vietnam War, 1945–1965.* New York: Cambridge University Press, 2007.

Nagl, John A. *Counterinsurgency Lessons From Malaya and Vietnam: Learning to Eat Soup with a Knife.* Westport, CT: Praeger Publishers, 2002.

O'Ballance, Edgar. *The Wars in Vietnam 1954–1980.* New York: Hippocrene Books, 1981.

O'Neill, Bard E. *Insurgency & Terrorism: From Revolution to Apocalypse* (2d edition). Washington, DC: Potomac Books, 2005.

Olson, James S., and Randy Roberts. *Where the Dominos Fell: America and Vietnam, 1945–1990.* New York: St. Martin's Press, 1991.

Overy, R.J. *The Air War 1939–1945.* New York: Stein and Day, 1981.

———. *Why the Allies Won.* New York: W.W. Norton, 1996.

Owen, Colonel Robert C. *The Balkans Air Campaign Study.* Maxwell AFB, AL: Air University Press, 2000.

Packer, George. *The Assassin's Gate: America in Iraq.* New York: Farrar, Straus, and Giroux, 2005.

Palmer, Lt Col David. "American Strategy Reconsidered." In *Military History of the American Revolution,* edited by S.J. Underal. Washington, DC: U.S. Government Printing Office, 1976.

Parrish, Thomas, ed. *The Simon and Schuster Encyclopedia of World War II.* New York: Simon and Schuster, 1978.

Peters, Ralph. *Fighting for the Future: Will America Triumph?* Harrisburg, PA: Stackpole Books, 2005.

Pike, Douglas. *PAVN—People's Army of Vietnam.* Novato, CA: Presidio Press, 1986.

Pollack, Kenneth. *The Threatening Storm: The Case for Invading Iraq.* New York: Random House, 2002.

Porter, Gareth, ed. *Vietnam: A History in Documents.* New York: New American Library, 1981.

Preston, Richard A., and Sydney F. Wise. *Men in Arms.* New York: Praeger, 1970.

Record, Jeffrey. *Dark Victory: America's Second War Against Iraq.* Annapolis, MD: Naval Institute Press, 2004.

———. *Hollow Victory: A Contrary View of the Gulf War.* Washington, DC: Brassey's, 1993.

Remini, Robert V. *The Battle of New Orleans: Andrew Jackson and America's First Military Victory.* New York: Viking, 1999.

Richards, Denis. *The Hardest Victory: RAF Bomber Command in the Second World War.* New York: W.W. Norton, 1994.

Ricks, Thomas E. *Fiasco: The American Military Adventure in Iraq.* New York: Books, 2006.

Ropp, Theodore. *War in the Modern World.* New York: Collier Books, 1974.

Sandler, Stanley. *The Korean War: No Victors, No Vanquished.* Lexington: University Press of Kentucky, 1999.

Scales, Robert H., Jr. *Certain Victory: The U.S. Army in the Gulf War.* Washington, DC: Brassey's, 1994.

Schelling, Thomas C. *Arms and Influence.* New Haven: Yale University Press, 1966.

———. *The Strategy of Conflict.* Cambridge: Harvard University Press, 1960.

Sharp, U.S.G. *Strategy for Defeat: Vietnam in Retrospect.* San Rafael, CA: Presidio Press, 1978.

Singletary, Otis A. *The Mexican War.* Chicago: University of Chicago Press, 1964.

Snow, Donald M. *What After Iraq?* New York: Pearson Longman, 2009.

———. *When America Fights: The Uses of American Military Force.* Washington, DC: CQ Press, 2000.

Sorley, Lewis. *A Better War: The Unexamined Victories and Final Tragedy of America's Last Years in Vietnam.* New York: Harvest Books, 2007.

Spector, Ronald H. *Advice and Support: The Early Years of the United States Army in Vietnam, 1941–1960.* New York: Free Press, 1985.

———. *After Tet: The Bloodiest Year in Vietnam.* New York: Free Press, 1993.

Stevenson, Jonathan. *Losing Mogadishu: Testing U.S. Policy in Somalia.* Annapolis, MD: Naval Institute Press, 1995.

Stewart, Richard W. *The United States Army in Afghanistan: Operation Enduring Freedom, October 2001–March 2002.* U.S. Army Center of Military History, 2004.

Stoessinger, John G. *Why Nations Go to War,* 7th edition. New York: St. Martin's Press, 1998.

Stokesbury, James L. *A Short History of World War I.* New York: William Morrow, 1981.

Stueck, William. *The Korean War: An International History.* Princeton: Princeton University Press, 1995.

Summers, Harry G., Jr. *On Strategy: A Critical Analysis of the Vietnam War.* Novato, CA: Presidio Press, 1982.

Sun Tzu, *The Art of War.* Translated by Samuel B. Griffith. Oxford, UK: Oxford University Press, 1963.

Tanner, Robert G. *Stonewall in the Valley: Thomas J. Stonewall Jackson's Shenandoah Valley Campaign Spring 1862.* Garden City, NY: Doubleday, 1976.

Trassok, Leonid, and Claude Blaiz, eds. *The Complete Encyclopedia of Arms and Weapons.* New York: Simon and Schuster, 1982.

Traxel, David. *1898: The Birth of the American Century.* New York: Random House, 1998.

Thompson, Sir Robert. *Make for the Hills: Memories of Far Eastern Wars.* London: Lee Cooper, 1989.

Ulanoff, Stanley M., ed. *American Wars and Heroes: Revolutionary War Through Vietnam.* New York: Arco, 1985.

United States–Vietnam Relations 1945–1967. Washington, DC: U.S. Government Printing Office, 1971.

US Air Forces in Europe. *The Air War over Serbia: Aerospace Power in Operation Allied Force—Initial Report.* Washington, DC: Headquarters, United States Air Force, September 1999.

Van Creveld, Martin. *The Transformation of War.* New York: Free Press, 1991.

Ward, Christopher. *The War of the Revolution.* New York: Macmillan, 1952.

Weigley, Russell F. *The American Way of War.* New York: Macmillan, 1973.

Werrell, Kenneth P. *Blankets of Fire: U.S. Bombers over Japan during World War II.* Washington, DC: Smithsonian Institution Press, 1996.

Westmoreland, William C. *A Soldier Reports.* New York: Dell, 1980.

Williams, T. Harry. *A History of American Wars: From Colonial Times to World War II.* New York: Alfred A. Knopf, 1981.

Winders, Richard B. *Mr. Polk's Army: The American Experience in the Mexican War.* College Station: Texas A&M Press, 1997.

Wright, Quincy. *A Study of War: An Analysis of the Causes, Nature, and Control of War.* Abridged edition. Chicago: University of Chicago Press, 1964.

Young, Peter. *Atlas of the Second World War.* New York: Berkley, 1977.

Zook, David H., Jr., and Robin Higham. *A Short History of Warfare.* New York: Twayne, 1966.

Index

Colonel Dennis M. Drew (USAF, retired) is professor emeritus at the USAF Air University where he taught for more than thirty years both in and out of uniform and served as the Director of the Airpower Research Institute and as the Dean of the University's elite School of Advanced Air and Space Studies. He has authored or coauthored many books, book chapters, monographs, and journal articles concerning military theory and history published in the United States, Latin America, Europe, and Asia. He holds an undergraduate degree from Willamette University and graduate degrees from the University of Wyoming and the University of Alabama. Among his many honors, in 2002 Queen Beatrix of the Netherlands dubbed him a Knight in the Order of Orange-Nassau for his efforts to improve professional education in the Royal Netherlands Air Force.

Dr. Donald M. Snow is professor emeritus of political science at the University of Alabama, where he taught from 1969 to 2006, specializing in foreign and national security policy and international relations. He has held visiting faculty appointments at the U.S. Air Command and Staff College and the U.S. Army, Naval, and Air War Colleges and has served as president of the Section on Military Studies of the International Studies Association and the Section on International Security and Arms Control of the American Political Science Association. He is the author of over forty professional books and approximately forty articles. His most recent books include *What After Iraq?* (2008), *Cases in International Relations,* 4th edition (2009), and *National Security for a New Era,* 3rd edition (2007). He currently serves on the Board of Directors of the World Affairs Council of Hilton Head, South Carolina. He holds the BA and MA from the University of Colorado and the PhD from Indiana University.